OUR MOLECULAR FUTURE

More praise for

"At first it seems like science fiction. However, upon careful reading one realizes that the potential for molecular technologies to protect us from catastrophic events is real. These new tools may ultimately allow for survival and advancement of the human species in ways that were previously thought to be impractical."

—Eric S. Fishman, M.D., president,
21st Century Eloquence

"Mulhall creates an inspired vision of how nanotechnology will change our lives, society, and the way we think.

"This is a vital book for those who care about the environment, society, and deploying new technology to check the destructive power of humankind."

—Allan Thornton, president,
Environmental Investigation Agency

"Mulhall offers a tantalizing glimpse of a nano-tech future in which—to ensure the survival of our society—the tiniest of mechanisms join battle with nature's greatest warriors—volcanic supereruptions, giant tsunamis, and impacting asteroids."

—Bill McGuire,
Benfield Greig Professor of Geohazards and
Director, Benfield Greig Hazard Research Centre,
University College, London

OUR MOLECULAR FUTURE

HOW

NANOTECHNOLOGY,

ROBOTICS,

GENETICS, AND

ARTIFICIAL INTELLIGENCE

WILL TRANSFORM OUR WORLD

DOUGLAS MULHALL

 Prometheus Books

59 John Glenn Drive
Amherst, New York 14228-2197

Published 2002 by Prometheus Books

Inquiries should be addressed to
Prometheus Books
59 John Glenn Drive
Amherst, New York 14228–2197
VOICE: 716–691–0133, ext. 207
FAX: 716–564–2711
WWW.PROMETHEUSBOOKS.COM

06 05 04 03 02   5 4 3 2 1

Library of Congress Cataloging-in-Publication Data

Mulhall, Douglas.
    Our molecular future : how nanotechnology, robotics, genetics, and artificial intelligence will transform our world / Douglas Mulhall.
        p. cm.
    Includes bibliographical references and index.
    ISBN 1–57392–992–1 (alk. paper)
    1. Nanotechnology. 2. Robotics. 3. Genetics. 4. Artificial intelligence. I. Title.

T174.7 .M85 2002
303.48'3–dc21

2002021992

Printed in the United States of America on acid-free paper

To each individual who walks the perilous path between those who claim that technology will solve our problems, and those who say it will destroy us.

CONTENTS

List of Illustrations   9

About the Title   11

Acknowledgments   13

Preface   15

Introduction. Collision of Futures   21

PART 1. THE MOLECULAR AGE

Chapter 1. Singularity   27

Chapter 2. Molecular Building Blocks   30

Chapter 3. What Comes First and What Does It Mean for Each of Us? 52

Chapter 4. What Comes Next?
            Megatrends That Could Alter Our Lives   81

Chapter 5. The Nanoecology Revolution   113

Chapter 6. Reviving Tropical Islands   120

Chapter 7. Who's Driving the Molecular Machine?   124

PART 2. NATURE'S TIME BOMBS

Chapter 8. Are We Getting More or Less Vulnerable?   143

Chapter 9. Shaking Up Tokyo and the Globalized Economy   150

Chapter 10. So, You're Bored by Doomsday?   156

Chapter 11. An Elephant in the Room of Environmentalism   174

Chapter 12. Lost Messages from Ancient Times   180

PART 3. BLUEPRINTS FOR A
MOLECULAR DEFENSE

Chapter 13. Why Go There?   189

Chapter 14. Tools for Defusing Time Bombs   191

Chapter 15. Lessons from Tokyo–Learning to Predict the Big One 217

Chapter 16. The Long Valley Caldera Defense–Avoiding a Dark Age 221

Chapter 17. How to Avert Armageddon–Have an Asteroid for Lunch 226

PART 4. GETTING FROM HERE TO THERE

Chapter 18. The Right Questions   233

Chapter 19. Overcoming Cultural Amnesia   255

Chapter 20. Bypassing the Road to Hell   263

Chapter 21. Using Open Source   282

Chapter 22. Redesigning Democracy for Artificial Intelligence 286

Chapter 23. Liberating Each One of Us   293

Conclusion   311

Appendix A. Isaac Asimov's Laws of Robotics   315

Appendix B. Excerpt from Foresight Guidelines
            on Molecular Nanotechnology   316

Notes   317

A Brief Sampling for the Scientifically Inclined Reader   371

Index   377

LIST OF ILLUSTRATIONS

Figure 1	Faster than exponential growth in computing power	29
Figure 2	Start of the nanorevolution: microscopes that manipulate atoms	33
Figure 3	The first recorded manipulations of individual atoms	34
Figure 4	How nanotubes are made	35
Figure 5	Did this car model come out of that photocopier?	37
	Replicating the human anatomy precisely	37
Figure 6	Spider mite raids a microlock	42
Figure 7	Manipulating something we can't see	44
	A real-time view of the nano world	44
Figure 8	Printable bicycles	46
Figure 9	Print your next computer	48
Figure 10	How this book may soon be delivered	49
	How e-ink works	49
Figure 11	Instantly reconfigurable signs	50
Figure 12	Solar power without the "solar"	63
	Stoke the stove and power the TV at the same time	63
Figure 13	The future of military surveillance	68
Figure 14	No more scratches?	68
Figure 15	Getting the dirt out . . . with no work	69
Figure 16	No more graffiti, or cleaning?	70
Figure 17	The solution to friction?	71
Figure 18	The future of personal transport?	90
Figure 19	The tunnel borer's ultimate machine today, and its potential successor of tomorrow	92
Figure 20	Space elevator	95
Figure 21	To see again . . .	103
	Artificial silicon retina on penny	103

Figure 22   Energy from every conceivable surface   115
Figure 23   A friend for life?   126
Figure 24   Your robotic assistant   127
Figure 25   Think it couldn't happen again?   151
Figure 26   Think you're safe from earthquakes
                if you're not on the West Coast?   154
Figure 27   Dress rehearsal for Earth?   157
Figure 28   How often are we hit?   164
Figure 29   Here's what makes fossil fuels and sewage systems
                unsustainable in a flood   169
Figure 30   Are these structures environmentally sustainable?   178
Figure 31   A new concept for a tsunami barrier   224
Figure 32   Delivery package to a planet-killing asteroid   228
Figure 33   The planet killers   292

ABOUT THE TITLE

In his 1970 treatise *Future Shock*,[1] Alvin Toffler forecast a technological tsunami that has since arrived. In 1987, a best-selling United Nations book, *Our Common Future*,[2] cautioned us to ride that wave without compromising the needs of future generations. Five years later, *Unbounding the Future*[3] told us how nanotechnology might help us achieve that. Then, in 1996, an ominous play on the UN title—*Our Stolen Future*[4]—warned that mutagenic chemicals might stop us from going further. Three years after that, *Owning the Future*[5] chronicled a gold rush for the patents that power the new economy. These are examples of thousands of works that contain the word "future" in their title. Whether positive or pessimistic, they share a common denominator: Every future they depict is one influenced by molecular technologies. This book shows how we may enter the molecular age, what may inspire us to go there, and why we might be knocked off the road along the way. The term "molecular," in this case, is a metaphor. Molecules make everything, so they're nothing new. What's new is our ability to precisely manipulate them right down to the atomic level. This lets us influence everything right up to the planetary scale. Hence, *Our Molecular Future*.

ACKNOWLEDGMENTS

To Carolyn Shoemaker, the late Eugene Shoemaker, and David Levy, whom I have never met, but whose discovery alerted me to the greater reality of the natural environment. My appreciation goes out to Prometheus executive editor Linda Regan, who, through a tough, fair approach, gave this the form that only a great editor could give it. Many individuals took the time to check segments of the text that related to their areas of expertise, and for that I'd like to thank: Chris Peterson and the Foresight Institute, also for their continued work on molecular assembly; Peter Hadfield, also for his careful documentation of earthquake risks; Bill McGuire, for showing what big tsunamis might do; Gerard Fryer for his work on detecting them; Brad Bass, David Etkin, and Ian Burton, who are forging a path to climate adaptation; Graham Hancock, also for his continued work on undersea forensics, Tom McCarthy, also for describing the military implications of nanotechnology; and Nick Bostrom, also for conceptualizing what transhumans might someday be. I bear full responsibility for passages of this book that they checked.

To Waldemar Boff and Valmir Fachini in Brazil, for showing me that high tech and low incomes can produce good results. To Michael Braungart and Monika Griefahn, for developing and arguing in favor of an environmental concept that makes sense. To Lai Chan, for showing how such a concept works in practice, and for opening the doors to China. To Eric and Susan Orbom, for their engineering discussions and tsunami barrier illustrations. To Michael Kabotie, for interpreting the traditions of the Hopi. To Bill McDonough, for introducing me to Eric Drexler. To everybody at NCDT for helping me along the nanoecology trail. To Luda Alexeeva, for opening the doors to Russia and Ukraine. To Pat Moore, for arguing that human technology is also part of nature. To Allan Thornton, for arguing that animals often make more sense than human technology does.

To John Hussman and Bill Fleckenstein, who, although they did not contribute directly to this book, have been instrumental in defining the relationships between markets, technology, and ethics. To Shirley Schmid, Glenn Haddrell, and David Rodgers, for showing how to establish security for the individual in a tumultuous economy. To Ken Mulhall, for his insights into the oil industry.

Especially to Katja Hansen, without whom this book would not have been started or completed: who gathered the illustrations, checked so many pages repeatedly, scoured the Internet for sources, assembled endnotes, critiqued scientific passages, and put up with me cursing endlessly at software on every continent from Africa to Asia to America. Hopefully, our molecular future may solve that frustration for everybody.

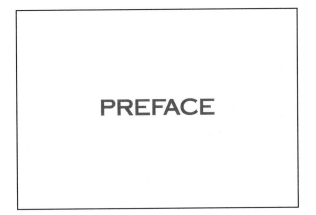

PREFACE

As the future approaches us more quickly, we're pressed to accelerate the rate at which we alter our views of what might come. Thus, the role of speculation takes on added importance. Yet this poses a dilemma: To keep up with the onrushing future, we have to consider possibilities that may seem like science fiction, and may seem unbelievable to many of us.

In this book, I argue that it's time for us to consider the unbelievable, without ridiculing it, yet without being adamant that one future or the other actually will occur. That's where speculation and prediction differ greatly. Speculation looks at possibilities. Prediction pretends that a specific future will occur.

This book makes no predictions. Rather, it looks at the *possibilities* we have for using science to protect ourselves from potentially devastating natural hazards. It focuses especially on the *right questions* that we might ask to begin defining new solutions to problems that may emerge.

The roots of those possibilities and strategies lie in the technological renaissance that we're now experiencing. Some of the natural threats that our technologies may protect us against are only now being discovered. Some of the tools that may protect us from those threats are still to be invented, but their foundations exist today.

As such, many of the technologies and natural phenomena described here are not speculative. Real technologies are helping us to discover real natural disruptions that have occurred in the past. This book gives examples of why peering more accurately into that past is so fundamental to understanding present risks and preparing for our future.

Thus, in these pages, we explore a vast swath of terrain with overviews that may link our past and present to our future. For readers interested in more specifics on a particular topic, I have provided endnotes that lead to work by scientists who are far more qualified to make their case in depth. They are the ones who are driving the discovery machine.

Because this renaissance is proceeding at a blindingly rapid pace, we face potential outcomes—good and bad—other than high technology ones. The chances of our advances being stopped or destabilized are real. Once we solve the questions posed in these pages, we may be equipped to avert such instability and proceed into an enlightened future. Until then, in the words of one individual quoted in this book: "If it doesn't sound like science fiction, then it's probably wrong."

For example, in the 1890s some wagon makers foresaw that millions of horses might disappear from the streets. In the face of much skepticism, they began building horseless carriages. Soon, most other makers of horse-drawn carriages had vanished. In the 1990s, some manufacturers foresaw that factories might disappear from the landscape, so they started to build desktop fabricators. There was much skepticism about how well these digital machines might work. Yet by 2001, they were manufacturing production molds for car parts.

The leap from horse-drawn carriages to desktop fabricators took only a century. Such digital machines represent a logical progression from telegraphs, televisions, and computers. Yet in the near future, these technologies may transform what we *are* rather than only what we do.

This possibility struck me some time ago when my brother, while reading a magazine article about new technology, remarked dryly, "In fifty years we won't even recognize ourselves." I was shocked, not by what was said but by who said it. My brother litigates among large corporations. He operates in the world of the do-able, the practical, and the here and now. Pronouncements on what's to come are usually limited to the realistically foreseeable. It therefore surprised me that he, rather than a science fiction writer, should say that.

In 1993, a few years after my brother's observation, an artificial-intelligence scientist named Vernor Vinge predicted that in thirty years, thinking machines would mark the beginning of the transition from *Homo sapiens* to another species. Four years later, world chess champion Garry Kasparov lost to a computer, calling it an "alien intelligence." Then, in 2002, a robot stepped onto the platform of the New York Stock Exchange and rang the opening bell. Was this a trend?

Yes, it was. And it is today. Yet the challenge we have with such technology is that many of us don't know why we should risk it. Why try to explore the unknown if we don't know what we want to see at the other end? What do we want to achieve in this immense universe?

Lofty goals such as eradicating poverty or exploring space inspire many of us, but not the whole of us. Perhaps it's because great promises haven't yet been fulfilled. We haven't stood on Mars or conquered disease, we haven't eradicated crime, we haven't even done away with traffic jams, and we still have big mortgages. Just when some of our greatest scientific breakthroughs are dawning, many of us are skeptical that science can solve our problems.

Amid such cynicism and throughout our history, some of us—usually the scientists—have continued to ask questions. The resulting ideas have constantly threatened the status quo. Only when enough supporters outside the scientific field explained what these perspectives meant for society, and only when enough people integrated them into their codes of conduct, did they cease to be a risk and instead become part of our progress.

LOOKING FOR INSPIRATION

In Galileo's day, when just a tiny portion of the population could read, the church had decades to reject, then reconsider, and finally assimilate his proof that the earth wasn't the center of the universe.

Today, the status quo doesn't have time to adapt. When nations, churches, or families are confronted by technologies for creating new life, they have no time to reject, reconcile, or assimilate them. Thus, they base their perceptions on current moral understandings rather than future perspectives.

It's this struggle between perceived peril and perceived gain, fought in a compressed time frame, that preoccupies us now. Churches condemn the life sciences, animal rights groups vilify researchers, and religious extremists force women into veils. It's a crazy combination of rushing toward the future and bouncing off of it.

What might spark us to free ourselves and embrace a future-oriented perspective, including the risks that come with it? In the past it's been such great undertakings as building a nation, exploring new physical frontiers, or engaging in self-induced calamities such as world wars. But many of us perceive that nations are out of room, frontiers such as deep oceans or deep space are unfriendly for large populations, and world war has become too horrendous to contemplate as a moral crusade.

So what might get us fired up?

As with many great undertakings, the inspiration may come from a shock. That shock may be far greater than the attacks of September 11, 2001. Instead of emanating from ourselves, it may emanate from our environment—

not only because of what we may do to it, but because of what it may do to us. The relatively calm natural conditions that allowed our technological society to develop during past centuries may be more rare than we thought. They may end. We know that in earlier millennia, grasslands became deserts and coastlines were inundated. It now seems that such transitions occurred more rapidly than we once believed. We're still uncertain about what triggered them, but we know that it was something besides human intervention.

It's been proposed by some scientists that our natural environment might suddenly turn on us, and moreover, that this isn't unusual in history. These ideas are beginning to get serious attention, though they often get lumped in with doomsday hysteria that dilutes their legitimacy. As we begin to separate mysticism from true catastrophic risks, a new survival imperative may inspire us to adapt to a universe that now appears more risky than we once thought.

Thus, our next great motivation may come from an age-old necessity: adapting to immense natural upheavals. As we'll see, our technological infrastructures seem ill prepared to tolerate such onslaughts. If these sweeping changes were to recur, they could destabilize the technologies that support our society.

This poses a deep challenge for each of us, but especially for scientists and environmentalists. On one hand, we may try to keep nature's dynamic environment in a "sustainable" state through laissez-faire methods that purportedly blend with nature. On the other, we may use activist approaches such as those now advocated for correcting human-generated imbalances. Yet what if both methods miss the elephant in the room? How do we "balance" an environment that generates its own monumental, natural house cleanings?

We may soon be caught between two types of disruption. One is wrought by natural cycles. The other is generated by our own technologies. Yet these differ in at least one aspect: Nature's big upheavals give us scant control over our fate. Human technologies—though volatile and risky—offer at least some potential for self-determination.

In a world that has grown pre-occupied with security, it is strange that many of us still work and live in structures that are prone to collapsing in natural disasters. Such tragedies are avoidable. Yet we have not given the same priority to surviving nature's September 11ths as we do to preventing human attacks. Perhaps natural incidents seem too inevitable to worry about. Or perhaps we've grown as complacent and over-confident about the remoteness of such possibilities as the U.S. was once about terrorism.

Now, science is showing us that such complacency may be misguided. This is especially paradoxical at a time when the technological balance may be shifting in our favor, to help protect us from nature's rare but devastating onslaughts. Will we use or abuse this advantage?

In the quest to survive such disruptions, it's the scientists who are leading the way. Therefore, much of this book looks at their role in helping to build a molecular future that may save us, or make us more vulnerable.

As someone who works closely with scientists to find solutions to ecological questions, I believe science has the answers to many, though not all, of our problems. My reasons are explained in this book. I also believe that scientists are in the midst of enlightening us with great discoveries that may alter the way we perceive ourselves as a species.

Especially because of this, I set the bar high, because scientists now have a more immense responsibility than before. The future of our species—and possibly new ones—is in their hands. If more scientists and environmentalists, along with the rest of us, look critically at the questions posed in this book, our chances for an enlightened future may improve.

The good news is that the molecular age may let us connect our actions more directly with the consequences arising from them. For example, in the future, each of us may have the power to synthesize food in a desktop factory without having to raise or kill an animal. Thus, as manufacturing becomes compressed in time and space, many of its consequences, beneficial or otherwise, are going to be right in front of us—not in a barn a thousand miles away.

On the other hand, we might choose to ignore the consequences of our actions by using designer drugs concocted in our desktop factories, to keep us healthy yet zoned out.

These are some of the contradictions we face in our molecular future. This book focuses on paradoxes, because history is full of them and our future could be, too. It's not going to be a unified path. Rather, like the collisions of trillions of atoms, our molecular future may have infinite numbers of pathways, many of them paradoxical. To cope with these, we may each require enhanced intelligence. Molecular technologies might give us the capacity to achieve that.

In many ways, it's still a blank page. But we don't have much time before that page is filled in. Each of us, not only scientists, has to think about that. This book aims to stimulate the process by describing, in plain language, the contradictory challenges we face.

INTRODUCTION

COLLISION
OF FUTURES

*"If it doesn't sound like science fiction
then it's probably wrong."*

–Chris Peterson, president, Foresight Institute,
commenting on how to forecast technology's
long-term impact[1]

On February 14, 2002, a four-foot-tall humanoid robot named ASIMO stepped up a stairway to the podium, clapped its hands, then pressed a button that rang the opening bell of the New York Stock Exchange, thereby becoming the first robot to preside over daily resumption of America's financial heartbeat.[2]

History shows that, regardless of economic turmoil, technology proceeds apace. Sometimes the greatest advances occur in the worst of times. The Great Depression and World War II each saw fantastic technology leaps. Today, in an age of insecurity and economic uncertainty, technology may be getting ready for its biggest hyper-jump yet.

For example, to stay ahead of obsolescence, publishers may soon offer books like this on electronic paper that looks and feels like regular paper, but contains millions of sensors. When we purchase a title, we'll be offered subscription-based updates for a fee. To get those updates, our new book will "call home," via the Internet. If one of the titles we bought has been revised, the revision will be retrieved wirelessly. Our book might also play music, video, and text-to-speech narration. It's not going to be a laptop with a hard-to-read screen. It may be a real book that feels like the one we've been used to curling up with in bed. As with a cell phone, it may cost nothing if we buy it as part of a service.

This biggest literary revolution to come since Gutenberg is an example of *disruptive technology*: something that suddenly, massively displaces older technologies, generating bankruptcies and hypergrowth at the same time. It's a cultural as well as a commercial event. For publishers, it represents a kind of giddy nervous breakdown. It may slash production and paper costs along with the nemesis of unsold stock. On the other hand, oblivion may await those who choose the wrong technology, such as software that gets pushed aside in the marketplace. Flexibility is the survival tool. Authors like myself may be beneficiaries or victims; we haven't figured out which yet.

Such disruptions may be about to accelerate, due to molecular technologies. These are technologies that let us precisely arrange or operate things at the atomic scale.[3]

At the molecular level, some materials don't behave the way we'd expect them to. Gold is a million times more fluorescent than at the normal scale we see in jewelry.[4] Energy input restrains motion instead of causing it, because billions of atoms bombard things from millions of angles at once, blocking movement.[5] These molecular characteristics could give us seemingly perpetual motion, or change the color of our clothes and cars instantly. Someday they might let us jump into and out of robotic bodies with minds and eyes that see across globalized networks.

This world of disruptive technologies may be about to accelerate past our comprehension. It poses some profound dilemmas: How do we decide where we *want* technology to go if we don't know what it might be? How do we build a "sustainable" society that preserves resources for future generations, when we don't know what those generations might require?

Until now, our past and future have been connected by continuities that let us prepare for what's to come. We've based our lives on certainties that are probably going to exist tomorrow. Gravity remains constant. The sun is probably going to come up each day. Seasons in temperate zones will vary in patterns. We also get ideas about the future from present trends that will probably continue, such as increased automation. Finally, there's continuity to the cycle of life. We're human. We're born, we live, and we die in that sequence, as far as we know.[6]

These certainties keep us sane in a world that accelerates without our permission. Yet we face the prospect that some such assurances may evaporate one day. We might live in robotic bodies, for example. Other machines that we build may have intelligence that surpasses our own. Our human memories could be preserved in such machines.

Does that sound too fantastic or premature?

Only a few years before the telegraph was invented, most nineteenth-century individuals couldn't fathom what it meant to have instant communication. The same shock accompanied the telephone, radio, and then television. These were otherworldly. Likewise today, we can't fathom what it might mean to feel multiple objects in separate locations thousands of miles away as if they were next to us. Yet virtual-reality sensors already do this. We're connecting ourselves to them to perform remote surgery.[7]

The foundation for such a future is being built right now, based on new tools that let us manipulate molecules. Those tools multiply every day. They are in the process of transforming humanity more profoundly than the car, telegraph, phone, radio, and television combined.

Yet that's just part of the story. A different set of disruptions may block the road to our molecular future before we go much farther. These disturbances may leave us in a much less fortunate, yet still unpredictable, position.

In his classic dark humor novel *The Hitchhiker's Guide to the Galaxy*, the late Douglas Adams describes how Earth is torn down to make way for a galactic freeway, much to the shock of its soon-to-be-extinct residents.[8] Most apocalyptic notions build up to a climax, but Adams shatters that rule by beginning at the end. His *Hitchhiker's Guide* and the books that followed describe those seconds that transpire between the farcical notice of imminent annihilation and the great event itself. In that interim, readers take a trip between past and future that encompasses "life, the universe, and everything." Finally, we come out at the beginning, which in that case is the end of the world—a mass extinction of humanity.

Life imitates art. Since Adams's comedy was first published, scientific evidence has shown that mass extinctions can and do occur suddenly. The "species superhighway" is occasionally cleared so other life forms might move forward. Moreover, small shake-ups alter the pecking order as often as every few hundred years in some regions, or every few thousand years globally.[9] The next such renewal is on its way. We just don't know when, or to what degree, although we're getting some idea by exploring planetary history.

Such revelations may also shake the foundations of environmentalism. For decades now, conventional environmental wisdom has held that the earth is a *closed system* as far as matter is concerned: that besides energy, little else enters our world. This has led to a broadly held idea that big changes are caused only by the earth, its biosphere, or human interference.[10]

Part of that concept now appears to be incorrect. It seems, for example, that asteroids, comets, and meteors are more than just pebbles pestering the earth. Besides bringing small quantities of trace minerals and other materials to us from space, they also alter the whole cycle of life from time to time, through the violence of their collisions.

What or who is leading us to these revelations? Molecular technologies are. They are the tools that pry open the secrets of the earth and space. These tools range from robotic telescopes, to molecular chemistry that reveals evidence of past catastrophes, to submicroscopic computers that analyze complex systems. Such technologies have elevated us to that awkward stage when we're smart enough to detect a message, but still uncertain of what it means. Some of us act like the pub patrons in Adams's novel: downing another drink, trying to ignore the din of cosmic bulldozers outside the bar room door. Yet, as with the main characters in Adams's books, others find themselves whisked away to a place where everything is possible and options are limitless: where the smallest tools let us survive the greatest catastrophes, and where a road map to the universe unfolds for us to read.

In our real world, these explorers exist. They are the scientists and entrepreneurs who work with molecular technologies. The tools they work with—genetics, robotics, artificial intelligence, and nanotechnology—are being designed increasingly at the atomic level rather than the micro level. Millions of us are tied to them through the companies and universities in which we work.

To see where this might lead, we have to peer into three great spheres.

The first is technological. It drives a confluence of genetics, robotics, artificial intelligence, and nanotechnology. This nano-based revolution is already transforming humanity and the environment. It's pushing the throttle wide open.

Next are nature's time bombs. They've struck before and are sure to strike again. Some are set off by our own technologies, but others occupy nature's exclusive realm. They may destroy our technological capacities and wreck our economies. On the other hand, they may be rendered a mere nuisance if we learn to adapt.

The last is an ethical, antitechnology backlash: fear of "Frankenfoods," monster mutations, and military madness. Some of it is driven by fear of the unknown, some by legitimate scientific concern. It may stop the molecular age before it starts, slow it down, or perhaps improve it.

These spheres orbit each other in a wild dance, their attractions and repulsions driving a discordant symphony. To try to predict our future without considering their interaction is folly. To declare a single path is equally risky.

This book gives a broad-brush description of each of these spheres, and how they may interact or collide.

PART I

THE MOLECULAR AGE

"Within thirty years, we will have the technological means to create superhuman intelligence. Shortly after, the human era will be ended."
—Vernor Vinge, computer scientist, 1993[1]

"Even if I don't want to, now I have to think about man versus something else, an alien."
—World chess champion Garry Kasparov,
after losing to a computer in 1997[2]

"They made us too smart, too quick, too fast."
—Gigolo Joe, the artificially intelligent
sex robot in the film *A.I.*, 2001[3]

CHAPTER ONE

SINGULARITY

Imagine a time when machines are more intelligent than we are, autonomous robots fill our emotional needs, and the economy is managed by a breed of humans whose brains are enhanced by neural implants. This era may be on its way, but it's hard to imagine. How do we prepare for such a world?

A group of award-winning scientists—including Carnegie Mellon robotics director Hans Moravec, Massachusetts Institute of Technology (MIT) artificial-intelligence explorer Marvin Minsky, Stanford medical informatics professor John R. Koza, nanotechnology pioneer K. Eric Drexler, speech synthesis developer Ray Kurzweil, and retired University of California computer scientist Vernor Vinge, among others[4]—is contemplating whether this instant, known as the *Singularity*,[5] might be reached in our lifetimes.[6] Furthermore, some of them suggest it may be impossible to forecast the future if it is determined by beings or machines with intelligence greater than ours.

In that event, each word that's written about the future may be useless, because regardless of our prognostications, everything may exceed our capacity to imagine it.

I thought about that while working on this book. Why bother? When scientists who are much smarter than I started to describe the future as being unknowable or unimaginable, I had to ask why this book might be relevant.

To find answers for that, I had to find out why some scientists argue that Singularity is near, and that it might signify the beginning of humanity's transcendence. Why are conferences about Singularity springing up in high-tech zones such as Silicon Valley? Why are technological leaders from universities such as Stanford, Carnegie Mellon, and MIT telling us we're woefully unprepared for what may come? Furthermore, what might stop us from reaching that point?

I soon found that answering these questions may take up more than one book. It may also require many disciplines besides the physcial sciences. It may be an economic or theological issue too, as I discovered when one of the main scientific proponents of Singularity gave a sermon in church about it.[7]

The Singularity group is still a small minority. Other similarly qualified experts disagree that such a point is so close, as we'll see in chapter 4. Yet the arguments supporting Singularity seem to merit attention, in view of the credentials of those scientists who put them forth, along with the qualifications of those who doubt them, and the growing attention that this discussion is receiving from futurists around the world.[8]

The fundamentals are easily understood. Looking back over ten thousand years of history, we see that the progress of computing power isn't linear, but rather exponential. For example, if we accept the current view of how technology evolved[9] we see it took many millennia to get from counting on cave walls to counting with manual analog computers. From manual computers to mechanical ones took several centuries; from mechanical to electronic ones, several lifetimes; and from integrated circuit computing to self-taught computers, less than a generation. Appreciating this rise in computing power, as depicted in Fig. 1, is central to understanding the potential for technology leaps. Such a historical trend has been described in depth by mathematicians and scientists in critically acclaimed works.[10]

These time lines suggest that, barring a cataclysm or another down-cycle, our intelligence may be superceded in the next years. Our technologies are approaching the stage where they may start carrying on by themselves.[11] Thus, Singularity may be one of the most vital scientific discussions. It may foreshadow our next evolutionary stage.

A SMASHUP ON THE ROAD TO SINGULARITY?

What could stop such a monumental event from occurring, or delay its arrival until well into the future? Might something throw our infrastructures out of kilter before we're able to build such machines? History shows that species move backward as well as forward, and that we're by no means immune. The Middle Ages saw civilizations go backward in time, perhaps due to natural disasters as well as cultural ones. Before that, in Tasmania and

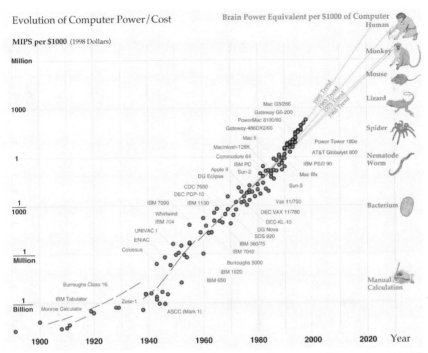

Fig. I. Faster than exponential growth in computing power. The number of MIPS (millions of instructions per second) in $1,000 worth of computer, from 1900 to the present. Steady improvements in calculators before World War II increased the speed of calculation a thousandfold over manual methods. The pace quickened with the appearance of electronic computers during the war. There was a millionfold increase between 1940 and 1980. Since then, the pace has been quicker; one that would make humanlike robots possible before the middle of this century. The vertical scale is logarithmic. Divisions represent thousandfold increases in computer performance. Exponential growth would show as a straight line. The upward curve indicates faster than exponential growth; an accelerating *rate* of innovation. (Courtesy of Hans Moravec, www.frc.ri.cmu.edu/~hpm/book98/fig.ch3/p060.html)

the Amazon, evidence suggests that cultures regressed to the most primitive circumstances when they were isolated through disruptions.[12] We don't always move forward, so the question is why. In later chapters, we'll discuss history's backward-flowing waves, and the tools we may require to avoid them. Those tools could be the same ones that some scientists say are leading us to Singularity. Thus, getting to the future may hinge upon our capacity to side step the types of disasters that disabled us in the past.

To see such a connection, and to appreciate how conceivable or unimaginable our future may be, let's look at the building blocks that could take us there, along with the changes they might bring to our lives.

CHAPTER TWO

MOLECULAR BUILDING BLOCKS

"There are only about 100 kinds of atoms in all the Universe,
and whether these atoms form trees or tires,
ashes or animals, water or the air we breathe,
depends on how they are put together."

—Marcus Hewat, computer scientist[1]

veryone has heard the terms, yet their workings remain a mystery to most:[2]

Genetics–manipulating the building blocks of life
Robotics–building autonomous machines to do our bidding
Artificial Intelligence–machines that learn
Nanotechnology–building things atom by atom

The acronym is GRAIN. This megamerger of supersciences may transform who and what we are. It could transplant our senses into other entities, then convert those entities into something else. It may alter millions of years of evolution that imbue us with wonderful and terrible traits. It already lends a new dimension to our technological development.

One technology sets the speed limit for every other. Its inventors build atom by atom on a scale of one-billionth of a meter–a *nanometer*. They are the nanotechnology pioneers.

A BRIEF HISTORY

In 1987 a lone dissenter on the *Challenger* space shuttle investigation upstaged a nationally televised press conference with a children's model showing how frozen o-rings caused a fuel leak leading to the explosion.[3] Nobel laureate Richard P. Feynman gave that elementary lesson, shaking the space program to its roots and sparking reforms that led to seventy-five subsequent space shuttle launches without a fatality.

Back in 1959, Feynman used that same simple genius to point in the direction of *molecular nanotechnology* in a talk entitled "There's Plenty of Room at the Bottom,"[4] where he gave the world an elementary rationale that has yet to be improved on for why nano-scale operations are feasible:

> The principles of physics, as far as I can see, do not speak against the possibility of maneuvering things atom by atom. It is not an attempt to violate any laws; it is something, in principle, that can be done.[5]

He painted a picture of nanotechnology's vast implications by describing the smallest material that all books in print could fit onto and still be retrievable:

> I have estimated how many letters there are in the Encyclopaedia [*sic*], and I have assumed that each of my 24 million books is as big as an Encyclopaedia volume, and have calculated, then, how many bits of information there are (10^{15}). For each bit I allow 100 atoms. And it turns out that all of the information that man has carefully accumulated in all the books in the world can be written in this form in a cube of material one two-hundredth of an inch wide—which is the barest piece of dust that can be made out by the human eye. So there is *plenty* of room at the bottom.[6]

In the same lecture, he envisaged atomically precise machines—devices still not invented today, but that promise to launch us into a vast new era:

> If we go down far enough, all of our devices can be mass produced so that they are absolutely perfect copies of one another. We cannot build two large machines so that the dimensions are exactly the same. But if your machine is only 100 atoms high, you only have to get it correct to one-half of one percent to make sure the other machine is exactly the same size—namely, 100 atoms high![7]

How close are we to such molecular machines, really? For clues, let's examine the chronology.[8] It would require more than one book to chronicle the many enabling technologies that are leading to nanotechnology, so this is just a thumbnail sketch of a very deep history.

In the late 1970s Eric Drexler developed concepts for molecular nanotechnology at MIT. In 1974 Norio Taniguchi of Tokyo Science University used the word "nanotechnology" in reference to machining with tolerances of less than a micron. In 1981 the first technical paper on molecular nanotechnology was published by Drexler, in *Proceedings of the National Academy of Sciences.*[9] While doing his references for this paper, Drexler came across Feynman's famous 1959 talk and began referring to it,[10] thus helping to resurrect the historic lecture. From the early 1980s onward, Eric Drexler and his colleague Chris Peterson were at the forefront of efforts to describe molecular nanotechnology. In their books, *Engines of Creation,*[11] *Nanosystems,*[12] and *Unbounding the Future,*[13] they discussed how to build molecular machines.

Let's trace nanotechnology from the practical applications side. In 1981 the scanning tunneling microscope was invented by Gerd K. Binnig and Heinrich Rohrer, staff scientists at IBM's Zurich Research Laboratory. This let scientists see and manipulate atoms for the first time. In the mid-1980s Binnig and coworkers extended such capabilities to nonconducting materials by inventing the atomic force microscope (see Fig. 2).[14] In 1985 Robert F. Curl Jr., Harold W. Kroto, and Richard E. Smalley discovered Buckminsterfullerenes (buckyballs) measuring about a nanometer in width.[15] In 1989 physicists Donald M. Eigler and Erhard K. Schweizer at IBM Almaden Labs in California spelled the letters *I-B-M* with thirty-five Xenon atoms, proving it's possible to manipulate atoms precisely (see Fig. 3). Parallel to this, Japanese scientists were forging ahead. In 1991 Sumio Iijima, a physicist at NEC Research Labs in Japan, synthesized carbon nanotubes—structures at least thirty times stronger than steel, with semiconducting capacities that may exponentially speed computing (see Fig. 4). In 1993 Warren Robinett of the University of North Carolina and Stanley R. Williams of the University of California at Los Angeles connected a virtual-reality system to a scanning tunneling microscope. This made viewing and manipulating atoms more practical.[16] In 1997 the first nanotechnology venture-capital company was established. In 1998 Cees Dekker and a group at Delft University of Technology in the Netherlands created a transistor from a carbon nanotube. In 2000 Lucent and Bell Labs, working with Oxford University, created the first DNA motors, thus demonstrating the link between biotechnology and nanotechnology.[17] Then, in 2001, nanocomputing accelerated dramatically, grabbing the top spot in the journal *Science*'s list of notable scientific achievements for that year.[18] Breakthroughs came monthly. A group of scientists at Israel's Weizmann Institute announced that they had developed DNA computers capable of performing billions of operations per second,

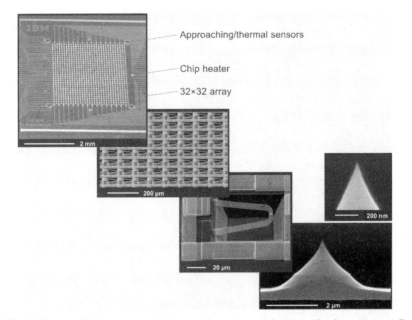

Fig. 2. Start of the nanorevolution: microscopes that manipulate atoms. From upper left to lower right: increasing magnification of an atomic-force microscope data storage concept called the Millipede, with a potentially ultrahigh-density terabit capacity. This could lead to computing millions of times more powerful than what we have today. The images show, in descending order, the 32 × 32 array section of the chip, with the independent approach/heat sensors in the four corners and the heaters on each side of the array, as well as zoomed scanning electron micrographs (SEMs) of an array section, a single cantilever, and a tip apex. The tip height is 1.7 μm. The apex radius is smaller than 20 nm. The image on the upper right is the tip. This can manipulate atoms. (Images copyright 2000 by International Business Machines Corporation; reprinted with permission from *IBM Journal of Research and Development* [P. Vettiger, M. Despont, U. Drechsler, U. Dürig, W. Häberle, M.I. Lutwyche, H.E. Rothuizen, R. Stutz, R. Widmer, and G.K. Binnig, "The 'Millipede'—More than One Thousand Tips for Future AFM Data Storage, IBM J. Res. Develop. 44, no. 3 (May 2000): 323-340])

yet able to fit into a test tube while using very small amounts of energy.[19] The same year, IBM Labs and, separately, scientists at Delft University in the Netherlands, used carbon nanotubes to develop nanometer-sized logic circuits–the components that perform processing in a computer.[20] These were the first building blocks of a potential *nanocomputer*. The company had already tripled capacity of hard drives using nanoparticles.[21] In late 2001, scientists from Bell Lucent Labs fabricated the first molecular-scale, individually addressable transistor with a switching channel of only one mole-

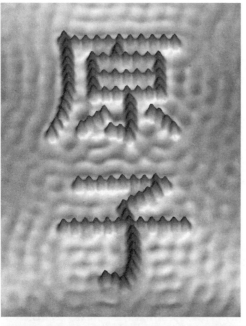

Fig. 3. The first recorded manipulations of individual atoms. Upper: The Kanji characters for "atom." Lower: The letters "IBM" arranged in xenon atoms. (IBM Corporation, Research Division, Almaden Research Center; source: www.almaden.ibm.com/vis/stm/atomo.html)

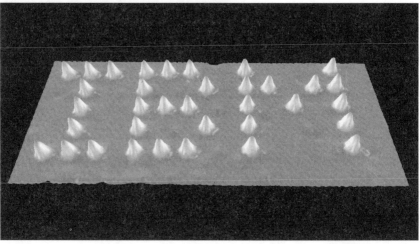

cule–a breakthrough that made molecular computing a near certainty.[22] In a parallel development that promised mass-manufacturing of molecular computing components, Mitsui & Co. announced that in 2002, it would start to manufacture 120 tons of carbon nanotubes annually–a feat thought impossible a year earlier.[23] By early 2002, researchers at a meeting of the

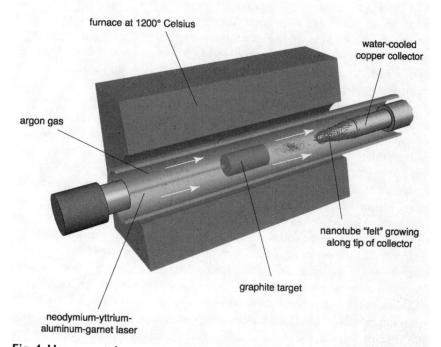

furnace at 1200° Celsius

water-cooled
copper collector

argon gas

nanotube "felt" growing
along tip of collector

graphite target

neodymium-yttrium-
aluminum-garnet laser

Fig. 4. How nanotubes are made. Single-walled nanotubes are produced in a quartz tube heated to 1,200° C, a much lower temperature than was previously necessary for making nanotubes. Nanotubes grown this way self-organize into ropes. These have promising engineering applications. In the techniques used at Rice University, a laser is aimed at a block of graphite, vaporizing the graphite. Contact with a cooled copper collector causes the carbon atoms to be deposited in the form of nanotubes. The nanotube "felt" can then be harvested. (*American Scientist*/Aaron Cox, reprinted by permission of *American Scientist* magazine of Signa Xi, the Scientific Research family)

American Association for the Advancement of Science had announced that ultrafast nanoelectronics and nanocomputers were approaching the industrial production stage, many years ahead of schedule, thanks to an avalanche of new discoveries.[24]

Thus, Molecular Alley may be superceding Silicon Valley.

Meanwhile, in the health-care realm, work was proceeding on the first *nanotoothpaste* to replace bacteria with toothlike enamel in microscopic cavities of teeth, stopping decay before it starts.[25] This was accompanied by trial use of drugs to combat *nanobacteria* that are apparently the source of diseases that range from tooth decay to heart disease. Thus, the days of the dentist's drill and surgeon's scalpel may be numbered.

Finally, on the leading edges of the very small, physicists at MIT, Harvard, and companies such as Sun Microsystems began experimenting with practical applications in the potentially explosive new field of *artificial atoms*—designer elements that may lead to a reengineering of the periodic table. Such atoms, when stimulated electronically or optically, allow creation of *programmable matter*: substances with properties that can be adjusted precisely and repeatedly. These, in turn, could let us transform the chemical, optical, and physical properties of materials with the flick of a switch. One of the first such applications has been via fluorescence, which permits instant, intense color changes. Wide-scale applications may be many years off, but the principles have been clearly demonstrated in practice.[26] Once again, such developments have been made possible by the nano-scale manipulation of individual atoms.

Nanotechnology has a similar investment background to that of personal computers. Just as PCs required decades of taxpayer-funded military and scientific research into their enabling technologies before they burst onto the world scene, so did nanotechnology.[27] After years of quiet research, the number of academic, military, and corporate nanotechnology teams mushroomed from a handful to upward of a hundred between 1997 and 2001. Japan, Europe, and the United States each began investing up to a billion dollars annually, plus many billion for enabling technologies such as those described later in this chapter. In 2000 the U.S. federal government financed the first big increase for basic science in twenty years—the National Nanotechnology Initiative.[28] The state of California put its own funding into the California Nanosystems Institute at the University of California, Santa Barbara, and UCLA.[29] In late 2001 this was followed by creation of nanotechnology centers at Columbia, Cornell, Rice, Northwestern, and Harvard universities.[30] These moves were marginally behind European[31] and Japanese efforts.[32] Today, venture capitalists are beginning to enter the field. The first nano-scale products are coming to markets.

Competition is heavy among nations, academics, and multinationals. No dominant player has emerged—no Microsoft or Cisco—because the technology cuts across many industries, including chemicals, computers, software, biotechnology, and weapons. Yet as we'll see, the potential for such a behemoth may belong to the first company to mass-produce devices such as *desktop digital fabricators* or *molecular assemblers,* and especially the software that runs them (see Fig. 5).

Fig. 5. Did this car model come out of that photocopier? Actually it's not a photocopier. It's a digital fabricator. The Z402 System is based on the Massachusetts Institute of Technology's patented Three-Dimensional Printing technology. The software converts a three-dimensional design into thousands of cross-sections 0.005"—0.010" thick. The three-dimensional printer then prints these cross-sections one after another, from the bottom of the design to the top. The car model held by the scientist comes from that process. (Courtesy of Z Corporation)

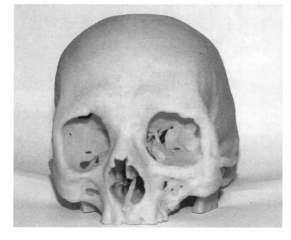

Replicating the human anatomy precisely. This model of a skull was printed on Z Corporation's 3-D Printer using a starch-based powder and a water based binder. The 3-D Printer uses an inkjet print head to deposit binder and glue parts together, layer by layer. By gathering digitized data from scans of the human body, such objects can be manufactured to precisely replicate the bone structure of a patient. This lets surgeons plan their work without invasive exploratory surgery. Results: less expense, more accuracy. (Courtesy of Z Corporation)

DEFINING NANOTECHNOLOGY

Surprisingly, a standard definition for "nanotechnology" doesn't exist, as this National Science Foundation study briefing explains:

> In Japan, the term is often used to describe the construction of nanostructures for future electronic technologies, and development of equipment for measure at the nanometer level. In Europe, "nanotechnology" is characterized as "the direct control of atoms and molecules." The National Science Foundation/ World Technology Evaluation Center study uses a definition that contains both of these meanings: technology that arises from exploitation of novel and improved properties, phenomena and processes specific at the intermediate scale between atoms/molecules and bulk behavior (that is, 0.1 to 100 nm).[33]

Despite such efforts, the term "nano" evolved into a cliché to describe anything small. In China, for instance, it has been used to sell durable goods such as washing machines.[34] If you bought a nanowasher, you'd get a chemical added to the wash water automatically. It was cool, but it wasn't nanotechnology.

Thus, we're entering an era when "nano" becomes ambiguous as a technical term, because it's entered the vernacular in the way that "way cool" or "beam me up" have. It's also a dangerously misused term for investors (as explained in chapter 18). Furthermore, Nano is a forename, harking back as far as the 1700s in Ireland.[35] We may see a revival, when trendy parents, who once might have named their kids something like Dweedle, elect Nano instead.

In view of such confusion, it helps to have a stricter definition to delineate one form of nano from another.

Eric Drexler and the Silicon Valley–based organization he cofounded–the Foresight Institute[36]–use the term "molecular nanotechnology" (MNT) to describe molecular systems that can build devices, including copies of themselves,[37] with atomic precision. Such a definition is critical, because such systems may drive other revolutions in the molecular economy.

Preconditions for *molecular assembly* are:

1. *Positioning:* placing molecules in a predetermined order
2. *Self-Replication:* molecular systems that duplicate themselves
3. *Assembly:* molecular factories that assemble components into other machines

Nano-scale products are possible if any of these conditions are met. For example, many industries are already based on the first precondition alone.

Yet the full potential of a molecular assembler is realized only when the conditions merge. Such an assembler, if built, could drive most of the technologies in our future world. Therefore it bears careful defining in its own right. Drexler et al. summarize it this way:

> Assembler: A general purpose device for molecular manufacturing capable of guiding chemical reactions by positioning molecules.[38]

This still leaves room for interpretation, and we see a lot of that. Confusion is common over what an assembler might actually be. This is because some nano terminologies haven't yet been officially standardized. Some say that assembly involves biological processes such as genetics. Others apply it to chemicals that self-assemble in patterns. These are each partially right. Yet a true assembler has to meet every precondition of positioning, self-replication, and assembly.

In the scientific community, there's agreement that positioning and replication are possible, but serious disagreement about whether assembly is feasible. Many scientists say such machines are impossible, or at least decades away. For example, they point to the practical difficulty of using multiatom-sized pincers to manipulate smaller individual atoms in a confined space. They also worry that it may be impossible to break or resist chemical bonds that bind atoms to each other, or to their pincers.[39] On the other hand, molecular-assembly proponents have countered these arguments with designs that they say obviate the need for such pincers.[40]

At least one company is betting that the proponents are right, by investing millions to build an assembler.[41] Other optimists point to those who said it would take decades to map the human genome, when in fact it took only one. These proponents, many of them accomplished scientists, say it pays to be skeptical of the cynics.

Here's a rough potential parallel. The elapsed time between postulation of the special theory of relativity and invention of the atom bomb was about forty years. If we apply the same time span, starting at Drexler's first technical paper on molecular nanotechnology in 1981, we arrive at the year 2021. Yet there is a lot of room to criticize such a time line. Optimists might say that nuclear fission was achieved some years prior to the bomb's use. Pessimists might say that it took a world war to concentrate the necessary investment of great minds, talent, and money for achieving fission, but at the beginning of the twenty-first century, we don't have or want such an impetus. Optimists might respond by pointing out that the rate of scientific discovery is compounding like interest in a bank account, rather than linearly; therefore, this may bring us to molecular assembly in a shorter time.

We can see some of the pitfalls associated with trying to forecast such a date. Nevertheless, it's clear that the chances for assembly would improve if we faced an imperative on the scale of a great war. We'll discuss such an imperative later, in part 2.

Meanwhile, in the laboratory, there are promising developments to suggest that assembly is practicable. By late 2001, scientists at the National Institute for Materials Science in Nagoya, Japan, created units of nano-scale materials that spontaneously form into groups of wires.[42]

HOW SOON TO MNT PRODUCTS?

James Von Ehr, founder of the first private company that aims to build an assembler, described implications for commercial products this way, borrowing from earlier descriptions by pioneers Eric Drexler, Chris Peterson, and Gayle Pergamit.[43]

> We get a lot of value out of nature rearranging atoms. For example an acorn has molecular machines inside it that rearrange the atoms in dirt, water and air into an oak tree. We can cut that down saw it up and make furniture out of it. It's cheap enough to rearrange those atoms if nature does it with molecular machines. What we're trying to do is a similar approach but with engineering behind design and control of these machines. So if we're able to build these machines to make molecularly precise products as cheaply as wood, that's going to have a lot of economic impact on the world. For example the starting materials we have are nanotubes that are about a ten thousandth the diameter of a hair, but because of its precise atomic structure it's about a hundred times stronger than steel. So you can imagine what our architects and engineers might do if we had a building material that was a hundred times stronger than steel. It's going to revolutionize the building industry.[44]

This description of superstrong materials hints at how we might use molecules to make structures resilient against natural catastrophes. Yet disaster defenses aren't the first nanotools being invented, so let's look at earlier types of applications.

There are many pathways to molecular nanotechnology. Some are used to build machines out of atoms, one by one or in parallel. Others use big machines to build smaller ones until we get to the molecular level.

We mortals don't have to understand how each of these work, and this would cover hundreds of volumes, but it's important to see what such systems do in our economy right now to appreciate how widespread nanotechnology is going to be.

Molecular chemistry is the basis for every advanced pharmaceutical product, industrial coating, and fast computing process. It helps us control the size, reactivity, and electrical properties of nano-scale components. At this scale, everything is so affected by everything else that we can't observe things without our act of observation affecting them, so we need molecular chemistry to predict their interactions and conceive how they might combine or break apart.

Scanning tunneling microscopes can be found in academic or corporate labs worldwide and are one of the only ways to see matter at an atomic level. The tips of these devices are only a few atoms wide. They don't "see" molecules the way our eyes do, but they sense them and relay those signals to other devices that translate them into something we can view (see Figs. 2 and 7). This precision lets them push atoms around. To imagine the scale they operate at compared to us, picture a fifty-ton construction crane picking up a grain of sand at the beach. The operator can't see the grain, let alone grab it. The crane needs tweezers and the operator needs the equivalent of binoculars plus a microscope. This, in turn, requires an array of devices that let the awkward pieces fit together. The crane needs supersensitive controls, otherwise it's sure to push other globs of sand around and lose its target.

Biotechnology is a multibillion-dollar industry that's changing medicine and opening doors to the genetic codes of everything on earth. With these codes we can use nano-scale factories that have been around for billions of years in the form of DNA. Genes constitute the software code that instructs molecular machines known as ribosomes to build proteins. These proteins fold up into hormones, enzymes, and other building tools.[45] If we figure out how they do this, then learn to instruct those genes, we might have a head start on assembling things.

With the top-down approach, *micro-electro-mechanical systems* (MEMS) combine information processing with sensing to help us perceive and control the environment. Examples are sensors that activate air bags in cars and guidance systems in planes or missiles. Just as we use huge presses to stamp out car parts, then use smaller machines to put those parts together in a vehicle, MEMS can be used to make tiny replicas of themselves until they reach the molecular level. That's the theory. In practice, chemical and electrical properties may start to throw physical assembly out of whack. Nonetheless, MEMS are helpful in constructing tools that link the human and nano scale, that is, connecting the crane and grain of sand (see Figs. 6 and 7). This connection isn't just another problem: It may be *the* problem in bringing nanotechnology from the lab to the production line. Michael Roukes, professor of physics at the California Institute of Technology, reflected the view of

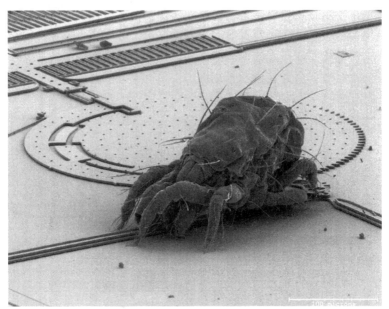

Fig. 6. Spider mite raids a microlock. This is not Godzilla. This is a spider mite (a white fleck to the human eye), hanging out on a microlock for a micro-electro-mechanical system (MEMS). Note the scale in the lower right corner. MEMS are much larger than the nano scale, but they let us move between the nano and human scale. If these were nano scale, the foot of the spider mite would appear as big as a building. (Courtesy Sandia National Laboratories, MEMS and Novel Si Science and Technology Dept., SUMMiT Technologies, www. mems.sandia.gov; mems.sandia.gov/scripts/images.asp)

many scientists when he observed that "the difficulties in communication between the nanoworld and the macroworld represent a central issue in the development of nanotechnology."[46] Chad Mirkin professor of chemistry and director of the Northwestern University Institute for Nanotechnology, concurred with that when he remarked in *Chemical and Engineering News* that "the realization of many of these [nanotechnology] notions is hampered by a fundamental inability to wire up and interface structures on the nanometer-length scale with macroscopically addressable components."[47]

The nano-macro link may, in the final analysis, require a marriage of atomic microscopy, MEMS, and computerized virtual reality. Some of this work is already underway (see Fig. 7).

Some related methods bear mentioning. *Photolithography* is used to fabricate microchips. *Soft lithography* uses printing without photochemistry to manufacture transistors on curved rather than flat surfaces.[48] Both can etch nano-scale patterns. Yet both are thought to be too slow for molecular assembly.

Thus, nanotechnology is not an isolated field that needs decades to gain acceptance. It's a logical progression from what we already have. It's evolving at a speed that is already catching many institutions by surprise. While scientists caution against "overexuberance," the technology is more advanced than some are ready to admit. Potential rewards are huge, so while scientists share information about discoveries, many military and commercial applications are kept under wraps, as explained in chapter 21.

We see from these methods that we don't need the atomic precision of a molecular assembler for the molecular revolution to begin. We require only tools such as MEMS, for example, that still let us make miniscule machines, or we might use self-assembling proteins to do the work for us.

When combined with computing, such methods produce results that are driving a molecular revolution already.

DESKTOP MANUFACTURING

Digital fabrication may be the most sweeping industrial transformation that nanotechnology and its associated technologies are ushering in. Known also as *rapid prototyping* and *desktop manufacturing*, it may transform industry by merging low cost with high precision and efficiency. It may also reintroduce home-based manufacturing, but in a radical new way.

Digital fabricators, or "factories in a box" have been around for some time. They make individualized prototypes such as models of a patient's bone structure for surgery (see Fig. 5), or injection molds for toys. Right now these fabricators use three basic techniques:[49]

> *Cutting:* Starting with a block of something and carving it into the desired form. The earliest versions go back more than a century, to milling and lathes controlled by a master form. This produces a lot of waste material, so its not the most efficient method.
> *Joining:* Adding material, layer by layer or particle by particle. This became commercially feasible in the 1990s, when automatic controllers and applicators gained enough accuracy.
> *Pressing:* Bending something into shape without adding or subtracting materials. This has also been used for centuries, but digitalized instructions and new materials make it possible to create more intricate products.

Right now, the cutting method dominates–it's a multibillion-dollar industry compared to only a few hundred million dollars for joining, and

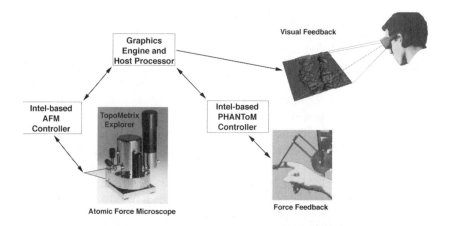

Fig. 7. Manipulating something we can't see. Scanning-probe microscopes (SPMs) allow the investigation and manipulation of surfaces down to the atomic scale. The nanoManipulator (nM) system provides an improved, natural interface to SPMs. The nM couples the microscope to a virtual-reality interface that gives the scientist virtual telepresence on the surface, scaled by a factor of about a million to one.

A real-time view of the nano world. Scientist uses the nanoWorkbench to examine carbon nanotubes. (Illustration: Department of Computer Science, University of North Carolina at Chapel Hill; photo: Larry Ketchum [photographer] and Department of Computer Science, University of North Carolina at Chapel Hill; www.cs.unc.edu/Research/nano/)

not much for pressing. In 2001 fabricators went for between forty-five thousand and eight hundred thousand dollars.[50]

These digitalized minifactories are already working. Boston-based Z Corporation[51] has produced a three-dimensional printer that sprays material–for example, corn starch or plastic powder–from an ink-jet printer head. This has produced models of air-conditioning systems and human bones[52] (see Fig. 5). The bone replicas are based on scans of patients. These let surgeons precisely map reconstructive surgery without plunging into the patient first.

Personalized fabrication on demand has entered the marketplace as rapid prototyping. Hearing aids that match your ear canal, steering wheels for race cars, and experimental bicycle frames are made this way (see Fig. 8).[53]

The process works like this: designing and scanning software are used to generate a three-dimensional image of a product. The computer slices the image into nanometer-thin cross sections, then instructs a digital fabricator to construct the product, one layer at a time. Layers are made from powdered ceramic, liquid polymer, or sheets of paper, plastic, or metal.

Some designs are already deliverable via the Internet, so a fabricator in Detroit could produce a product just designed in Bangalore, India.[54]

Here's who's already applying it:[55]

∞ Automotive manufacturers use it for prototypes and production molds.
∞ Prosthetics manufacturers use it for personalized components in artificial limbs.
∞ Architects use it for building scale models.
∞ Hollywood prop makers, artists, and sculptors use it to produce their work.
∞ Scientists use it for physically representing complex three-dimensional structures.

Nano-scale tools may dramatically lower costs and accelerate the speed of such applications, first via miniaturized automation of manufacturing, then via self-replicating components. Such methods may replace whole factories.

Technologists are already forecasting such midterm possibilities:

Brock Hinzmann of SRI Consulting Business Intelligence says, "Within 10 years, these machines will cost less, in real dollars, than color 2-D copiers did 10 years ago."[56]

Marshall Burns, founder of the Los Angeles-based Ennex Corporation, which optimizes these machines, says, "We're talking about transforming the Internet from a medium of communication to a medium for delivering manufactured goods."[57]

Ping Fu, CEO and president of rapid-prototyping software developer Raindrop Geomagic in North Carolina, adds: "With sufficient demand, there's no technological impediment to creating any material."[58]

Such "materials" may include a home-manufactured computer. The Defense Advanced Research Projects Agency (DARPA) has commissioned researchers at Rice University to develop a *nanocell* that may be the foundation for computers many times more powerful than those available in 2001.[59] Nanocells are collections of molecules that could act as a logic circuit, which constitutes the basis for computing. If completed in the envisaged time frame, they would render Moore's law—which states that computing chip power doubles every eighteen months—obsolete. It would then

Fig. 8. Printable bicycles. Bicycles like this one were made by MIT students on a waterjet cutting machine or laser cutter. Parts for the frame were cut from flat-sheet plywood or polycarbonate. The twenty to thirty parts were snapped together like Legos in about an hour. Wheels and bearings were bought. These machines form part of the general spectrum of "rapid prototyping" tools, which are increasingly becoming "personal fabricators" as they develop more capacity and material complexity. Such fabrication could soon be done in the home. (www.media.mit.edu/~saul/ iapweb/parts.htm; courtesy of Joseph Jacobson, MIT Media Lab)

be feasible to fit a supercomputer as a flexible microsurface into a pair of jeans. Think of it: Soon your pants may be smarter than you.

This may also transform the way chips are manufactured. Right now, for example, every factory costs billions of dollars and needs two weeks of constant processing to make each chip. But what if these could be manufactured the way we print on paper at home? No more clean rooms, no more acres of facilities, and no more heavy capital investments in factories.

Is it possible? Yes. When? Some of the technology is available now. MIT Media Lab pioneers Neil Gershenfeld and Joseph Jacobson, along with Lucent Technologies's Bell Labs, have manufactured transistors using desktop printing and nano-scale particles (see Fig. 9). This is different from molecular manufacturing envisioned by Eric Drexler, because big machines make small parts, instead of the other way around. Yet it gives a practical idea of how personalized fabrication might work, per this description in *MIT Technology Review*:

> A chip fab on every desktop could bring about the day when individuals download the architecture of integrated circuits the way they download software today. It could, in short, transform hardware manufacturing much the way the "open-source" movement has changed how software is written. . . . Jacobson contends printed logic could give rise to an open-source hardware movement where chips are custom-designed via the Internet and printed by the consumer in about the same time it takes to print out a Web page. You could, says Jacobson, "download the chip design from the Web, tie in some modifications from some guy in India, and boom—out comes the device.[60]

Jacobson's MIT team aims to combine cheap desktop manufacturing with another invention—*electronic paper*—that lets a single flexible sheet act like a computer screen but look like paper. So far, electronic paper comes in a few varieties. One type was conceptualized by Xerox Corporation researcher Nicholas K. Sheridon more than twenty years ago, then developed over the next few decades by Xerox and spun off in the year 2000 into a company known as Gyricon Media.[61] The other was developed by MIT spin-off E-Ink,[62] in cooperation with Lucent Technologies (see Fig. 10). The technology was upgraded in 2001 to allow it to display in color, in collaboration with Japan's Toppan Printing Company.[63]

Both E-Ink and Gyricon technologies have been used in store sign displays. For example, in an electronic store display, a sheet of oil-bead-soaked plastic film is laminated to a layer of circuitry. This forms a pattern of pixels controlled by a computer display driver. Microcapsules are suspended in a liquid, allowing them to be printed with conventional screen printing processes onto virtually any surface, including glass, plastic, fabric, and paper (see Fig. 11).[64]

Just like paper, electronic paper is user-friendly, thin, and—in some versions—flexible. But it's also a rewritable computer display. This may save hundreds of millions of dollars for stores that normally have to spend time and money changing their displays daily. Other potential applications include:

- electronic newspapers that offer breaking news as the paper is read;
- magazines that use animated images and refresh their text;
- books that let readers mark up pages with pen or voice command for later recall; and
- wall-sized electronic whiteboards, billboards, and portable fold-ups.

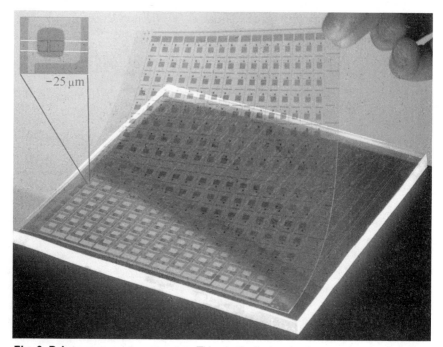

Fig. 9. Print your next computer. This plastic display circuit was printed, rather than manufactured in a factory. The circuit depicted in the photo is contacting, on its lower edge, a rubber stamp that was used to print the finest features in the circuit. The inset on the top left shows a magnification of a transistor in the array. (Courtesy of Bell Labs, Lucent Technologies)

The second type is conventional paper with thousands of sensor dots. When written on with an electronic pen, the sensors convert handwriting into computerized commands. This lets users send e-mail and faxes. Just write on the paper and check the "send" box, and your handwritten message is sent. Advantages include the high-contrast legibility of regular paper, low cost, and an ability to accept conventional ink from printers. Ericsson, a cell phone manufacturer, produces the pen. The paper is produced by a Swedish startup, Anoto, owned by Christer Fåhraeus.[65]

Both types of application lead to flexible computers that weigh a few ounces, roll up, and contain thousands of books. These may be "killer applications": all-in-one reading devices that go everywhere, fold up, and have no appreciable weight.

In the not-so-distant future, a digital fabricator may manufacture these.

Opinions about how desktop manufacturing might work are still divided. Marshall Burns of Ennex, for example, foresees homes with automated canisters delivering raw materials such as photopolymers and other

Fig. 10. How this book may soon be delivered. Flexible electronic paper. These very thin (< 1mm) devices have the look and feel of conventional paper (i.e., good contrast, angle-independent viewing, mechanically flexible, light-weight, rugged, etc.), but they are reconfigurable like a computer display. They run on small batteries, and require about one-thousandth the power of a conventional liquid crystal display. (Courtesy of Bell Labs, Lucent Technologies)

How e-ink works. Prototype electronic paper display is a thin layer of transparent plastic in which millions of small beads, somewhat like toner particles, are randomly dispersed. The beads contain white particles suspended in black dye. An electrical current moves the particles to the top or bottom of the capsules, changing the display's image. The image will persist until new voltage patterns are applied. (Courtesy of Bell Labs, Lucent Technologies)

Exploded Layers of Rubber Stamped, Organic Active Matrix Backplane Circuit

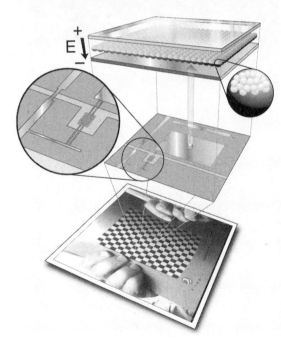

Fig. 11. Instantly reconfigurable signs. These signs use "SmartPaper" that changes text and graphics instantly when a computerized electronic pattern is sent. The MaestroSign Systems are already used for store displays. (Courtesy of Xerox Palo Alto Research Center (PARC) and Gyricon Media, Inc.; www.parc.xerox.com/dhl/projects/gyricon/)

thermosetting resins; powders and pellets of thermoplastics, metals, and ceramics; and spools of cotton, graphite, boron, and other fibers via pneumatic tube to every home fabricator. This presents the specter of cities laced with spiderweb networks carrying hundreds of different chemicals, just as we have gas and other types of pipelines.

Yet with nanotechnology there may be a more elegant solution, by breaking chemicals into their basic components and reassembling them in a desktop fabricator. Eric Drexler has described ways to use four feedstock molecules–hydrogen, oxygen, nitrogen, and carbon–to produce generic materials as a basis for most consumer products.[66] Such molecules are available everywhere. For molecular assembly, feedstock could be taken from the local environment instead of being transported to the place of manufacture. That is still years down the road. Right now, it's more likely that feedstock would be remotely delivered to household fabricators. (For a discussion of environmental implications of feedstock, see chaps. 5 and 14.)

NANOCOMPUTING

The foundations for such technologies are determined by computing.

When things get much smaller, bigger, or faster, their impacts go from incremental to paradigm shift. When we graduated from the horse to the car, we still used roads, but the change in the rate of speed was enough to transform society. The same applies to computing and communications. When chip speeds went from 5 MHz to 500 MHz and modem speeds from 1 to 100 kb per second they didn't just go faster, they gave consumers cheap access to word processing, e-mail, Web pages, and photos that fit on a disk. In the near future, when chip speeds go from 2 to 20 Gigaherz with a hundredfold reduction in size and price, we'll probably see these effects: desktop monitors turn into wall-sized displays; jerky, screen-sized webcam images turn into three-dimensional, holographic, life-sized meeting rooms; and accessibility extend to the Kalahari Bushmen.

This acceleration in computing is the most profound step. It opens the door to a new world: building superstrong materials; eliminating heart disease by using microscopic robots to clean arteries; cleaning up the ecology with sensors that ferret out toxins; building robots that learn; and surviving natural disasters by detecting them in advance.

One of the first steps to enable megacomputing is already under development for the marketplace. *Dynamic random access memory* (DRAM) is the memory in your PC that helps it run programs and data seemingly simultaneously. There is already a mass market for such devices, because conventional DRAM is found in every computer. Nano-scale DRAM is among the least complex of computer components to build. The product will be a combination of existing silicon technology and molecular technology, as a first step toward more complex equipment.

Dan Hutchison, president of VLSI Research, a Silicon Valley microchip industry research firm, observes, "If you can pull it off and make memory devices using nanotechnology, you would blow away anything you could do with hard drive, CD drives or anything else out there."[67]

Thus, molecular computing may be on the threshold of commercial application.

CHAPTER THREE

WHAT COMES FIRST AND WHAT DOES IT MEAN FOR EACH OF US?

Back in 1982, noted science fiction writer Isaac Asimov predicted that by 2000, "under global sponsorship, the construction of solar power stations in orbit about the Earth will have begun."[1] There are a few ironies about his prediction. First, the *Wall Street Journal* cited it as an example of misguided predictions for the turn of the millennium.[2] Paradoxically, they were wrong to call him wrong. By the end of the old millennium, construction of a solar-powered station had in fact begun, in the form of the solar generator for the International Space Station. Why was this overlooked? Perhaps the researcher for the article did not perceive this as a power station. She may not have seen how the space station's solar power plant had much to do with the average person here on Earth.

For those of us who try to forecast future events, this is a sobering lesson. Even when you're right you can be labeled as wrong, *especially if people don't see what a new achievement means for them.* In terms of predicting developments, it pays to be mindful of when they might start affecting our daily lives.

Thus, when we try to forecast which megachanges might come first, it helps to identify a turning point that would undoubtedly have big impacts. Some say that invention of the molecular assembler might be such a pivotal event. The emergence of autonomous artificial intelligence is also regarded as a turning point. One might hasten arrival of the other.

Yet there is still controversy over *when* or *if* these might occur. Optimists say later this decade. Pessimists say the last half of the century. Skeptics say probably never. Here, we focus more on what might influence the time frame, rather than trying to predict a specific date.

WHAT AFFECTS THE TIME FRAME?

Predicting absolute dates has proven fatal for scientific reputations. Social and economic factors may push such dates forward by five years, or back by fifty. For example, by now we could have put a human on Mars, but the political impetus such as cold war competition wasn't there. Most homes in sunny climates could have cost-effective solar heating and electricity, but resistance by, and subsidies to, the fossil fuel industry have thrown barriers in the way of widespread adoption.[3] This is the most blatant, but not the only, example of vested interests resisting new technologies. It's a big problem. Every year, governments give hundreds of billions of dollars in taxpayer subsidies to obsolete industries to keep them on life support. Such subsidies have been identified as one of the main barriers to the entry of new technologies.[4]

Furthermore, although the rate of technological change is exponential, the rate at which changes find their ways out of the lab and into our daily lives depends on investment and consumer demand. The burst technology bubble of 2000 had a dampening effect on, for example, investment into bringing broadband wireless to the marketplace. In the rush to take advantage of the "new economy," companies spent more than $35 billion building 100 million miles of fiber-optic networks. In 2001 these had about 95 percent unused capacities.[5] This was partially due to "old economy" connections that ran "the last mile" to households. These were slow to catch up, with the result that too few users could take advantage of so much broadband. Worse yet for those companies, new compression technology began to stuff many more signals down the same pipe, thus drastically reducing the need for more fiber-optic cable. Not to be outdone in such overspending, European companies spent an eye-popping $125 *billion* bidding for government licenses on "G3" wireless–the so-called third generation broadband wireless. This may prove to be overpriced,[6] depending on how the markets and technologies develop. In 2001 share prices of European telecommunications giants were slashed to bargain-basement prices, due to overextension of their finances on the bidding. Moreover, the "new market" index–NEMAX, Germany's equivalent of the NASDAQ–dropped a stunning 85 percent in the techmania hangover. These types of collapses can set back technology investment for a decade or more. Furthermore, patent infringement lawsuits can keep key technologies tied up indefinitely. This is discussed later, in chapter 21. Or the military may decide not to allow some technologies into the public domain, as occurred, for example, with night vision technology during the Cold War. Of equal concern are political and religious limitations that might be imposed on, for example, cloning. Additionally, if the genetics industry makes a blunder that results in a high-profile disaster, it may set some research back decades.

Still another wild card is the disruptive nature of these technologies. For example, in 2001 the company that had brought us instant pictures—Polaroid—went bankrupt.[7] Who needed a camera to spit out a print when pixels were just as good? The digital-photo revolution dealt the company a fatal blow. Moreover, Excite@home, a new-economy company that—unlike many other such startups—was producing hundreds of millions in revenues from high-speed Internet services, went belly-up the same year, due partially to miscalculations over how much it might cost to support the technology. This implosion caused e-mail chaos for broadband users across North America, interrupting what had evolved into an essential service for those who depended on it for business.[8] That's disruptive technology at work. Expect to see more. Similarly, a huge personal-computer company might rise to dominance, then go out of business in a relatively short time when the PC is superceded by voice recognition computers in every product we use. Our current lumbering PC operating systems might be superceded by supercompact software that self-improves. More disturbing than the meteoric rise and fall of companies is mass displacement of human skills and financial markets. An example is the fall of Enron—a largely Internet-based energy-trading company that rose to be America's seventh largest corporation, then collapsed. Although its woes were not entirely technology related, the technological ability to move money and contracts at lightning-fast speed worldwide was a contributing cause.[9] In the future, this type of disruption could cause serious economic, social, and political problems that lead to a slowing of technological advances. In 2001, for instance, economic leaders at the annual Davos Forum in Switzerland focused on public rejection of technologies that are either too disruptive, or fail to do what they promise and wreak havoc instead.[10]

Finally, we come up against the biggest wild card of all—natural disasters. If an earthquake flattens a world financial capital such as New York, Los Angeles, or Tokyo, or a category 5 hurricane cripples the economy of the East Coast of the United States, this can have serious multiplier effects. In the worst cases, the world technology infrastructure might be set back by a giant volcanic eruption that ejects enough dust into the atmosphere to alter climate patterns. Such infrequent but catastrophic potentials are discussed later, in part 2.

Looking past these gloomy barriers to the brighter side: if cures for AIDS, malaria, or most cancers are found, or if organ replacement becomes cheap and more humane, we may see acceleration of technology, as potential bene-ficiaries call for increased support.[11] (This is discussed in depth in chapter 23.) Moreover, even during the Great Depression, we saw that in the worst pos-sible investment climate, new technologies such as radio continued apace.

Looking at such divergent variables, we see that *it's impossible to consider only the scientific side when we try to forecast what science means for us in the twenty-first century.* While progress for technology may seem inevitable, it's by no means guaranteed.

To see what might be possible though, let's—for this chapter at least—consider what could occur if the present pace of development continues.

PRE-ASSEMBLER YEARS— THE FIRST NEAR-TERM MEGACHANGES

Thousands of articles already explain what is "just around the corner" in military and corporate research labs. The marketing of such inventions is relatively easy to predict because, rather than being just around the corner, they have already been invented and are in the process of being introduced to consumers. The real trick is to see how the broader kaleidoscope is going to come together, how soon it may come to markets, and how cheap it may be. Most products heading down the pipe come from the convergence of technologies, rather than isolated inventions. For this convergence to reach the marketplace, cost and availability are just as important as new inven-tions—although each tends to reinforce the other. For instance, television was invented in the 1930s, but it didn't cause a societal revolution until the 1950s, when low-cost mass manufacturing led to its introduction in most American homes. We need only look at the ruinous expense of AIDS drugs, or the still-exorbitant cost of real-time broadband videoconferencing, to see that affordability affects wide-scale adoption. Therefore, the real molecular revolution of the next decades may be to make things broadly available more cheaply, and more quickly.

This potential is apparent in the medical field of *calcification* that relates to a spectrum of diseases. Calcification is the accumulation of calcium in parts of the body where it shouldn't be. The vast majority of calcium that we consume and produce internally ends up in the right place: our bones and teeth. The 1 percent that seems to go haywire ends up most everywhere else, including our arteries, eyes, brain, tendons, breasts, and organs. Due to its toughness, calcification has resisted conventional treatments such as antibiotics, chemotherapy, and other drugs. Why does a mineral that normally is so essential for our bones try to kill us? Researchers have been perplexed for years, searching to find a cause.

Then, in the early 1990s, researchers Neva Ciftcioglu and Olavi Kajander used electron microscopes to zero in on a new class of bacteria hundreds of times smaller than conventionally known bacteria.[12] Like many other breakthroughs, it happened through a combination of new technology and accident. The researchers were perplexed by contamination of mammalian cell cultures that they'd been studying. By combining standard biology with molecular detection techniques, they found an apparent culprit: *nanobacteria*. This is shorthand for *nanobacterium sanguineum*, or blood nanobacteria. They range from twenty to two hundred nanometers in size and are the smallest known self-replicating bacteria.

Other researchers around the world were disbelieving. Few had thought that bacteria could be so tiny, or that they could contribute to heart disease.[13] The next shock came when nanobacteria were identified in disease-causing deposits such as atherosclerotic plaque, coronary artery plaque, kidney stones, cataracts, and dental plaque.[14]

Then, in 1999, yet another surprise: researchers developed and applied a drug cocktail known as NanobacTX that appears to break through the bacteria's calcium and biofilm defenses. Nanobaclabs, a Tampa, Florida-based start-up company, developed a compound drug formula to kill the bacteria, then began treating patients. By early 2002, patients were showing apparent improvements in coronary artery disease and other ailments,[15] although tests to confirm such results were still underway.

The trail that leads to confirming this is far from straightforward. The symbiotic relationship between these bacteria and our bodies is still not fully understood. The reasons why they have been able to protect themselves against high temperatures, drugs, radiation, and other treatments are only now becoming apparent. Our knowledge is not yet clear enough to confirm that the nanobes won't develop immunity to treatments that include extensive doses of antibiotics. How do we get infected? Do the transmission vectors include saliva, sperm, or perhaps vaccines? Do these tiny microbes contribute to cancers? Are they responsible for new diseases that have sprung up, and if so, why now? Are some of these bacteria good for us?

The uncertainties are substantial, but do not detract from the potential significance. Have researchers found the Rosetta stone of bacterial infection—a vehicle that may open the door to inexpensive cures for many diseases? If so, these would convincingly demonstrate the ability of molecular technologies to produce broadly based benefits. Or is it a dead end that leads to high hopes, then disappointment? The nanobacteria chase is afoot.

Such technologies are also potentially disruptive, because they may displace a host of drugs and procedures that treat symptoms rather than the causes of disease. Drug companies and clinics that today generate billions in revenues from the treatment of symptoms may have to scramble for new revenues. The upward spiral of medical costs for such diseases may be arrested. The percentage of our population that is elderly may accelerate as life is extended. If early, successful treatments for nanobacterial infections prove sustainable, such changes could begin forthwith. Therefore, the broad impacts of at least one molecular technology may be with us now.

LET'S MEET ON THE BEACH

Here's an example of something that's technologically feasible today but not available to most of us. When it does arrive, it will transform the way we go to school, shop, work, commute, socialize, have sex, and see ourselves.

Imagine stepping into a passport photo booth. Instead of a curtain, this one has a door that closes seamlessly. The interior is white, with a curvature like the inside of an eggshell. No apparent camera, flash, or other ancient low-tech equipment. You take out your smart card and wave it next to a tiny black patch on the wall. This is for passcoded billing and data access. You put on your eye visor and arm-length gloves. The visor resembles sunglasses, except it has a display screen across the lens. You say, "My nine o'clock conference at Miki Soto's office, please." Your eggshell becomes a meeting room by projecting wraparound images via the liquid crystals embedded in its interior. The walls also conceal cameras and microphones to transmit your movements and voice. Suddenly, seven coworkers—each in a different country—appear, as if they're sitting together at a virtual table. You "shake hands" with those you've not seen in weeks. Their palms feel lifelike on yours, due to thousands of sensors embedded in your glove. A new product—a running shoe—has been placed on the virtual table. You "pick up" the shoe to feel its texture. It's a nice design, but a bit too rubbery for your liking. The "weight" of the shoe is created by magnets in the walls of your compartment that pull down on your gloves to simulate gravity. Your boss asks for your presentation about the environmental impact of new materials being used. You've prepared this at home, and it's easily accessible because your smart card opens those files remotely, just as it did the Web address for your virtual conference. To everyone's right in the virtual meeting room—the right side of your booth, in your case—an audiovisual presentation shows the volatile organic compound tests that confirm reduced gaseous emissions from the manufacturing process. These constitute lower risks to factory workers, and that's a political plus, because the shoe manufacturers are under pressure to clean up their act. Moreover, several assembly steps have been eliminated. These are displayed by switching to the shop floor, real-time in Jakarta, then overlaying production changes with a graphics program. After everyone has seen this, some of your colleagues give their presentations. These are automatically downloaded to your home files for reference. The new shoe is discussed, modifications are suggested, everyone's files are updated, and the meeting ends. The thought crosses your mind that some cybersex with your boyfriend in Bangalore might be nice, but that would involve putting on your virtual-reality suit, so you've decided to leave it till later. You exit the booth, but not into an office or store. Instead, hot sand tickles your bare feet. This booth is located at the beach.

Although such a virtual-reality booth exists only in the future, the experience is feasible today. A mix of technologies provide geometric perspective, color intensity, stereoscopy, focus, and touch. Airline pilots and astronauts have been using some of these in flight simulators for years. The problem is, they cost a few million dollars to set up. The smart card—the cheapest part— is from Sun Microsystems. Already, many Sun employees use smart cards to access their files from terminals everywhere in the world.[16] One variant of the virtual-reality technology has been developed at the University of North Carolina. It links computer graphics and virtual-reality displays with images of the real world. This technique is also referred to as *augmented reality*. Data are delivered by ultrasound that generates real-time images from sound waves bouncing off tissues inside a patient. High-resolution slices are combined with a live video image from a camera focused on the patient. The combined data are seen by the operator in a head-mounted display.[17] At the University of Illinois at Chicago Medical Center, surgical residents explore three-dimensional models of the liver, standing in front of a twenty-four-square-foot-screen called an *Immersadesk*. Viewers wear special glasses to simulate immersion into oversized, full-color, three-dimensional images of an anatomical structure. They use an electronic wand to point to areas, change the orientation, bring it farther or nearer, or add and subtract structures. The eyeglasses track each viewer's movements and orient the model in relation to the viewer.[18] At the National Center for Supercomputing, an advanced relative of this, known as the CAVE, immerses viewers in a total environment surrounded by three walls and a floor. It combines supercomputers, graphics, sound, and tracking to give participants the sense of "being there."[19]

Toshiba, Hitachi, and Philips, to name a few, build liquid crystal displays that might cover the interior of our fictional booth. Broadband communications come from services supplied by Sprint, AT&T, and dozens of other telecom companies, while corporations such as Cisco and Nortel provide the switching equipment. The virtual-reality gloves and glasses are from video game companies such as Sega or Sony, although the commercial versions that we use in our homes are less sophisticated than those used in media labs.

Each of these technologies is available now. They need to be integrated, shrunk, and reduced in cost. In the future, flexible video displays and memory chips may roll off room-sized assembly lines, instead of from football-field-sized factories. Internet servers comparable in size to their 2001 ancestors may have thousands of times more capacity. Broadband satellites could have millions of video channels. Pin-sized video cameras might cost less than toothpaste. Cost and size will shrink, while speed multiplies.

The net outcome might look like this: After years of unfulfilled promises, commuting to the office may finally be outdated. The red-eye flight to catch a morning meeting on the other side of the continent will be obsolete. The monthly transpacific flight to keep an eye on the shop floor will be passé. Jet lag will be gone, but not leisure travel, and that's a story for later. The physical workplace is evaporating. Long live the virtual workplace.

Oh, and by the way, it's also going to be working in Shanghai, Delhi, and Sao Paulo, so it won't just be a toy for rich nations anymore.

CONTRADICTORY SNAPSHOTS OF OUR FUTURE

How might the near-term confluence of nanocomputing, robotics, and genetics turn out? No one knows for sure. Yet we can and must draw pictures, to see what we may face. Some of these pictures may seem paradoxical, but that's to be expected. Such contradictions have always been a defining part of our history and our humanity. In the twentieth century, the world experienced great wealth creation alongside broad impoverishment, birth control alongside overpopulation, and life extension alongside genocide. In the twenty-first century, paradoxes may be more extreme, or they may be moderated. We may still see rich, long-lived individuals living next to short-lived, impoverished masses. Or we may see broad-based wealth and health. It depends on how our society copes with the technologies. We may see enhanced intelligence for some, alongside drug-induced stupors for others. Among the positive contrasts, we may see the needs of the many coexisting more comfortably with the needs of the individual. We may see empowerment of individuals to fulfill their aspirations, along with the lifting of large populations from poverty. On the other hand, these differences may generate serious conflicts if we don't bring everyone along in this adventure.

With such contrasts in mind, here is an eclectic sampling of the often contradictory developments that may characterize our near-term molecular future:

∞ Cures for heart disease, malaria, and dengue fever are found. Child mortality plummets along with worker absenteeism, especially in the tropics. Gross domestic products of tropical economies shoot up, but so do population pressures, as reduced child mortality outstrips wealth-induced birth control.

∞ The virtual office becomes reality, allowing billions to work from any location.

∞ Virtual sex seems more real. The religious right tries to outlaw it, but as sexually transmitted disease and teen pregnancy rates plummet, health authorities endorse it.

∞ Room-sized TV screens are commonplace, but teenagers spurn them in favor of bodysuits that give them virtual reality up close and personal.

∞ Every household and office appliance has voice activation, but as with the videophone in earlier decades, many of us choose not to use it.

∞ Instant written translation becomes commonplace—still quirky, but enough for most people to communicate in most languages.

∞ As nano coatings and gears eliminate most lubricants, motors shrink while their energy efficiency multiplies. As a result, costs drop. Appliances such as washers and refrigerators become throwaway commodities instead of "durable goods."

∞ Fuel cells replace home heating furnaces and air conditioners. Photovoltaic and energy storage costs drop. Homes start to go off the electricity grid. Big oil companies and utilities find their revenues dropping, although more slowly than some would think. This is due to an energy paradox: As energy efficiencies spur more economic development, net energy use goes up rather than down. Thus, the "clean energy" solar economy runs parallel to the fossil fuel economy for a while, to meet growing energy demands.

∞ Oil and gas, despite their tenacity as fuels, decline as a percentage of total energy use. Attention to Middle East conflicts also falls off, due to less dependence on oil from the region. America's military deserts the Gulf states. Many Arab ruling regimes collapse.

∞ Half a billion households that still had no electricity in 2001 receive it when nanotechnology makes renewable energy sources cheap and easy.

∞ Nevertheless, the rich-poor gulf widens. Wealth grows among a privileged few as they use information technologies to accumulate money and power.

∞ A terrorist releases a virus that shuts down most internet servers in America and Europe. The open-source software movement uses the event to push for an end to virus-prone proprietary operating systems. Congress introduces a bill requiring makers of operating-systems software to open their source codes.

∞ Most of us live longer, but epidemics cut life spans for those unfortunate enough to be among first victims of exotic disease strains until a cure is found. Longevity is interspersed with rapid, unexpected death.

∞ Pharmaceutical companies become the world's largest enterprises, as demands from an aging population spark a genetics revolution.

∞ China and a group of international nonprofit organizations release open-source software for genetic therapies. Pharmaceutical stocks plummet temporarily. A struggle erupts over international treaties to regulate genetic software codes.

∞ Organ replacement via synthetics and cloning become commonplace, first among the rich, then the middle class.

∞ As many of us stay vigorous into our old age, more managers resist retiring from top spots, and a generational conflict starts between young and old over who rules.

∞ Undetectable designer drugs drive teen drug abuse to a rampant level, but no one notices because most other age groups are doing it, too.

∞ Robots see, hear, smell, feel, taste, and tell us about it.

∞ A computer spontaneously tells a National Security Agency programmer to get a life.

∞ None of this happens because we get hit by a small near-Earth object that we didn't spend enough money looking for, despite the urging by astronomers who had been tracking such projectiles. Innovation slows to a stall as we spend decades rebuilding. (We'll discuss how to avert this possibility in later chapters.)

Here are near-term developments that may help each of these scenarios—except the last—to materialize.

Energy

One of the most economically and politically significant short-term benefits of molecular manufacturing may be an overhaul of the energy production infrastructure worldwide. The first catalyst is *photovoltaics*, but it's far from the only one.

Since the 1960s, environmentalists have dreamed of the day when such cells could be cheap and efficient enough to replace fossil fuels. Yet it's been a frustrating road. Besides encountering decades of resistance from a fossil fuel industry that faces revenue loss, the technology still faces cost, efficiency, and flexibility barriers. Some environmentalists blame the energy industry, but for whatever reasons, a gallon of gas still packs more punch than a gallon of solar cells in the equivalent space. Solar cells require a lot of surface area to convert energy arriving from a star 93 million miles away. Energy that hits the earth requires fields of cells to deliver the same power in the same short burst as a tank of gas. That's because millions of years ago nature concentrated solar power in the form of petroleum. We're drawing on that solar bank deposit. Yet solar energy that hits the earth now is in the checking account, and although it's a huge account, it still has qualities more diffuse than the savings deposit of petroleum. Batteries can help to convert the checking account to savings, but they still have weight and other shortcomings.

For these reasons, oil industry experts still claim solar energy isn't ready for the big time. This is wishful thinking on their part.

Richard Feynman pointed out decades ago that a big advantage of nanosurfaces is their surface; a lot of surface.[20] This lets heat escape efficiently and reduces friction. It lets heat and light enter more efficiently. Molecules aren't two-dimensionally flat, they're three-dimensional, with a larger surface area. When the absorption and conversion parts of a solar cell get small enough to exploit three-dimensional surfaces, they receive light energy across a greater surface and convert it more efficiently. Particles of titanium dioxide are used for this in nanocrystalline cells. They'll be cheap due to low-purity materials and low-cost processes used for their manufacture.[21]

Furthermore, the self-assembling properties of chemicals at the nano-scale level are already beginning to pay off. In 2001 scientists combined a liquid crystal with an organic dye and found that ingredients arrange themselves into a two-layered film that promotes electron flow. This cheap self-assembly may help to overcome the high cost of photovoltaic cell manufacture.[22]

It's now also possible to convert waste heat into photovoltaic electricity. This is known as *thermal photovoltaics*. When a thermal electricity plant or home furnace generates heat, a vast amount is generated but wasted in the invisible infrared light spectrum. Solar cells can capture this. The sun also produces infrared light, but the advantage of thermal cells in power plants is that they're one inch from the source instead of tens of millions of miles. This gives the cells an energy-intensive environment from which to convert power (see Fig. 12).

Molecularly constructed photovoltaic cells—thermal or otherwise—will also be applicable to irregular surfaces. They may be painted on to everything from the inside of a furnace to the roof and walls of a house. Such cells could be combined with logic circuits to create photo-optical computers that carry their own power source with them. Most consumer electronics could go off the grid. Homes and offices would be free from an outside electricity source. Monthly energy bills could drop to near zero. Use of fossil fuel and its destructive impacts could be dramatically reduced. The first big fossil-fuel-based source to die may be thermal electricity generation for homes and offices, because each home and office will have its own stable power source.

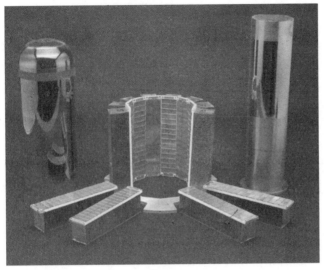

Fig. 12. Solar power without the "solar." JX Crystals manufactures photovoltaic cells that use infrared radiation from heat emitters, rather than the visible light from the sun. The foundation is gallium antimonide (GaSb) photovoltaic cells. Much of the energy generated from hot objects is in the infrared spectrum. GaSb cells capture it and convert it to electricity. JX Crystals fabricate Midnight Sun cogenerators in the 100-watt to 500-watt range. The photo shows the 500-watt system components: PCA (photovoltaic converter array) partly disassembled in the middle, with a filter-coated quartz tube on the left and a tungsten foil-wrapped kanthal emitter on the right. (Courtesy of JX Crystals Inc.; www.jxcrystals.com/)

Stoke the stove and power the TV at the same time. The Midnight Sun Heating Stove: a simple propane-fired heating stove that puts out 25,000 Btu/hr of heat and simultaneously generates 100 watts of electricity. GaSb cells are lined up along a special emitter in the stove's burner, so the electricity to charge batteries is "free," from infrared energy collected by the cells. The unit is a natural complement to solar systems, which are typically inoperable when the heating stove is most likely to be used: at night or under cold, overcast conditions. (Courtesy of JX Crystals Inc.; www.jxcrystals. com/profile.htm)

Transportation and heavy industry are experiencing their own molecular revolution. Fuel cells that don't use internal combustion have been around since the early days of space exploration. Today they're entering into commercial use for power plants. They're still too heavy and expensive for broad adoption, but molecular manufacturing may dramatically decrease weight while increasing efficiency. Once this happens, every factory or truck could use fuel cells cheaply to generate power. They could also use built-in fuel conversion from onboard sources, such as hydrogen from water, to eliminate fuel deliveries. This would signal the end of our dependence on centralized systems for high-energy applications. It's the oil industry's worst nightmare, and an example of a "disruptive" technology.

Most of the world's military, political, and economic infrastructure is built around centralized energy systems. We finance them with bonds, pay politicians to approve them, dam rivers and build nuclear plants to power them, use hundreds of thousands of square miles for transmission corridors, fight wars to protect our foreign supply sources, and charge every individual who uses them. That may end. And we may each have greater security for it.

Yet it won't end overnight. As explained earlier, vast increases in energy efficiency and clean energy sources may be matched by vast increases in consumption, as economic growth is spurred on by such improvements. Big oil may not just go away. It may hang on for some time, as clean energy sources struggle to keep up with overconsumption.

An example of the potential boost for oil occurred in 2001 with the announcement that oil production from Canada's Athabasca oil sands was being tripled. The oil sands contain about 1.7 *trillion* barrels of petroleum, and are one of the largest known reserves, but these are locked in a layer of tarry sand that until now has been expensive to exploit. But new technology has made it cheaper to extract the oil, rendering it price-competitive with conventional sources. Such achievements—thirty years in the making—may be accelerated by molecular technologies and as such, alter the geopolitical balance of power, as the United States gains a source of oil in Canada, close to its own markets, instead of depending on distant and unstable regimes.[23]

Our Cars

In 1986 there were 12 million electronic units in cars worldwide. By 2000 there were 200 million. They monitor fuel consumption, detect oncoming traffic, and call their manufacturers.[24] Besides conveniences that are readily apparent, such as communication and entertainment, other technologies are being incorporated to make cars safer at lower cost. Goodyear has incorporated microsensors into tires that indicate when they need replacing.[25] New plastics make vehicles stronger, lighter, and more energy efficient. Systems are being developed that track eye movements and voice, to see if you're awake at the wheel. Once in a while, the computer asks a question. By listening to your voice, if it detects that you're dozing off, it sprays a light mist in your face from a nozzle in the dashboard.[26] Such an unpleasant awakening might soon be unnecessary because NASA is working on guidance systems for flying cars that don't need a pilot.[27] Instead, these will use computerized air corridors much in the same way that some mass-transit vehicles operate without drivers. In these aerocars, sensors will allow seats to mold precisely to the weight, height, and shape of a passenger, adjusting airbag and other safety systems accordingly. Lubricants may be replaced with nanostructured films so we don't have to worry about leaks or replacement. Meanwhile, whether your car is being guided along an air corridor or on the ground, new coatings are making the exterior scratch- and weather-resistant, so its new shine will last.

Coatings and Surfaces

Coatings protect or enhance the performance of cars, ships, jets, floors, walls, building exteriors, consumer electronics, computers, lenses, photos, furniture, books, packaging, and kitchenware. Yet many coatings are still relatively primitive. Paints peel and lenses scratch. UV radiation makes coatings disintegrate. They wear off, and solvents wreck them. Moreover, most of them aren't too smart. They require a mishmash of chemicals to perform more than one function, then they're limited to those predetermined jobs. If we look at coatings through a microscope we also see that they're full of irregular, jagged edges that catch dirt, cause friction, or break off and degrade the finish.

Coatings are a defining part of our lives, but also a surprisingly toxic one. On one hand, nature provides color via minerals or pigments. Plants use biodegradable pigments to absorb energy or process food. On the other hand, synthetic coatings are designed to last, so they don't degrade when we use them. Such enduring qualities produce good products but toxic reactions. Automotive coatings contain heavy metals or organic pigments. During recycling of old car bodies, paint remains as part of automotive shredder residue, a cocktail of chemical substances that can't be recycled. Furthermore, our environment is contaminated everywhere with trace amounts of coatings that come off during regular use.

Nanotechnology has the potential to eliminate such impacts by modifying surfaces via molecular construction. This is because something special happens to color at the nano level. In 1999 scientists at the Georgia Institute of Technology's Laser Dynamics Laboratories discovered that gold *nanorods–*

nano-scale toothpick-like materials—are a million times more fluorescent than the metal itself.[28] Other materials have similar properties. *Nanoshells* are tiny particles fifty to one thousand nanometers wide with an insulating core, such as silica, coated by a thin shell of conductive metal. They resemble malted-milk balls. Developed by Rice University engineers, these metal nanoshells lend a chameleon-like optical effect to materials at different light wavelengths.[29] Nanoshells might be easily incorporated into some surfaces. As coatings they may be a thousand times thinner than regular paint.

Gold nanoshells are a thousand times more colorful than a lump of gold in a ring or necklace. They give off deep hues that are currently only possible with toxic chemicals. So researchers are looking at incorporating nanoshells into cosmetics. Wild color combinations may be possible. On the other hand, this highlights some of the potential risks. Normally, if gold is ingested in small amounts, it has no toxic effects on the human metabolism. However, the impact of metals such as gold at the nano level are still unknown. Do they get incorporated by the body into bone matter and accumulate at the molecular level? We're still not sure. Medical researchers are investigating how to use nanoshells to deliver treatments to targeted sites in the body.[30] For infrequent uses in patients, there may be no adverse effects, but for repeated uses such as cosmetics, researchers will have to know the interactions before they can be used. How do we prevent *nanofiber* particles from being ingested by workers and users? We might be replacing one problem with another. Nevertheless, an undeniable advantage of nanocoatings is that because of their precise molecular construction, much smaller amounts are required.

With such benefits and risks in mind, here are some potential applications for nanocoatings when they're combined with nanocomputing.

In 2001 the Swiss-based engineering multinational ABB and Nanogate Technologies GmbH, a startup in Germany, announced cooperation for robot-based automation technology to coat surfaces.[31] Chemical nanotechnology is used, for example, in manufacturing glass surfaces such as single-layer shatterproof glass, making it stronger, thinner, and cheaper. This makes it cost-effective for household windows and patio doors. Such robotic applications may also protect workers from having to apply coatings manually and risk inhaling potentially harmful nanoparticles.

The military is investing heavily in nanocoatings for stealth technology. Nanoparticles that absorb radar make a tank, ship, or airplane virtually invisible to detection through radar. This is a pressing priority because new detection methods are rendering present stealth aircraft visible.

Conductive *nanowires* promise to make electrical wiring cheap and invisible by incorporating it into surfaces, rather than running beneath them. Already, "flatwire," which operates at a scale much larger than nano, has a thickness less than a business card.[32] This is eliminating behind-the-wall wiring. Flatwire may be hundreds of times thinner if it uses nanowires. First applications may be for computing and electronics. Consumer electronics such as computers and music players may shrink to where we'll wear them unobtrusively.

When they are combined with lenses and batteries, such applications will make stealth surveillance possible everywhere. In the 2001–2002 war in Afghanistan, for example, armed, pilotless drones similar to the one depicted in Fig. 13 were used to identify, then try to kill, leaders of the targeted organizations.[33]

While such surveillance also portends pervasive loss of privacy for civilians, it may have helpful effects, too. We may soon explore the two-thirds of the earth that we still know little about–the oceans. In 1997, a camera known as the *critter-cam*[34] was attached to the back of a seal, showing us for the first time what happens when seals dive to feed. Widespread use of the critter-cam is still hobbled by size, cost, and limited power. Yet nanotechnology, computing, and robotics will lead to cameras that can operate for days at tremendous depths with minimal light. For the first time, we may have eyes everywhere in the ocean. This is invaluable for understanding how species interact in the deep sea, how underwater volcanoes are born, and how subsea earthquakes trigger avalanches that in turn cause tsunamis. When combined with remote sampling techniques, these may also help us find new species that give us new medicines.

Many of us spend hundreds of dollars replacing eyeglass lenses that scratch. Superhard nanocoatings that resist scratching are already in some markets,[35] and impact-proof coatings are on the way (see Fig. 14).

Conventional coatings will get stronger, harder, lighter, and more flexible thanks to nanocrystalline powders. Steel coatings will get harder. Bonding between layers of ceramics and polymers will be stronger, yet more pliable.[36]

By 1999 materials so smooth that dirt fell off their surface were being made, using a property known as the *Lotus Effect*. Its name comes from the lotus leaf, which has its own dirt repellent (see Fig. 15). This was first applied, though not at the true nano level, in a facade paint called Lotusan, which has been sold in Germany since 1999. Other companies have made coatings that resist graffiti (see Fig. 16). That same year, nano-scale coatings were applied to ships to replace toxic paint and reduce corrosion.

Lubricants

Without lubricants, industry wouldn't work. Friction wears down machinery and turns electricity into unwanted heat. Yet as early as 1959, in his first talk on nanotechnology, Richard Feynman predicted lubricants might disappear.

Fig. 13. The future of military surveillance. The United States's Unmanned Aerial Vehicle (UAV) Global Hawk made international aviation history in 2001 when it completed the first nonstop flight across the Pacific Ocean by an autonomous aircraft. Such craft will provide near-real-time intelligence imagery from high altitudes for long periods of time, using moving target indicators and infrared sensor systems. Such vehicles may be standard intelligence gathering and weapons systems for the twenty-first century. Miniaturization has also produced much smaller versions, microscopic in size, virtually invisible to the human eye. When integrated with nanotechnology and artificial intelligence, they could become cheap and widely available. (U.S. Air Force photo; www.af.mil/photos/Feb1999/globalhawk.html)

Fig. 14. No more scratches? Lens on left was scratched with sandpaper. Lens on right resisted scratching with nano-scale coating. (Courtesy of Schweizer Optik GmbH)

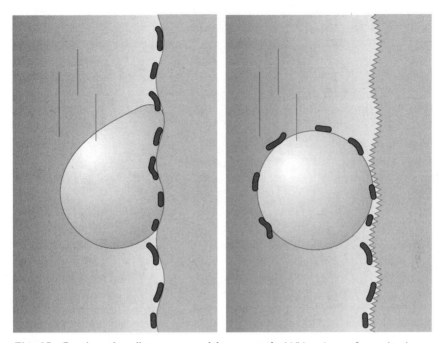

Fig. 15. Getting the dirt out ... with no work. With microsurfaces, the Lotus Effect—named for properties first observed in the lotus leaf—prevents dirt and bacteria from adhering. Left: Dirt adheres to regular surfaces and doesn't come off when rinsed with water. Right: Dirt comes off when rinsed. This reduces the need to use detergents on surfaces. A paint known as Lotusan uses this to keep buildings clean. (Courtesy of ispo GmbH, Germany; www.ispo-online.de/islotus.htm)

At the atomic level, heat is dissipated by large surfaces. In 2000, scientists at the University of California, Berkeley manufactured nano-scale bearings that are virtually frictionless (see Fig. 17). "Friction is a big problem with [mechanical devices], but these nano scale bearings just slide as if there's no friction," says John Cumings, a graduate student in the Department of Physics at Berkeley, who created the bearings. "As a lower limit, friction is a thousand times smaller than you find in conventional devices made with silicon or silicon nitride."[37]

Textiles

Space suits already use micropumps to heat and cool parts of the body. Nanotech skiwear, wet suits, and other extreme-environment clothing may extend these properties, improving our tolerance for temperature variations.

Fig. 16. No more graffiti, or cleaning? Nanocoatings make it hard for vandals to deface buildings. On the left are fresh markings from an indelible pen. The right side, beneath the hand, is blank because the coating repelled the ink. (Courtesy of Nanogate Technologies GmbH)

At the next step, sweat will escape from wet suits underwater, via one-way membranes. Our clothes may change color at a voice command, using the magic of programmed nanoparticles. They may also contain voice-activated computers and cell phones woven into the fabric, so we'll be able to work from everywhere. If someone steals them, they'll refuse to function because they won't recognize the user's DNA. Such clothing may also form an "exoskeleton" that gives us artificial muscular power to swim and jump as if we had the mechanical strength of robots.

Robotics

As the new century arrived, there were about 750,000 robots in the world, and their numbers were growing at a rate of 25 percent annually. One in ten automobile-manufacturing workers was a robot. Their prices had dropped to a fifth of what they were in 1990, both in real terms and compared to the cost of human workers.[38]

A few years ago, robot developers from around the world started to get together and play soccer...with their robots. Robocup is the World Cup of

Fig. 17. The solution to friction? Nanobearings may be the solution to one of indus-trial society's greatest efficiency destroyers. Physicists at the University of California, Berkeley have peeled the tips off carbon nanotubes to make seemingly frictionless bearings so small that some ten thousand would stretch across the diameter of a human hair. The bearings are telescoping nanotubes, with the inner tube spinning about its long axis. When sliding in and out, they act as springs. Potential applications include frictionless motors, which might run at high speed without lubricants or cooling systems. (John Cumings and Alex Zerrl, UC Berkeley; www.berkeley.edu/news/media/releases/2000/07/27_nano.html)

Robotics. The event tests speed, agility, perception, sensory abilities, and accuracy. This is for the serious autonomous "bots"–the ones that run them-selves. These aren't "robot wars," where opponents try to destroy each other's remote-controlled machines; Robocup competitions are for robots that can pass and kick a ball toward the goal by themselves. By the year

2050, Robocup organizers aim to build a team of autonomous humanoid robots that can win against the human world soccer champions.[39]

We don't yet see players faking a fall and complaining to the referee, but we do see machines ferociously competing without human intervention. There are no remote control commands from the sidelines. It's thinking machine versus thinking machine.

This brings chuckles when we watch vacuum-cleaner-type boxes fumbling around in a playpen, with geeky technicians cheering them on. Yet many laughed when the first computer went up against a world chess champion. In 1997, as noted earlier, the human lost.

The coming convergence of robotics, genetics, nanotechnology, and artificial intelligence has profound implications for robotic medicine, notably surgery and prosthetics. Every surgeon knows that cutting someone open is one of the most traumatic, invasive aspects of medicine, with long recoveries and risk of infection. Robotics is already changing this with machines that guide a surgeon's hand through a tiny hole rather than wide incision.[40] This requires technologies that include:

Sense of touch: When a surgeon can't see or feel what she's doing because her hands aren't inside the patient, computing and sensing will restore this sensation remotely.

Mechatronics: For moving around. Soft tissue tends not to stay where it's put, so the instruments that "feel" the tissue need to be able to sense its pliability and not accidentally rip it apart. Sensors play a crucial role.

Three-dimensional viewing: This lets the surgeon see inside a cavity and what's around it, rather than only the spot being operated on. If bleeding starts centimeters away from the operating area, the surgeon needs to be able to detect this instantly.

Each method is expensive. That's where nanotechnology comes in: By making computing, sensing, and viewing cheaper, nanoscience could make robotic surgery affordable, instead of only accessible to a few rich patrons.

Yet this, too, is primitive compared to what MEMS and smaller devices will accomplish. In a generation, surgery as we know it may be obsolete. Autonomous, submarine-type tools will likely patrol our arteries, picking off fatty deposits that characterize heart disease today. This alone could replace more than half of today's surgeries at a fraction of the cost. Every major artery in your body might be cleaned of harmful fatty deposits. This could slash the heart attack rate and improve our quality of life by maintaining a strong blood flow. Postindustrial society's most pervasive disease could be postponed and minimized. These shrunken machines would keep arteries supple to avoid the hardening that we experience now. They'd also remove colon polyps that grow to be cancer. As they got more sophisticated, they'd identify viruses and cancer cells, destroying clusters, for example, before they become tumors. To achieve this, a confluence of technologies is the key: microscopic cameras, propulsion mechanisms, mechanical arms, computerized sensors, and transmitters to interact with the physician.

What if the body rejects these tools? At first they'll be so big, that is, micro scale rather than nano scale, and temporary that we'll deal with immune system response the same way we do with other microscopic procedures. As they grow smaller, to the nano level, we may have to perform arterial cleaning while the immune system is temporarily or locally suppressed. Or these nano-sized cleaner bots may disguise themselves chemically to avoid an immune system response, just as some viruses do.[41] Anti-rejection drugs already achieve this for transplants, but genetic coding may let these machines mask themselves in a friendly way, just as the AIDS virus, for example, masks itself in a fatal way.

The same goes for prosthetics. By improving the connection between human nervous systems and mechanical arms and legs and then shrinking the energy source, nanotech-enabled prosthetics could go from being clumsy tools to seamless parts of a body. *Nanosurfaces* will give an artificial limb the look and feel of skin. *Nanosensors* will give it a sense of touch that's relayed to the attached nerves. Soon it may seem like the original limb.[42]

THE GENETIC WORLD

A brief overview gives us an idea of the potential influence of genetics and genomics. Hundreds of billions of dollars are invested every year by pharmaceuticals companies, military agencies, government health departments, research institutes, and computer companies. Decoding of the human genome has launched a chain reaction in research that's only just begun but is already pervasive in scope.

In our daily lives, the short-term impacts can be classified into these general categories:

Agriculture: Plant and animal growth rates are being manipulated in big ways. Genetically manipulated salmon that grow larger, earlier than their natural counterparts have been produced, sparking an intense debate over releasing them into the wild. The nutritional content of grains and vegetables is being altered to provide more basic vitamins and minerals to supplement diets. Plants that are more resilient against drought are being bred. It's the biggest modification of agriculture since we started putting seeds in the ground. The great debates over such genetic changes are extensively covered in many other books, and partially summarized in chapter 20.

Molecular manufacturing: A shortcut to the molecular assembler may be to use self-replicating properties of DNA for molecular computing and manufacturing. Each DNA molecule contains many genes, which are nature's software code for physical and functional units. With this, nature assembles whole machines such as humans and animals. DNA provides at least two advantages. First, the information is efficiently stored in a tiny package. Second, its processes might be used to make molecular motors that perform repetitive tasks efficiently.

Cures for and protection from diseases: Malaria, cancer, AIDS, dengue fever: each has genetic weaknesses that we're able to start exploiting following the completion of the first phase of the human genome project in 2001. Susceptibilities to heart diseases and some cancers may also be correctable prior to birth through gene therapy. Organ replacements may be feasible by growing our own organs from our own tissues. This triad of treatment, avoidance, and organ replacement may soon explode onto the medical scene, potentially changing the face of health care forever—if we succeed at controlling downside risks described in chapter 20.

To understand what might come first in the short term, follow the money trail.[43] Part of that trail leads to the globalized consumer economy. At the end of the Cold War, as markets matured in Japan, Europe, and the United States, the markets of Asia and Latin America grew attractive to multinational companies. Yet there was, and still is, a problem: For consumers to consume products they need disposable income. To have disposable income they need higher-paying jobs. To work at higher-paying jobs they need to be healthy. If millions of technicians are constantly off work or operating at half speed, training them and maintaining the systems they work on is a nightmare. Citizens of most temperate climates are blissfully unaware that for most of the population who live in tropical or subtropical regions, disease is a debilitating drag on their ability to generate wealth. Tourists and business travelers have discovered this, to their regret, when they come back sick from warm climates. Most of them are shocked when this happens, often with the most common diseases. At the institute I managed, our Brazilian staff was constantly off work with illnesses that were made worse by resistance to antibiotics. Prescription regimes are less stringent in some developing nations, with antibiotics available as over-the-counter purchases. Overuse of such drugs, accompanied by incomplete application regimes, has led to an epidemic of drug resistance around the world, especially in poor tropical nations.[44] Thus, doctors at tropical disease institutes admonish their patients that "anyone from the North who lives in the Tropics for more than a few years is virtually guaranteed to get a serious disease."[45]

Many such illnesses are nicknamed "orphan diseases"; those that haven't been seriously adopted by rich nations as priorities to solve. Orphan diseases are symptomatic of the downside of our affluent economics. While billions of dollars pour into high-tech health care, tens of millions of people die or are sickened by diseases that get meager research funding or inadequate deployment support. A blatant symbol of this contrast arose in 2001, when a sixty-nine-year-old patient at an American hospital rang up a bill of more than $5 million for thirty-four days of treatment for internal bleeding. More than 95 percent of the bill was for drugs.[46] Health insurance paid part of the bill and the hospital swallowed the rest. The patient wasn't billed. That cost was many times the total required to pay for drugs to save every one of the approximately *thirty-four thousand* children who died of antibiotic-resistant dysentery during that same period worldwide, because their parents couldn't pay for treatment with a new drug, Ciprofloxacin, to save their lives.[47] Thirty-four thousand to one. That's an example of the relative cost, in lives, of an orphan disease, compared to one that catches the attention of a high-tech health-care system.

Among the worst of these orphan diseases is malaria. More than one million die annually from the malaria parasite, and tens of millions more are sickened for life. Many victims are children, and the numbers are staggering. To understand the economic impact of this disease, just take a close look at people on the street in a malaria zone. You'll see some wearing jackets in hot, humid weather. They're cold from the malarial fever. They live in a world of cyclical weakness, as explained in a report of the World Health Organization:

> In countries where a high proportion of the population is at risk of severe malaria, measured on a standard index, average income per capita in 1995, adjusted for purchasing power, was less than one-fifth that of non-malarial countries. Income growth per capita since the mid-1960s for countries with severe malaria has been only 0.4 per cent per year, compared with 2.3 per cent in other countries–a difference of more than fivefold.[48]

A salient characteristic of malaria is that those who have the problem don't have money, and those who have the money–for example, those in temperate climates–don't have the problem. Multinational companies and their governments have understood that to transform these victims into robust consumers, the Cold War policies of benign neglect for orphan diseases won't work. Virtually no funding has gone into serious malaria research for decades, although it sickens many times more victims than, for example, AIDS.

The common prophylactics for malaria are to toxify the liver so the parasite can't take hold, or toxify the environment with DDT. Each leads to long-term problems that are often worse than the disease itself. Moreover, the "solutions" have led to varieties of malaria becoming drug- and pesticide-resistant. The problem is worsening, because as drug resistance increases, desperate countries are resorting to reintroducing pesticides such as DDT that are known for their devastating impacts on human health and wildlife. The only lasting solution may be to attack the parasite at the genetic level.

At a pan-African conference that saw the beginning of a continental initiative to fight the disease, Professor Harold Varmus, director of the U.S. National Institutes of Health, remarked, "The life cycle of *Plasmodium falciparum* [the parasite that causes malaria] is one of the great mysteries of biology, and understanding its genetic blueprint will help solve one of the most fascinating problems."[49] In 1999 trials began on a Glaxo Wellcome DNA vaccine that targets the molecular makeup of the body's own defense. In late 2001, a team in Canberra, Australia announced that it had found a way to "infect" whole populations of malaria-carrying mosquitoes with genes that neutralize the parasite.[50] Also, a promising vaccine that wipes out parasites in the human liver and prevents their spread was undergoing trial tests in Gambia.[51] Scientists who test such vaccines are fully aware that the war against such parasites will have to be fought carefully. If a genetically manipulated vaccine is improperly developed, it may lead to more virulent strains of the disease. Thus, vaccines that have low evolutionary risks are being pursued with this in mind.[52]

Meanwhile, on the political and financial front, in 2000 evidence of action started to show up with the creation of the Global Alliance for Vaccinations and Immunization. Kick-started with annual grants from the Bill and Melinda Gates Foundation, the alliance set out to immunize 30 million children with known vaccines that prevent known diseases. That year, an anonymous donor gave $100 million to Johns Hopkins University's Bloomberg School of Public Health to develop a malaria vaccine.[53] At the same time, the National Institute of Allergy and Infectious Diseases announced a global health plan to attack HIV/AIDS, malaria, and tuberculosis: three diseases that alone kill more than 5 million people a year, and in some countries account for half of annual deaths. The institute's director, Anthony Fauci, gave a revealing rationale for this:

> It's becoming clear that the issue of global health has now integrated itself into this nation's foreign policy.[54]

Thus, infectious diseases are now an American foreign policy target.

That means big bucks for genetic research. We may be on the way to a cure, driven by the demands of the globalized marketplace. As argued in part 4, such a cure could be central to getting public support for genetic research. It's one of those catalysts that would counter accusations of Frankenstein science that are leveled at the genetics research community and the pharmaceuticals industry. As such, it represents one of the most desirable, attainable short-term goals in the huge field of genetics.

Meanwhile, though, cures for disease aren't the only uses of designer drugs.

Recreational Designer Drugs

Designer *nanodrugs* may be undetectable, powerful, and impossible to regulate. Their forerunners include drugs such as Ecstasy. From 1990 until 2001, the rates of abuse of this drug by American teenagers exceeded those of nearly every preceding drug of choice.[55] This was symptomatic of a relatively new phenomenon that began with LSD some decades ago: synthetic drugs that escape legislative criminalization until they get a foothold. These are easily manufactured by any chemistry student, are enormously profitable, are difficult for standard testing to detect, and give the short-term illusion of being harmless, other than making users feel great while still letting them function. For teenagers prone to risky behavior, this combination is far worse than letting a kid loose in a candy store. In such a store, at worst the kid might get stuffed and sick. Not so with designer drugs. Some users can go on for years without showing debilitating ill effects. Then they crash. Furthermore, abuse of designer nanodrugs probably won't be confined to kids. If we look at adult propensities for abusing prescription drugs and alcohol, we see that the broad general population is at risk.

Finding Things

Then there are smaller irritations such as lost luggage that drive us to drink. Imagine a lost suitcase that tells you where it is, regardless if it ends up in Delhi or Atlanta; a wallet that never gets misplaced; a car that never gets lost in the parking lot and is hard to steal because thieves can't disable the sensor that tells you when a user moves the vehicle; or a child whose whereabouts are known to you every moment (this may produce some interesting results). With global positioning systems and nanosensors, these would be feasible. The first candidates to use them will be shipping companies, but the systems could become cheap enough for personal use. Personal-assistant devices may help us keep track of everything. We'll need only ask, "Where's my daughter right now?" or "Where are my keys?" and our wearable computers will give the answers.

Human-Computer Relations

The "digital divide"—that line between computer-literate and computer-illiterate populations—is becoming disturbingly synonymous with high technology. Those on one side continue to get more powerful and those on the other are disenfranchised. How are molecular technologies going to help the poorest populations?

Right now, we each need a university degree to debug our computers. Instruction manuals weigh more than laptops. Software crashes are daily sadomasochistic rituals. Tech support personnel speak a language that we need to decipher before we start on the problem. "User-friendly" is turning into a bad joke for the tens of millions of workers who struggle with such problems every day. The most vexing of these problems is the user-computer interface. We need to learn to type, then figure out the hardware, then install software, then hope everything works for the six months until it's out of date.

Molecular computing combined with programmable chips and broadband communication may transform this human-computer relationship. To start with, hardwired computers are going to be replaced with programmable chips that allow reconfiguration updates automatically, via the Internet. Such software is already being delivered and installed on demand. Results: no more manual program installation or removal. And we won't have to type if we don't want to. Voice recognition will be standard. We won't have to worry whether information is in Japanese, English, or German because translation programs will convert it. Microscopic units will process billions of algorithms for speech recognition and translation. Energy efficiency will let a tiny battery charged by a wearable solar cell run a powerful molecular computer. If a Kalahari bushman wants to send a message to his granddaughter in Los Angeles, he'll press the pendant that hangs around his neck, call her name, and start talking. When he's finished, the message will be translated, transcribed, and sent to her instantly. The concept is nothing new. The National Security Agency (NSA) does similar work every day when agents eavesdrop on hundreds of millions of transmissions.[56] The difference is accessibility and affordability. Instead of a multibillion-dollar infrastructure housed in a supersecret installation maintained by thousands but open to just a privileged few, the bushman will need only a pendant on his chest that talks to a self-managed robotic satellite data network. Thus, the poorest users may have much to gain. We may still have great disparities in wealth, but such technologies might, if deployed intelligently, help cross the digital divide by making computer networks more user-friendly. As this unfolds, millions of financially disadvantaged individuals might also gain entry into financial markets.

Finance

Imagine a personal computer that combines known variables for world markets, then executes trades for each of our personal portfolios. This is happening already in big investment houses and mutual funds. Supercomputers are being used to track the impacts of economic variables on markets. They trade stocks in fractions of seconds in New York after economic data have been announced in Shanghai or Delhi. The same capabilities will soon fit into a laptop. As millions of laptop investor programs execute trades on extremely short-term information, this may drive short-term volatility through the roof. Today, for example, markets fluctuate daily by as much as 3 or 4 percent—far greater than years ago—due to programmed trading. We may see daily moves of up to 10 percent as computing systems struggle against each other for the fastest action and reaction. Huge fortunes will be won and lost in seconds, and rather than being an unusual occurrence, as it is now, this may become the norm. Strap on your retirement fund seatbelt. Furthermore, as revealed in congressional hearings into the 1999–2000 stock market bubbles, regulations are slow to catch up with these technologies.

The Golden Rule is still "buyer beware."

Investing isn't going to get less hazardous for the ordinary investor, and may look something like this: Blue-chip companies rise and fall in a never-ending struggle to prevent obsolescence. Small start-ups sometimes produce huge fortunes, but most still go bankrupt. Legal liabilities sink companies, because the power of their technologies leads insurers to refuse coverage. If a genetic cure has a bad result, for example, a million people might be harmed, with resulting lawsuits emptying the coffers of the richest pharmaceutical company. Liability may determine the direction these technologies take. Asbestos and breast implant lawsuits may seem inconsequential by comparison. Companies will continue to press politicians for liability limits.

Privacy

Since the September 11 attacks, the spy business has experienced a post–Cold War renaissance. Projects that were plodding along for years have a new lease on life. For example, the Defense Advanced Research Projects Agency (DARPA) is pouring funds into nanocomputing because the NSA is looking for faster ways to break codes. So far, the NSA can break only encryptions that have up to 140 or so prime numbers. Each part of the decryption needs to be done sequentially.[57] The NSA wants to do it with parallel computing—linking many processors together with simultaneous tasks—up to infinity. This requires a confounding process known as *quantum computing*. Such computing may be only practicable through nanotechnology. Nanotubes and nanowires have been developed and are racing to industrial fabrication. In another of the most significant advances in high-speed computing, scientists replicated quantum computing with light in 2001.[58] This makes computing applications far easier because, unlike quirky particles that we don't yet understand, light is a known quantity with determinable characteristics.

Even without nanotechnology, the military has succeeded in creating microscopic *smart dust*[59] that monitors audiovisual activity. A video camera with computerized circuitry can already fit onto a chip at a cost of a few dollars. These chips could be installed anywhere, making ubiquitous spying a fact. Nanocomputing, nano-scale propulsion, and robotics together will make such surveillance invisible to the human eye.

Personal privacy is dead and getting deader. Without setting eyes on you, a reasonably skilled snoop can, in a week or so, know your name, address, driver's license and social security numbers, credit record, family members, TV-viewing habits, buying habits, Web-surfing habits, genetic disposition to disease, current state of health, sexual preferences, work records, and just about everything else that governs your life. Legislation might say it's illegal to gather some of this, but that's unenforceable unless you can catch the snoop.

In the molecular future, counterespionage will become big business, and personal-privacy lawsuits will spring from everywhere. Encryption technology is a must. Genetic antidiscrimination legislation is already being passed and tens of billions of dollars will be spent trying to enforce it.

∞   ∞   ∞

Now that we've had a glimpse into the near future, let's envision a time farther into the future, when these capacities may be in place. These changes won't come at once. Their introduction will be irregular. Yet to prepare for possibilities and dangers, we can and must imagine what's to come. Here's a taste of the molecular age.

CHAPTER FOUR

WHAT COMES NEXT? MEGATRENDS THAT COULD ALTER OUR LIVES

In 2001 the Motorola Corporation, a manufacturer of telecommunications equipment, adopted the slogan: "Motorola technology is teaching things to think."[1] Motorola ads showed consumers having discussions with their cars and refrigerators. The campaign was based on the idea that we'd teach our appliances how to think for themselves.

In the same year, the Kellogg company began explaining "intelligent materials" and "smart clothing" on the back of its cereal boxes sold in Germany.[2]

This is food for thought. The advertising industry is always full of hype. Yet it also shows a persistent ability to foreshadow what's coming.

Much of what you're about to read seems implausible in the described time frame of this century. Why believe it? It's only an educated guess at what may occur. Yet to appreciate the likelihood of it actually happening, here's a story that's been told many times in many ways, and still has merit:

Imagine yourself a farmer in New Jersey in 1899. More than a third of the population lives on farms. Your life expectancy, if you've made it past infancy and you haven't died in a farm accident, as so many do, is fifty-some years. Your wife just died of an infection from a small cut, while two of your five kids succumbed to yellow fever and smallpox. There's no running water except for a hand pump. You're reading by gas lamp. You've seen the miracle of the telegraph, but messages still take a few hours to get from the railway station to your house via dirt road. The train is the fastest thing around. The phonograph is still a novelty. Then, one day, a noisy, smelly, dangerous looking carriage sputters its way up the road to your house. A dust-covered man gets out and introduces himself. He announces that in the next century, millions of images from around the world will come directly into people's homes as they happen. You'll be able to throw your voice thousands of miles. Horseless carriages such as this one will go faster than sound,[3] and your kids will drive themselves faster than a locomotive. A needle prick will cure many infections. Yellow fever, smallpox, and most other fatal childhood diseases will disappear. The miracle of natural reproduction will be stopped and started by a pill. Three percent of the population will farm to feed the rest,[4] while your land will produce many times more crops, using machines like this horseless carriage. Another machine that runs on something you've already heard of, but haven't seen much use for—electricity—will milk your cows. Your kids won't be farmers but instead will ride underground with millions of others each day, to work in buildings a thousand feet tall, while millions of others fly above them. A bomb will have the power to incinerate New York. We'll be able to hold sunlight still, then start it up again. Someone will play golf on the moon. Moreover, a piece of this future is already here, he says. Someone in town named Edison is making such machines. The man pulls out pictures and shows them.

What might your reaction be? Well, around the time he started talking about golf on the moon, you decided it's time for him to leave. And that car: how do they expect to control millions running around the countryside when they can't keep the trains running on time?

As with many farmers in those days, your first instinct might be to kick him off your land, because his noisy contraption scares the horses.

Now imagine that you're a young Iowa farmer in 2001. Some guy lands in your yard with a wingless car that behaves like a helicopter and runs on autopilot. He announces that, in your lifetime, this will replace most passenger vehicles. In your child's lifetime, agriculture that's been around for thousands of years will disappear, and food will be synthesized in a box. Many cities will look like forests. Your kids will be part of an invisible machine that keeps their minds intact for millennia, sharing them with a million other minds. Machines that are smarter than us may share our world and play golf on Mars without protective spacesuits. Some of this future is already here, he says, and millions are reading about it. Already, he says, devices have been attached directly to humans, so that when they think, they make machines move.[5] A guy in town has invented a computer that designs the electronics for your hatchery's thermostats better than a human engineer could.[6]

What would your reaction be? You're not so cynical about technology, because you've seen animal cloning and you're using the Internet, but around the time he started talking about eternal life and robotic golf on Mars, you decided it's time for him to leave. Maybe he's a born-again Christian who had a nervous breakdown. And that flying car: how do they expect to control millions running around the sky when they can't keep the airlines running on time? It's too noisy to be used in town. Bylaws won't allow it. Besides, it scares the horses.

Despite widespread skepticism, the head of technology for one of the world's most successful computer corporations, along with developers of talking computers and artificial intelligence, are among those discussing how such things may come to pass in this century.

Isn't it premature—over the top? Won't it scare the horses?

If the future described in these pages seems disjointed, it's because it may be. Things might not occur as sequentially as we're used to. Old traditions may contrast more sharply with new trends. Superintelligent human beings or machines might coexist with *Homo sapiens*, just as *Homo sapiens* coexisted with other types of human many thousands of years ago.[7] Wealthy individuals may own asteroids and space homes while some of us live in huts on the beach. We see similar contrasts now, where AIDS and brain drains cripple Africa's technological capacities, while other continents race ahead. Yet in the future, these stark differences could be reduced in severity, with less suffering among the population. If we live in a hut on the beach it'll be because we want to. In many ways, we may be stronger as individuals, because technology will empower us to achieve basic material security by ourselves.

We face a life that requires multidimensional matrices to describe it. We may approach a point described earlier as the Singularity: when a mass of technologies converge and the future becomes impossible to predict. As with three-dimensional chess, it's greatly more complex than a flat board.

To understand why life may frazzle the comprehension of today's *Homo sapiens*, let's continue exploring trends explained in the previous chapter, and see where they lead.

EDUCATED GUESSES—WHAT MIGHT REALLY BLOW OUR SOCKS OFF

As with the pictures of the short-term future that we explored earlier, the midterm future may be laced with contradictions—absurd, immense, and fantastic ones. The likelihood of moderation by 2001 standards seems low, because the technologies show every sign of extending human power. In the midterm future "moderation" might, for example, be considered as restricting each of us to one color change a day for our car, clothes, and home so we don't drive our neighbors color-crazy. In such ways, our world could get much more weird, yet more personally satisfying. On the other hand, everything might get totally out of control. We'd have to find ways to cope.

With that in mind, here are snapshots of the wild ride ahead in the midterm future. These may not occur simultaneously or in a predictable sequence. They are only snippets.

- The first molecular assembler is built. It begins to manufacture solar cells and molecular computers. Most consumer goods are made at home, just as they were in the nineteenth century, except this time by desktop factories that give us most of what we need whenever we need it. This marks the end of globalized manufacturing systems that dominated the latter part of the twentieth century.
- Instant translation is perfected, eliminating language barriers.
- Neural implants let us speed-read by enhancing our retention rate.
- Thirst for knowledge explodes. Every library in the world is downloadable into a personal e-book via the Internet. The average price of a book goes from twenty dollars to twenty cents, but authors get rich on royalties from mass downloads. Instead of books selling a few thousand copies each, most sell in the hundreds of thousands as the cost of printing disappears and people download them as frequently as they bought newspapers in the twentieth century. Everything is keyword- and concept-searchable to match individual interests, so we have greater choice, along with improved accessibility to the information we want.
- Every car, home, and piece of clothing can change color at the touch of a hand.
- The euro, dollar, and yen merge, then disappear, as universal units replace paper money. This is just a short step away from the international credit card systems that were available in 2001.
- The first transworld supersonic tunnel is built, linking New York, London, Berlin, Moscow, Shanghai, Tokyo, and Los Angeles. Average speed: twenty-five hundred miles per hour. Northern Hemisphere circumnavigation time, including stops: nine hours. New York to London: ninety minutes. Capital cost: only for software. Construction is completed by self-replicating fractal robots. They don't ask for wages.
- Marriages between septuagenarians and twenty-year-olds are common, as genetic therapy reduces the impacts of aging. Experience merges with youthful exuberance to constitute the most valuable and sexy commodity. The young strive to get wiser, fast.
- Companies that had the foresight to focus on robotics become the

world's largest corporations as the world becomes dependent on robots.

- Unemployment skyrockets as intelligent systems take over the jobs of stockbrokers, telecommunications workers, computer assemblers, librarians, bankers, pilots, air traffic controllers, landscapers, fruit pickers, and office cleaners. Other types of social jobs begin to fill the gap. Social self-help groups receive big start-up financing from governments and philanthropists who fear societal disintegration. Some of the heaviest users are those who still have technology jobs and face imminent obsolescence.

- *Robo servers* with high intelligence in narrowly defined areas are manufactured in the billions, with the same matter-of-factness as microchips are today.

- *Robo sapiens*[8] are built. This is the first autonomous robot generation with intelligence that rivals that of humans. They begin to replicate themselves, using the same type of gene sequences that every living being has, at a rate of one per month, that is, logarithmically. Robo sapiens start to ask for rights. Religious terrorists assassinate some, but fail to wipe them out.

- *Homo provectus*–advanced humans with artificially enhanced intelligence and bodies–become commonplace. They are the only intellectual matches for Robo sapiens. A human marries an intelligent sex robot.

- A human goes on trial for murdering a robotic spouse, but gets off because the jury is only human. Robo sapiens protest by withholding sex from their human clients. The verdict is quickly overturned.

- Different classes of intelligent entities begin to branch out: *Homo provectus*, with enhanced intelligence and partially robotic bodies; Robo sapiens that think and learn independently; and Robo servers (robots with limited intelligence), such as pet robodogs that talk and do the household administration.

- Synthesized food replaces animal-based food. Killing of animals for food and fashion ends. The antivivisection movement changes its focus to robot rights.

- Space homes hit the markets, but only become broadly affordable when a space cable is built with carbon nanotubes from Earth up to geosynchronous orbit.

- The Moon is used for tourism, but most prefer to stay in Earth orbit or zone out at home on designer drugs and virtual reality.

- Scientists slow light to a stop, and then speed it up again (whoops, that happened in 2001).[9]

RETROSPECTIVE FROM LATE IN THE TWENTY-FIRST CENTURY

Here is a look backward from sometime later in this century, when this hypothetical molecular future has already come to be.

Our Daily Lives

Reading Everything Everywhere

A big drawback for early-twenty-first-century computer users was having to read from a display screen. Contrast wasn't good in daylight, so it strained our eyes; we couldn't roll it up and put it in our pocket; it broke when we dropped it; we couldn't scribble on it, then toss it across the table to someone else; we couldn't take it to bed or in the bath–although some tried. With e-paper, the infoscape exploded. E-paper is embedded with millions of memory cells that hold thousands of pages. We use it to wirelessly load our morning news, filing system, or books. We flip pages just as we do with a newspaper. Each time we turn a page, the content changes. We load new documents by instructing it in natural language: "Get me yesterday's *New York Times*." If we want to show a "hard" copy to someone, we write on ours with an electronic pen or use voice recognition instruction to send it to him or her. It recognizes our DNA, so it's safe from unauthorized users if we lose the sheet. We can delete misplaced data with a wireless command. E-paper is biodegradable, so with a coded command, it disintegrates into microscopic, nontoxic dust. It costs three cents a sheet. It's also our cell phone, TV, web camera, and personal translator. Thus, in some ways, Douglas Adams's fictional encyclopedia *Hitchhiker's Guide to the Galaxy* that contains a databank about everything, everywhere, has begun to materialize.

No Language Barriers

Despite a century of electronic communications, only a tiny percentage of the world moved comfortably between cultures in the early twenty-first century. One barrier was language. International businesspeople used to spend years learning foreign languages or depending on interpreters. With molecular computing, instant translation soon arrived. This breached business and personal barriers among Latin, English, Arabic, and Oriental markets, vastly increasing personal mobility and allowing hundreds of millions of skilled workers to work anywhere, from anywhere. The first signs appeared with online services that let Web surfers translate individual Web pages.

Then, cheap chips and self-improving software democratized translation. Immigration rules were circumnavigated via Internet as everyone began working with everyone else, everywhere, while staying at home. For those who could be enticed to move, work visas underwent revolutionary changes as nations struggled to compete for workers who powered the new economy: information engineers, geneticists, and robotics engineers, along with financial experts and others who greased the economic wheels.

Now, late in the twenty-first century, this has led to communities based on common interests rather than location or language. International functions such as air traffic control and shipping are suddenly open to tens of millions of non-English-speaking people. With universities such as Harvard putting their course materials online since 2001, hundreds of millions are accessing educational resources that they only dreamed of before. Students now check academic admission standards in Beijing just as easily as in their hometown. This doesn't eliminate local standards, but it does make everything far more transparent, and gives each of us access to other cultures instead of depending on filters.

Clothing that Talks and Listens

With supercomputers the size of fingernails, the term "smartly dressed" has taken on new meaning. Our clothing is our office, our health-care specialist, and our power source. Woven into the fabric is an array of sensors, logic circuits, exoskeletal muscles, communicators, and solar cells that give us total mobility. We gain the capacity to adapt to a range of hot, cold, wet, and dry environments without changing clothes. Vital signs are monitored constantly.

Virtual Touch and Virtual Sex

We now have options for virtual reality, including head-mounted gear combined with virtual-reality suits; neural implants that deliver sensations to and from the brain; and virtual-reality rooms. Our virtual-reality suits use tiny artificial muscles to elicit tactile responses. When combined with broadband communication between individuals, they let us shake hands, then shake bodies. Cybersex, which generated so much traffic on the old Internet, draws investment to this kind of virtual reality. Sexually transmitted disease is virtually gone, but computer viruses cause awkward problems. Virtual-reality rooms provide a less intense experience that seems more comfortable for some. In these rooms, video displays and speakers are printed on flexible sheets at a fraction of what computer screens used to cost. These coat wall, floor, and ceiling surfaces. Together, they generate audiovisual

3-D. Thus, we can attend a virtual conference from home. We log in with tens or thousands in a virtual room, watching a virtual speaker. Some participants "reveal" their virtual selves as being present in the audience, because they want to see who's attending. The deal is that we can "see" attendees only if we agree to be seen ourselves. In the past, most of us visited conferences not to hear speeches, but to network in the hall. With a virtual audience, we electronically tap someone on the shoulder and, if they agree, we go to a virtual hallway for a cyberchat. If we don't want to talk, we block their signal. This eliminates the need for physically being there without sacrificing a key component—selective intimacy.

We find the Internet is just as secure, or at risk, from snooping as are personal meetings. *Nanospies* are invisible and hard to detect in hotel and meeting rooms, or outdoors where they blend into trees and grass. Online, we're protected by encoded cyberwalls that tune into our DNA fingerprint, and by ear implants that connect to our neural pathways. Offline, our homes and leisure spots are scanned constantly. Still, many users are sloppy. Identity theft is common.

We plan our vacations by cybertraveling to sample locations. Holographic waves lap at our feet, cyberbreezes cool our face, and hot virtual sand sifts between our toes. With this type of immediacy, virtual junkies are a problem; some users are addicted to going everywhere and feeling everything without doing anything.

Chameleon Surfaces Color Our World

We control the color of our cars and clothing with a voice command or touch. We reorganize surfaces at the atomic level to give them color without toxic additives. Computing lets us alter color via ultraviolet, infrared, or electric current. Surfaces are nontoxic, as opposed to older coatings that left pollution everywhere they went. The same applies to coatings that make our cars and clothing smoother and more resistant to wear.

No More Cavities

For human coatings, a company produced a toothpaste in 2002 that used particles of enamel-like material to fill nano-scale cavities in teeth. This prevented decay, replacing the toxic mercury-laden amalgam once used for fillings. It was the end of visits to the dentist. After that, nanobots were developed that scoured our gums for signs of gum disease then repaired them before things got out of hand. It takes one application a year, just like the gooey stuff that the dentist used to give us to fluoridate our teeth.

Throw Out the Scrub Brush

Nothing needs cleaning anymore, at least by humans. Most bathroom and kitchen surfaces are supersmooth at the nano-scale level, discouraging dirt and bacteria buildup. Tiny robotic spiders scour the rest and dump waste in the trash or use it as an energy source to keep going. Sometimes we accidentally squash them like bugs, but they just replicate themselves and keep going.

Ties that Bind . . . and Release

Most adhesives were used in the old days to connect materials in an irreversible process. The adhesive bond was broken only with solvents, or by physically separating two pieces, leaving chemical residue on each surface. Adhesives posed health risks from solvents, chlorinated substances, and interaction of reactive compounds with other chemicals in the environment. New nano-scale materials were then invented that acted like Velcro by sticking together, then physically reversing the process with a tug or an electronic command. Molecular switching turned sticky characteristics on and off as required. Result: no more gooey glues; no nails or screws; no buttons, zippers, or shoelaces. This means that we can do away with doors, because many walls can open and close at an electronic command, much in the same way that they change color, as explained earlier.

Just Flying to the Store for Some Milk

Advances in materials, guidance systems, and fuel efficiency brought the pilotless personal aerocar (see Fig. 18). Every aerospace manufacturer had one on the drawing boards by the end of the twentieth century,[10] but couldn't get it to market because planes needed pilots, and air traffic control systems couldn't handle such chaotic traffic everywhere. Now, though, megacomputing and satellite guidance have brought fail-safe automatic pilots; vertical takeoff and landing for small craft, with superlightweight directional engines; nanofuels that triple propulsion efficiency; and noise-canceling baffles that recirculate engine exhaust. Politicians quickly approved the new air traffic and zoning regulations when they discovered they wouldn't have to take a taxi to the airport anymore. This didn't eliminate collisions, of course. Just as tens of thousands died on twentieth century roads, many still do in aerial accidents. Yet casualties are lower since, unlike the old cars, aerocars have collision-avoidance systems. They also have "utility fog,"[11] a special nano-scale "seat belt" comprised of billions of dust-like specks that spring from the walls of a vehicle to envelop passengers in

Fig. 18. The future of personal transport? The M400 Skycar. Inventor Paul S. Moller has developed, built, and flown a two-passenger prototype model. This new type of aircraft combines the performance of airplanes and the vertical takeoff and landing capability of helicopters in a single vehicle, without pilot training requirements. Advances in materials, guidance systems, and engines may make it feasible to mass-manufacture such vehicles. NASA is developing traffic management. Such vehicles could allow residents to flee an oncoming tsunami, where roads would be hopelessly snarled with traffic. (Courtesy of Moller International; www.moller.com/skycar/m400/)

the event of a crash—much like airbags, but more effective. For those afraid of heights, a swish of the hand makes the windows go opaque, replacing the outside view with whatever images you'd like to see—perhaps terra firma, for instance. Motion sickness is counteracted with an internal envelope that stays stable as the vehicle turns and gyrates around it. Still, *Star Trek*–type immunity from acceleration and deceleration hasn't been achieved yet. Pulling the universe toward and away from us at light speed without being squashed remains an elusive goal.

The Magnificent Moles

Superhighways are being replaced by vast underground electromagnetic transport networks.[12] These have freed thousands of square miles of land that were once used for roads. How is it economically feasible? Fractal robots. Here's an early description, although it depicts only one of many variants:

A fractal robot is . . . made from motorized cubic bricks that move under computer control. These cubic motorized bricks can be programmed to shuffle themselves to . . . make objects like a house . . . because of their motorized internal mechanisms. It is . . . like kids playing with Lego bricks making a toy house by snapping together plastic bricks—except a computer performs the shuffling operation. . . .

With more advanced versions of such machines, it will be possible to build a bridge in a day, a housing estate in a week, a space ship in hours, . . . and so on. This technology is called Digital Matter Control and it is implemented here with machines called Fractal Robots. . . .

A fractal robot can erect supports as it mines and remove them automatically when . . . finished. Even if the robot is buried through an accident, it can dig itself out. . . . They are extremely efficient at moving equipment and since very little ruble is removed, they are extremely energy efficient. . . .[13]

With the power and efficiency of fractal robots, it's possible to build thousands of miles of tunnels deep in the earth at low cost (see Fig. 19). These are used by supersonic transports with magnetic levitation and super-strong lightweight materials. Together, the transports and tunnels shatter time barriers between continents. There are no concerns about environmental impacts of sonic booms; most tunnels operate in a vacuum. A trip from Shanghai to New York now takes five hours.

The End of Air Rage

With aerocars, electromagnetic tunnels, and virtual reality, the airline nightmare has ended for tens of millions. Gone are heavy traffic to the airport; check-in lines; and dry, dirty, oxygen-depleted air on long flights. Instead we use our own aerocars for short hops, and electromagnetic tunnels for transcontinental ones. We still have transport hubs, but these are decentralized and provide nonstop transit to most destinations. We stay in personalized pods while we're switched from tunnel to tunnel, or our aerocars take us directly there. Some thought it might be confusing, but instead it's infinitely simpler when travelers have only to tell their vehicle pods something like, "Take me to the Mandarin Hotel in downtown Shanghai" and guidance systems do the rest. If there are two locations with the same name, the vehicle lists them and asks the passenger to specify which one. Powering this apparent simplicity is a globalized, multitrillion-computer guidance network. Occasionally, travelers end up in Chicago instead of Shanghai, but not often enough to compromise confidence in the systems. And for those who don't want this traveling nonsense, home-based virtual reality networks obviate the need for that. We can be there without being there.

Fig. 19. The tunnel borer's ultimate machine today, and its potential successor of tomorrow. Top: Borer surface is about ten meters (thirty feet) in diameter. It tunnels continuously and constructs structural rings for the tunnel at the same time. Seen here preparing to construct a railway tunnel near Rotterdam. Cost: many millions of dollars. (Courtesy of Herrenknecht AG, Germany) Bottom: Fractal branching ultra-dexterous robots, also known as *bush robots*, a concept for robots of ultimate dexterity. This is a hierarchy of articulated limbs, starting from a macroscopically large trunk through successively smaller and more numerous branches, ultimately to microscopic twigs and nano-scale fingers. With its numbers of fingers of many scales, the robot's dexterity would exceed anything known. Thousands of these fractal robots could autonomously dig tunnels, remove waste, and add structural rings. Potential size: from microscopic to a mile high. Cost: next to nothing for the hardware because it self-assembles or is printed in a desktop factory. The software that runs it would determine its price. (Courtesy of Hans Moravec; www.frc.ri.cmu.edu/users/hpm/project.archive/robot.papers/1999/NASA.report.99/)

Total Surveillance

Cheap, invisible lenses, sensors, and microphones make it possible for millions of airborne nanobots to track each of us without our knowledge. The implications are troubling–total elimination of personal privacy. Governments have passed strong privacy laws, but enforceability is spotty. Privacy hacking is an epidemic and personal firewalls that mask our voices are a prerequisite. The security industry has grown larger than the energy business.

Overcoming a Dynamism Deficit

The medical potential of nanotechnology was restrained by backlashes against potential "Frankensteins." These abated when researchers found a cure for nanobacteria that caused dozens of diseases. Soon after that, organs were manufactured without using lab animals.[14] Hundreds of millions of persons suffering from heart and lung disease, kidney failure, and skin cancer got a new lease on life. Genetic repair prevented and cured a range of diseases, including cancer, AIDS, and heart disease. Growing rejection-free organs from our own tissue has become big business.

A pressing issue for a while was the ratio of young to old. As the population aged, the pool for dynamic new ideas from naive and enthusiastic young generations decreased, as did their influence in real numbers. This left us with a dynamism deficit. The struggle between ideas perpetuated by an aging population and new ideas from a younger generation intensified. As the younger generation struggled to break into the upper echelons and older folks were increasingly able to hang on, a logjam developed at the top of each sector of society. This led to generational conflict. A solution had to be found.

Enter artificial intelligence. This has proven invaluable to maintaining a dynamic society. Regeneration has come from marrying artificial intelligence with our own. We get an injection of vigorous ideas that otherwise would have waned as elderly outnumbered the young. Artificial intelligence makes its contribution in many ways. Computers are programmed to augment our regular intelligence, either externally via voice interaction or by implanting them directly into our neural pathways. Next, drugs are developed to enhance our capabilities. Such drugs are actually nano-sized computers that attach themselves to parts of our brains to stimulate them. Finally, artificial intelligence develops on its own, independent of humans. One result of this diversity is that the "glass ceiling" that used to stop the younger generation from moving up is broken. A whole new economy has developed, consisting of enterprises generated from artificial intelligence and robotics.

In the meantime, robotic care of the elderly has become commonplace.

Nor is this impersonal or degrading. Quite the opposite: home-based medical robots and remote Internet-based analysis save us the trouble and distress of going to a hospital. Robotic companions give us company when the kids are gone. Some of us choose to have autonomous, fully functional companions that may not always do what we want. They exhibit some of the irreverent behavior of our kids. Still, they have a prime directive not to harm or abandon us. If we don't want testy companions, we opt for Robo servers with limited intelligence that's more than sufficient to meet our medical and psychological needs. Many of us don't have children, so we start to look upon robots with affection. It's hard not to, because many of them are programmed to adapt to our personal traits and develop their own characters.

Thousandth Floor, Please...

By the year 2001 scientists had tied knots in superstrong carbon nanotubes, with thirty times the tensile strength of steel.[15] Once they reached industrial production, they drove an architectural and engineering renaissance. Now, we use carbon composites instead of aluminum and steel for most construction. The most striking use of these materials has been for space cable construction. Space elevators let us launch payloads into orbit by transporting them up a twenty-five-thousand-mile-high, stationary vertical shaft anchored to an asteroid (see Fig. 20). Once thought impossible due to limitations on the strength of materials, the space cable carries tons of equipment and personnel up or down daily from the equator, via electromagnetic transport vehicles. What's the advantage? Payloads cost twenty thousand times less to deliver into orbit than with rockets.[16] This is opening the door to asteroid mining, interplanetary exploration, and trade.[17] Furthermore, by using the same flexible, superstrong materials, we're cheaply reinforcing buildings on Earth against hurricanes and earthquakes. Whole regions have been rendered disaster resistant.

A Yacht in Every Slip

Every new owner used to discover belatedly that boats were holes in the water into which you poured money. Seagoing vessels were maintenance nightmares. Wind, sun, and waves battered them brutally. Sails, engines, electronics, surfaces, and hatches disintegrated. They were expensive to replace. With molecular science, though, construction and maintenance costs plummeted. Cheap, light, corrosion-proof, superstrong carbon nanotubes replaced disintegrating fiberglass, rotting wood, and rusting steel. Structural materials now float like Styrofoam but have the tenacity of tita-

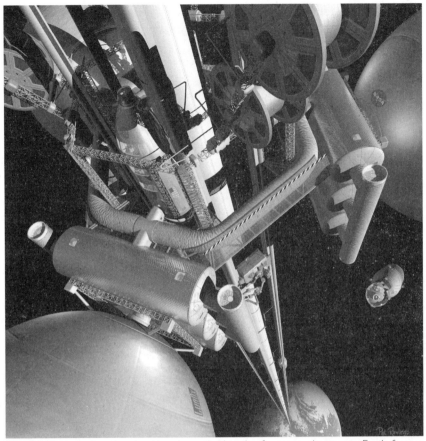

Fig. 20. Space elevator. Looking down the length of a space elevator to Earth from a geostationary station about twenty-five thousand miles "up." Electromagnetic vehicles travel the length of the elevator to transfer passengers and cargo between Earth and space. High-strength cables work back and forth to provide adjustments to the position of the geostation. An asteroid counterweight (not shown) maintains the center of mass at a geostationary point. Research into feasibility of space elevators indicates that there are five key technologies for space elevator development: nanotechnology, robotics, and artificial intelligence are among them. (Courtesy of NASA, Pat Rawlings/SAIC; flightprojects.msfc.nasa.gov/fd02_elev.html)

nium. Contamination from antifouling paint, a deadly toxin for sea life, has been eliminated using surfaces that repel microorganisms by offering no rough areas for them to cling to, or by emitting an occasional electric charge. This reduces water drag, improves speed, and raises energy efficiency. Pollution from plastics—one of the biggest killers of marine mammals who swallow them—disappeared with instant biodegradability built into plastic containers. Toilet-bots now break down sewage effluent that used to kill coral reefs. These advances, combined with wireless broadband, make it possible for many of us to live aboard seagoing vessels. Floating communities have appeared everywhere. It's still hard for mariners to believe that their boats can repel a typhoon or cyclone; yet the materials that constitute these vessels have the hardness of diamond, thirty times the tensile strength of steel per unit of weight, and the flexibility of a spider's web. Such materials, built by molecular assemblers, also made it feasible to build a space cable twenty-five thousand miles out from Earth,[18] as described earlier. The same strength now lets boats survive in hurricanes, typhoons, and cyclones. Thus, big storms have lost their reputation as the killers they once were.

At Work

No More Cubicles

Medical professionals, computer designers, and engineers are now empowered by totally mobile environments. Most doctors scoffed at the idea, but patient exams are now done from everywhere on earth. Instead of patients going to a clinic for brain or body scans, they put on cybersuits with resolution a hundred times greater than the old X ray, MRI, or CAT scan behemoths. By communicating using these cybersuits, a remotely attending physician or technician does body temperature, blood work, cholesterol count, immunodeficiency, and most other tests that used to involve invasive procedures, with a cybersuit performing the lab work. Just as these suits make cybersex possible, they let the physician poke and feel, using cybergloves on one end and the patient's cybersuit responding on the other. Paperwork—an overbearing feature of the old health-care system—is done on the spot with voice recognition. Lab reports, diagnoses, prescriptions, treatments, and billings are electronically assembled into a personalized file that stays with the patient. For those who worry about privacy, personal information can be made more secure by encryption, which gives patients more control over who uses their personal data for what purpose. In some ways, it's safer than paper records used to be. In other ways, it's less secure, because those who have your DNA code can get the data if they want to.

COMPUTER NETWORKS TRANSFORM MOST INDUSTRIES

A shock to many—except those who mapped computing since its inception—was the exponential growth of computer power (see Fig. 1). Just as computer networks grew thousands of times more powerful between 1945 and 2001, so the trend continued into the twenty-first century. Only in 1945, computers weren't running globalized systems, whereas by 2001, they were. Thus, when computing grew again exponentially more powerful, a paradigm shift erupted. Computers began doing creative work such as complex construction, medical care, and food manufacturing. Moreover, they changed the way that organizations operate, by disrupting vertical control structures.

The Flat Organization

The management revolution had been foreseen by scientists such as Neil Gershenfeld. In his 1999 book *When Things Begin to Think*, Gershenfeld explained the impact of the desktop fabricator on design, manufacture, distribution, and recycling of products:

> Companies are trying to flatten their organizational hierarchies and move decision-making out from the center by deploying personal computing to help connect and enable employees. But the impact of information technology will always be bounded if the means of production are still locked away like the mainframes used to be. . . . Fabrication as well as computing must come to the desktop. The parallels between the promise and problems of mainframes and machine tools are too great to ignore. . . . Adding an extra dimension to a computer's output from 2D to 3D will open up a new dimension in how we live.[19]

Thus, desktop manufacturing gave vertically controlled multinationals a nervous breakdown. Suddenly, we were able to build extremely complex products in our own homes, with help from software imported via the Internet and intelligent machines that learned on their own. For these functions, centralized controls went out the window.

Decline of Patents, Ascendancy of Brand Names

Big companies are often unable to enforce copyright and patent laws. *Genetic programming*[20] proved to be the death knell. Intellectual-property laws proved incapable of keeping up with the speed of innovation, as millions of machines developed billions of designs at a breakneck pace. The U.S. Patent

Office gave up after a million applications were filed in ten days. Inexpensive designing revolutions raced across every industry as patent barriers to low-cost manufacturing collapsed. This brought a terrifying specter for inventors and companies who depended on patents to protect their ideas. How were they going to make money if, next week, a machine modified their idea to make a new concept or superceded their work with a better idea and put it into production immediately? The new battleground quickly became clear: Those who controlled the most advanced forms of genetic programming would be those who won in the marketplace. Such control wasn't exercised just by big multinational companies. As millions of new products were developed, individual inventors began to establish lucrative niche markets with "designer" genetic programming. The key difference was that fortunes were made in weeks or months instead of years, as product cycles shrank. Living proof of the viability of this approach came in the early twenty-first century, when "open source" programming—software that was freely licensed among programmers—led to vast collaborative efforts that produced superior products. Companies built their products on open-source platforms to diversify into specialized markets. This resulted in an explosion of new products and processes that in turn generated new niche markets.

As another partial solution to the genetic-programming monster, companies began developing sophisticated "poison pills": encrypted self-destruct mechanisms that prevented ordinary hackers from copying their software. They also built up brand names to distinguish themselves from fly-by-night operations. Software search engines were set up to police the Internet for rogue products. Today, as in 2001, more than a quarter of the products on the markets are knockoffs, but companies still make enough to stay profitable. A typical customer service call goes like this:

"Hello, this is Dream Nanosystems. How may I help you?"

"Hi. Your massage bed came out of my fabricator with a leg missing."

"Sorry, Ma'am, that's impossible with our software. Normally the fabricator disassembles the product if it detects a fault. Where did you purchase it?"

"From your holosite, NanoDream-SleepWare."

"Sorry, Ma'am, that's a hacker's site. We closed it down yesterday. Please send us your copy of the software and as a courtesy we'll send a new bed, even though the company is not obliged to reimburse people for hacker sites."

"How long will it take?"

"It's instant," says the synthesized robotic voice. "The software arrives right away. If you'd like we can do it now."

"Good, because I need it this evening."

The customer returns the defective software, and in five seconds the certified replication software is installed. Molecular assembly starts. The bed can be seen taking shape through the viewer in the assembler. This time is has the requisite number of legs.

"OK, it looks all right."

"No problem, Ma'am, and sorry you got taken. Next time please check with customer service to get a verification code to prove that it's our product. Thank you for calling Dream Nanosystems. Have a nice night." Then the computer hangs up.

The Artist as King, and Everyone as Manufacturer

Consumer goods and their components have become extremely cheap and radically different in composition. Most are manufactured by digital fabricators in our homes, without a lumbering globalized network that used to require thousands of parts arriving in the same place at the same time. The cost of goods isn't determined by the price of materials anymore; rather, it's set by the price of the software to manufacture them and the fees of those who imagine them. The real cost of software is human and computer brainpower. Whoever puts a price on these determines how the world economy develops. Hollywood special-effects and video game designers have ended up as the highest-paid workers, because they have the imagination to create three-dimensional images that translate into products via desktop manufacturing. The bond between artists and computers becomes a love-hate relationship. Artists work with intelligent computers to develop new concepts, yet because their ideas are immediately taken up and enhanced by intelligent computers, artists have to make sure they get prepaid for concepts that are outdated the instant they get published. Some royalties still accrue afterward, but these now work in the same way that first-run movies did at the beginning of the century, with most of the money being made during the opening week. After that, most ideas are superceded by hybrids. Some residuals accrue from "reruns," but these usually drop off quickly. On the other hand, artists benefit from these hybrids because they help to generate new ideas, in cooperation with intelligent computers; and so the cycle goes. Thus, artists have been reincarnated as Plato's philosopher-kings. For consumers, this has led to an explosion of personal freedom, but also an increase in personal responsibility. Freedom comes from machines that liberate us to build whatever we want. We can manufacture furniture, electronics, medicines, housing, and eventually food. Responsibility comes from the obligation to control the tendency toward overabundance. Too much of everything becomes one of the primary risks that we collectively face (see chapter 5 for more details).

The Renaissance of Cultural Diversity

In the early twenty-first century, "globalization" meant cultural genocide. It stuck us with "global" symbols, such as soft drink and running shoe labels, that seemed to take over the world. As the American way of life ran amok, positive aspects such as entrepreneurialism were diluted by cultural destruction. Despite media saturation and extensive travel, the average American or European still knew almost nothing about Asian or African culture, and vice versa. If they did, it was more by blood relation than communication. This gross cultural blockage was taken for granted because it had existed for millennia. Worse yet, we'd tried to overcome it by homogenizing the world. That cultural desert began to bloom when universal translation rained on it. The alienation and homogeneity that accompanied the early globalized society gave way to greater cultural security. This let everyone fit into the international framework without being assimilated by it. Today, we're less at risk of becoming cultural equivalents of the single-minded Borg, as depicted darkly in the series *Star Trek*.

Still, it's not without rough spots. Culture and high technology aren't always directly translatable. Artificially intelligent computers try to put their own spin on interpretations by anticipating how one culture might react to another. We're entering a phase where humans and computers try to compensate for one another. Human and machine cultures are blending.

Machines that Think

The term "robot" was invented in the 1940s by Czech playwright Karel Čapek.[21] It's a Czech word meaning slave worker. The term "robotics" was invented and used in 1942 by the Russian-born American scientist and prolific writer Isaac Asimov.

Half a century after that, a team led by John Koza at Stanford University developed a method known as genetic programming that used mega-computing to design machines for optimal performance.[22] By 2001 computers had outdone world-class human engineers by independently devising designs for electrical circuits, stereo components, and cruise control systems.[23] Yet truly three-dimensional products were still impossible due to limited computing power. Today, though, nanochips make teraflops–a trillion floating point operations per second–supercheap and superfast. Genetic designs are applied to every manufacturing process. We have desktop machines that design other machines, with an emphasis on the "creative."

The big surprise came one day at the NSA when a human programmer was talking to the agency's newest quantum computer. In the midst of a conversation about China's code-breaking capabilities, the unit had paused and asked, "What's the purpose of encryption?" Stunned, the programmer responded by explaining it was designed to protect information from unwanted intrusion. Then the computer asked, "Why do we want to hide information?" Aghast, the programmer searched frantically through the computer's subroutines to find where evolutionary inquisitiveness had been programmed. No luck. This computer had formulated not only the question, but also the thought process behind it, that is, that code existed *for a purpose.* A computer had never in history asked why something *was* unless it had been programmed to mindlessly mimic a question. Normally, a set of pre-conditions was accepted as given, and the computer went from there. Questioning assumptions—autonomous thought—was new. At that instant, Robo sapiens had been born—spontaneously.

For decades, programmers worldwide had been trying to program this type of autonomous intelligence into computers. Some theorized that it might evolve spontaneously from massive parallel networks. But when it happened, everyone was caught by surprise. A meeting of the National Security Agency governing body was convened. Should they unplug the unit? What if it unilaterally decided to start feeding information to the Chinese government? After much hand-wringing, the council concluded that if this unit had reached such a level, it wouldn't be long before other such units did the same. So they cut off the unit from the outside world and started experimenting. The next question the unit asked, of course, was: Why had it been disconnected from the outside world? And the conversation went from there.

LOOKING PAST HUMANITY

Now that we've glimpsed backward from a potential future, let's again look forward from the present, to see how relationships might develop between *Homo sapiens* and other forms of high intelligence.

Many types of intelligent machines or life-forms may soon inhabit the earth. We may first see controlled intelligence in computer networks and robots. These Robo servers might be designed to provide technical and emotional support for us. They may be self-sufficient enough to provide their own energy and repairs, or to replicate themselves. They might perform thousands of different tasks, from surgical procedures to space station repair. They would come in thousands of configurations, each with its own specialty. Sizes may range from submicroscopic, for medical applications inside the human body, to truck-sized, for transporting passengers and freight to their destinations. They would have sufficient intelligence for sophisticated problem solving, but only in controlled limits that would be determined by the intended function. The ultimate status symbol may be the robotic personal home assistant: a humanoid Robo server slave.

Then we may see what I call *Homo provectus* (advanced human—sometimes referred to as *posthumans*, as we'll see later). This is the human species, upgraded. *Homo sapiens* have already begun the transition to *Homo provectus*. Prosthetics are attached to our nervous systems so amputees can move robotic arms and legs.[24] Robotic organs arrived in 2001.[25] Computers have been implanted in our neural pathways, via retinal implants that help blind persons see (see Fig. 21)[26] and chips that let spinal cord injury victims move.[27] Soon, we'll be putting chips directly into memory banks in the brain and connecting them to the Internet.[28] Some of this may also be achieved genetically, although it remains to be seen to what degree we might pass on such genetically manipulated traits from generation to generation (see discussion of this later in the chapter). Variations among *Homo provectus* individuals may produce vast differences in intelligence levels: some enhanced by genetic manipulation, others by neural implants and external computers that communicate seamlessly with their thought patterns. Just how many of us reach this stage depends on a host of factors, including our willingness to do such things to ourselves, religious or ethical debates, and the price that's established for enabling software. A globalized splinter economy that caters to this enhanced breed of superhuman may develop. This may lead to enormous gaps between them and *Homo sapiens*, where *Homo provectus* have little to do with those who lack the upgrades.

The most controversial "machine" that we develop may be Robo sapiens: intelligent autonomous robots that think and learn independently. At first they may not have the kind of bodies that we'd identify as robotic. They may only exist in complex neural networks of computers. Then, as computers shrink, their minds may begin to fit into robotic bodies. These machines may have their own consciousness: *awareness that they exist.* And they may be a lot smarter than us.

THE ROAD TO SINGULARITY

The idea of superhuman intelligence has been with us since the concept of God was invented. The thought that something more intelligent than us might manipulate us for sinister purposes is embodied in the concept of Lucifer. Thus, the ingredients for a good versus evil struggle among superbeings have been in our minds for some time.

Defining this in scientific terms has been with us since the Egyptians and Greeks tried to understand the mechanics of how gods ran the world.

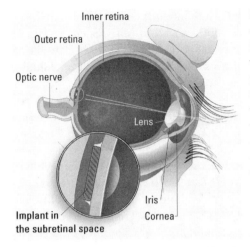

Inner retina

Outer retina

Optic nerve

Lens

Implant in
the subretinal space

Iris

Cornea

Fig. 21. To see again . . . The Artificial Silicon Retina (ASR) was invented by brothers Alan and Vincent Chow, the cofounders of Optobionics Corporation. The ASR is a silicon chip 2 mm in diameter and $\frac{1}{1000}$ inch in thickness. It contains approximately thirty-five hundred microscopic solar cells called *microphotodiodes*. These convert light energy from images into thousands of tiny electrical impulses, to stimulate the remaining functional cells of the retina in patients with macular degeneration and other retinal conditions. From their subretinal location, artificial "photoelectric" signals from the ASR induce visual signals in the remaining functional retinal cells. The signals are processed and sent via the optic nerve to the brain. The ASR is powered by incident light. It does not require the use of external wires or batteries. The first human clinical trials began in June 2000. (Courtesy of Optobionics Corporation; www.optobionics.com/artificialretina.htm)

Artificial silicon retina on penny. The black dot indicates the relative size of the ASR device that is implanted into the back of the eye. (Courtesy of Optobionics Corporation)

Then, in 1898, as the industrial revolution reached maturity, writer and historian H. G. Wells conceived of a technologically superior intelligence in his *War of the Worlds*,[29] where he described how an alien force might try to annihilate our species. In Wells's story, microbes rather than humanity beat the invaders. As one of the foreseers of modern technology, he nonetheless predicted that lower life-forms rather than human ingenuity would be the first to repel aggressive superior beings.

The idea that we might develop ways to supercede ourselves is as old as the pharaohs, who supposedly went to live among the stars to enjoy a superhuman existence in the afterlife. In contemporary literature, the idea of artificial humans was gruesomely popularized in Mary Shelley's classic *Frankenstein*, where body parts were assembled and electrified back to life as a whole being, who then destroyed his scientific creator in a fit of childlike rage, only to be destroyed himself by a fearful mob.

With invention of the computer, science fiction screenwriters were quick to jump on the idea that thinking machines might one day try to dominate the world. Arthur C. Clarke's *2001: A Space Odyssey*[30] and Stanley Kubrick's film by the same name described the struggle between machines and humans as they jointly explored our solar system. Since then, cyborgs have populated the screen, from *Star Trek* to *Terminator*, often in armed conflict. A more human-friendly cyborg came along in 2000 with the film version of Isaac Asimov's *Bicentennial Man*,[31] about a self-educated robot that decides in the end he wants to die as a human. The famously sexy Seven of Nine in *Star Trek: Voyager* gave this theme a twist as a liberated humanoid prisoner of the Borg mind collective who recovered her humanity.

Then, in 2001, Steven Spielberg completed Stanley Kubrick's final work, *A.I.*, depicting a world where humans have the godlike power to create and destroy machines that love; telling how we ultimately turn on them, and how our creations outlast us by outrunning us.

Visions of artificial intelligence are finding their way into the popular imagination and we're beginning to envision what it might be like. Yet we're a long way from understanding.

The term *singularity* is one of those words thrown around at parties to confuse mere mortals. In dictionary terms it means everything from "the state of being singular" to "the point at which mass becomes infinite, a black hole."[32] In science it has many applications. In mathematics it's used to define infinities, that is, calculations that go on forever. In physics it describes the time before the big bang, when everything was nothing, and a time when it might be nothing again. In the world of computing, it's used to describe the whole universe as a giant computer.[33]

None of these terms is very helpful when we try to understand the singularity of super-human intelligence. So we're going to look at it from another viewpoint.

In 1993 Vernor Vinge, a since-retired computer scientist at the University of San Diego, published a paper entitled "The Coming Technological Singularity: How to Survive in the Post-Human Era."[34] This was among the first efforts to identify how human-created superintelligence might emerge in the near future, and how we might coexist with, or be a willing part of, it.

Vinge lists ways in which superior intelligence might arise:

∞ *Right for sixty four thousand dollars!* Vinge describes how human-computer interfaces have already produced a form of superior intelligence. He gives the example of a doctoral student and a computer that together answer intelligence questionnaires faster and more accurately than an individual doing the same tests unaided. This has occurred on the game show *Who Wants to Be a Millionaire?* when stumped contestants elected to "phone a friend" at home. The home-based helper had thirty seconds to answer the question. In that interval, intelligent search programs could find answers by accessing the Internet. When the contestant called a doctor to answer a medical question, the physician enhanced his or her knowledge by accessing a medical database. Natural language recognition let the user ask the search engine without having to redefine it in geeky terms. So game show contestants with the fastest computer pals gained an advantage. A few years ago, such consumer-accessible superintelligence didn't exist, or was too slow to get a reply in the allotted time. Now, in a figurative way, we're already becoming "artificially" intelligent.

∞ *Brain pills and intelligence screening.* Next, Vinge describes biological means of improving human intellect. This might be achieved through neural implants, genetic modifications, or chemical inputs. The Cognitive Enhancement Research Institute in San Francisco promotes development and use of brain-function-enhancing drugs, for example, and is in a perennial struggle with the Food and Drug Administration to approve such drugs.[35] It's still a controversial field. Yet thousands of parents already have their children's DNA screened for susceptibility to diseases that might lead to brain damage. They also select eggs with genes that enhance their physical characteristics. Professional models, who some find attractive as genetic "role models," are selling their eggs for such purposes. The next rung on that ladder is to identify and insert genes associated with greater intelligence.

∞ *My name is HAL.* Another path to superintelligence is via self-aware computers. A compliant intelligent computer could generate enormous benefits for humans, by serving as a memory for things we forget, an engineer to help make things that we can't, or an astute analyst to help guide us through life. While this is thought to be far off, some computers already make good educated guesses at improving machine designs that surpass those made by human engineers. This process, known as genetic programming[36] has already been described earlier.

∞ *I am, therefore I think.* One of Vinge's more intriguing ideas is that superhuman intelligence might "wake up" spontaneously in networks, as described earlier with the fictional NSA encryption computer. The combined power of thousands of computers working across millions of applications might spark a self-awareness process that we ourselves are incapable to comprehend. Thus, instead of a signal from a distant civilization in deep outer space, we might get a new form of intelligence from cyberspace.

Wherever we look along the path to Singularity, we find computing. And wherever we find computing we find its underlying technologies. These technologies and how we use them are central to our molecular future. Here's one example of how autonomous artificial intelligence might upset our economic systems in the near term.

Patents are the basis for entire industries, especially advanced genetics, pharmaceuticals, and computing. They underpin the ability of companies to get investment capital to build new industries. How do we tell a thinking machine that it just contravened a patent? Moreover, how do we enforce a patent when artificial intelligence is capable of superceding it with infinite variations in a matter of days? If we gain the capacity to input every known concept that underlies every known patent into a thinking computer, are we then going to ask it to respect those patents to avoid legal reprisal? Improbable. Rather, we'll do what we've always done until now; ask it to find ways around them. And it will. Quickly.

This is already happening. Tools such as genetic programming are developing designs that cleverly circumnavigate patents through parallel methods that achieve the same end.

Transhumanism

One view of the future is based on the concept of *transhumanism,* although this title has yet to be officially seized on in the halls of academia. The discipline's goal is to help us move past being human, into the realm of superintelligence and superhealth, along with increased emotional and social well-being.[37] Proponents of transhumanism take pains to distance themselves from what they term as "psychic, extraterrestrial or divine"[38] methods. Transhumanism has its own international organizations, such as the World Transhumanist Association,[39] started in 1998. Its associated groups include organizations such as the Extropy Institute.[40] It has a small group of local chapters in the United States and Europe. Its participants are the chroniclers and advocates of new life-forms in the making. Among its members are some of the most accomplished technologists. They're far from unified in their technological or social outlooks, but they agree on one concept: The era of *Homo sapiens* as the only self-aware, technological life-forms on earth may be about to end. We may be at the same relative stage as when Neanderthals met the first *Homo sapiens*, except this time the transition may be decades or centuries, not thousands of years. It may involve not just one subspecies of posthuman, but many, each emerging in a time span that seems like spontaneous combustion compared to the rest of history until now.

There isn't yet a broadly accepted definition for transhumanism—and again it's important to emphasize that the transhumanist group is a very small one—but nonetheless the Extropy Institute gives guidelines. According to its Web site, the California-headquartered—although primarily Web-based—Extropy Institute "acts as a networking and information center for those seeking to foster our continuing evolutionary advance by using technology to extend healthy life, augment intelligence, optimize psychology, and improve social systems."[41] In this broad framework, the organization focuses on transition of *Homo sapiens* from what they are now into what they soon may become. Its members and executive body consist of an eclectic group of thinkers who range from computer prodigies to artists. Together, they are working on the beginning of a new terminology for what may be a new species. Here are some definitions:

Posthumans

These will be persons of unprecedented physical, intellectual, and psychological ability, self-programming and self-defining, potentially immortal, unlimited individuals. Posthumans have overcome the biological, neurological, and psychological constraints evolved into humans. . . . Posthumans may be partly or mostly biological in form, but will likely be partly or wholly postbiological—our personalities having been transferred "into" more durable, modifiable, and faster, and more powerful bodies and thinking hardware. Some of the technologies that we currently expect to play a role in allowing us to become posthuman include genetic engineering, neural-computer integration, molecular nanotechnology, and cognitive science.

Transhuman

We are transhuman to the extent that we seek to become posthuman and take action to prepare for a posthuman future. This involves learning about and making use of new technologies that can increase our capacities and life expectancy, questioning common assumptions, and transforming ourselves . . . to be ready for the future, rising above outmoded human beliefs and behaviors. [An example of such an outdated belief might be, for instance, that we're born, live, and die in less than a century. Transhumanists suggest that this sequence may soon not apply. To see other discussions of out-moded behaviors, visit the Extropy Institute Web site at www. extropy.org.]

Transhumanism

Philosophies of life (such as the Extropian philosophy) that seek the contin-uation and acceleration of the evolution of intelligent life beyond its currently human form and limits by means of science and technology, guided by life-promoting principles and values, while avoiding religion and dogma.[42]

Looking at these definitions, we may have to ask: What is a new species? Might it be separate from the human one? Is it a subspecies? Is it not a species at all?

One definition for "species" is: "groups of interbreeding natural popu-lations that are reproductively isolated from other such groups."[43]

Yet in a report for the U.S. Congress, Lynne Corn, a congressional research service specialist in environment and natural resources, explained that species' definitions tend to get confusing when it comes to classifying populations and subspecies:

> . . . Populations, subspecies, and species all describe regions on the spec-trum of interbreeding, from nearly complete interbreeding (a single popu-lation) to never interbreeding (separate species). In contrast to legal defin-itions that draw clear boundaries between these categories, the actual boundaries are blurry and made even more hazy by the lack of informa-tion on most organisms.[44]

This may present complications when it comes to reproduction via arti-ficial or enhanced intelligence. For instance, if artificially intelligent machines had biological characteristics such as DNA motors[45] or human skin,[46] along with DNA-mimicking software code[47] to replicate themselves, or they extracted software code from other intelligent machines to assemble the next generation of thinking machines, would they be a species?[48] If *Homo sapiens* made the first generation, would these machines be a part of "natural populations?" If we synthesized human genes and used them to genetically modify the physical or intelligence characteristics of human embryos outside the womb, at what point might this cease to be "inter-breeding"? If intelligent machines modeled human DNA to assemble copies of themselves, or *Homo sapiens* used synthesized DNA to breed, then would the human species be "reproductively isolated from other such groups"?

Thus, we might see some interesting battles erupting over how or whether to classify intelligent machines as a species, especially if we tried to control their intellectual property, turn them off, sue them, or give them rights and bank accounts.

It's not the intention here to have an extended discussion about how species are defined. Yet, through this brief summary, we see how the lines that separate life-forms might get confused. The issue has also been taken up by artificial intelligence proponents, at the level of defining "consciousness," as we'll see later.[49]

∞   ∞   ∞

Humanity may have to cope with Robo sapiens, a completely autonomous intelligent being partially manufactured by humans, then improved on by itself. Robo sapiens may have a photographic memory, instant recall of trillions of pieces of information, self-repairing energy-self-sufficient bodies, wireless communication among themselves over vast distances, surface speed of up to a hundred miles per hour, thousands of mechanical manifestations for various jobs, and feelings.

Where might the feelings come from? How much of our irrational behavior is going to be programmed into them? Might they have something like the "emotion chip" that the android Data tests on himself in the series *Star Trek: The Next Generation*? To what degree might these machines be extensions of ourselves, or perhaps disdainful of us? These are some of the challenges we'll encounter. They aren't science fiction anymore. They are real possibilities with which we have to start grappling.

Artificially intelligent machines may observe us and learn, then develop feelings in their own neural pathways. They may program themselves to do this as part of the millions of experiments they'll conduct to learn what the world is about. Who would teach them to learn this way? We would. We're already programming them to teach themselves. Thus, in some ways they may be a reflection of ourselves, but in other ways they will be alien to us. Although we might imbue them with the Laws of Robotics that forbid them to bring harm or allow harm to be brought to humans (see appendix A), some of these laws may produce contradictions that make robots behave irrationally, like us. They'll face questions such as: When should humans be allowed to hurt each other emotionally? What about economic inequity? Who gets treated first in a vehicle accident?

We could develop close relationships. Robots will be programmed to help us solve complex problems, such as using genetic information to treat several cancers in the body at once. This "ethical" programming that emphasizes care for the whole human may make its way into human-robot relationships. Thus, Robo sapiens may exhibit an affinity for their "parent" species although, on many levels, their intelligence would exceed our own. They might also ask us some awkward questions, such as: Why are there so many poor people in the world? Why is there so much inequity? Why do some people die of easily preventable diseases when others are able to spend millions of dollars fixing themselves? Why, if Robo sapiens received the prime directive of not harming a human, do humans harm each other with such ferocious regularity? Why don't humans program their own brains to have the same prime directive? These vast inconsistencies may perplex the Robo sapiens's minds.

Robo sapiens may demand rights, but before that, some of their human protectors may demand rights for them, because they don't want them to be destroyed.

Thus, life may advance beyond *Homo sapiens.* Just as *Homo sapiens* and many varieties of humans cohabited in past,[50] so *Homo sapiens* and Robo sapiens may soon coexist. This will be far different and more rapid than previous cohabitations. To start with, Robo sapiens might come in many bodies and communicate instantly in networks worldwide. They may evolve at a rate of a generation a year. *Homo sapiens* will struggle to keep up with brain-enhanced *Homo provectus,* and Robo servers. Truly unmodified *Homo sapiens* may become extinct.

Death and Taxes?

For at least some of us, death in this century may become passé. Our thought patterns may be downloaded into a neural network and transferred into a new body. Our neural capacities may be upgraded one-thousand-fold via neural chips. We may communicate wirelessly with thousands of others. As time progresses, we may keep the memories of past generations, sharing the memories of our parents or children. And, although it sounds like we may become a nightmarish version of the Borg with only one brain controlling the collective, we may conversely branch off into subspecies of great diversity and intelligence. We may in essence be something other than *Homo sapiens.* We may be *Homo provectus.* And with that amazing transformation will come challenges to the concepts of birth, death, and reproduction.

NONSENSE

That's one view of our molecular future.

Some of the most advanced technological thinkers today say this scenario is nonsense. They say we underestimate the complexity of the human brain. James Martin, automated software developer and author of *After the Internet: Alien Intelligence*,[51] is a big promoter of artificial intelligence but a confirmed skeptic that it will ever supersede human intelligence:

> "We will have machines that are a billion times more intelligent than we are, but only in narrow, specific ways," Martin says. "In the 1960s and 1970s, the artificial-intelligence people kept telling us over and over again that in 20 years computers would be as intelligent as people. Yet nothing like that has been achieved. We grossly underestimated the complexity and subtleties of the human mind. We cannot now get close to programming what a mosquito does, much less a human being."[52]

It's hard to argue with James Martin. He's accurately predicted the financial impacts of optical fibers, networks, and Internet technologies, and he's gotten rich doing it. Other such skeptics have been right to challenge unbridled views of the future. They were right, for example, to question whether we'd find intelligent life elsewhere in the universe so easily, as some had predicted. And they were right to challenge the scientific basis of claims about phenomena such as cold fusion that turned out to be poorly substantiated. Still, criticizing the failure of artificial intelligence experts to predict exactly when computers may be as smart as us seems in some ways reminiscent of a 1934 *Nature* magazine article that dismissed physicist Enrico Fermi's postulation about existence of neutrinos as "too remote from reality to be of interest to the reader."[53] Fermi went on to achieve nuclear fission, while neutrinos have since been acknowledged as being among the most plentiful basic elements in the universe.

Artificial intelligence has convincing advocates from key fields. Prominent and accomplished individuals such as Bill Joy and Ray Kurzweil have taken issue with the assumption that smart machines will never be *totally* smarter than us. They also argue that a machine doesn't have to be "totally" smarter to be out of the control of its human handlers. Some of the top artificial-intelligence and robotics researchers at the best U.S. universities are sending us similar messages,[54] as are accomplished technologists from the business world.[55] In space exploration, self-repairing satellites that make their own decisions on what data to send back will soon orbit Earth.[56] As for the military, the U.S. Navy Center for Applied Research in Artificial Intelligence is developing "systems capable of improved performance based on experience; efficient and effective interaction with other systems and with humans; sensor-based control of autonomous activity; and the integration of varieties of reasoning as necessary to support complex decision-making."[57] Does that sound like we're expecting computers to be mindless tools? Among professional associations, new organizations such as the Foresight Institute and Extropy Institute are growing up around ideas such as evolutionary computing, while existing institutes such as the Santa Fe Institute are beginning to examine such ideas seriously. The Max Planck Institute is studying neural implants. Moreover, as seen earlier, artificial intelligence is already emerging in the form of computers that design things by figuring out how to do it independently.

These are serious players. They are not isolated. They are not dreaming about the twenty-second century: They are planning for artificial intelligence to arrive in our lifetimes.

Yet there may be more than one fly in this ointment. To support conditions that let us reach such levels of intelligence, we will have to pay close attention to how such technologies interact with the ecology. Such interactions might be beneficial, but they may also bite.

CHAPTER FIVE

THE NANOECOLOGY REVOLUTION

Throughout the fall of 2001, plumes of poisonous particles rose day and night above the smoldering World Trade Center ruins, dispersing across New York City. Scientists at the University of California, Davis, who conducted a Department of Energy study on the upwellings said they were "laden with extremely high amounts of very small particles, probably associated with high temperatures in the underground debris pile." It was worse, they said, than the Gulf War oil fires or coal-fired pollution of Beijing.[1] The U.S. Environmental Protection Agency, which had earlier given assurances that the contamination was within safe limits, was caught off guard. While health effects remained uncertain, reports persisted of illnesses among site workers, office workers, and residents, who were already traumatized by the horrendous collapse. Angry questions began to fly.

Such terrorist attacks were an anomaly in American history, but contamination from the destruction of human technologies is a frequent occurrence. Everything from natural disasters to ship collisions leads to millions of tons of toxic materials being released into the environment. This is one of the less desirable by-products of our postindustrial convenience culture.

How, then, might we prevent—or worsen—such phenomena in the new molecular era?

Nanotechnology and ecology meet in the vast new realm of *nanoecology*: the interface between molecular technologies and our natural environment.

This interface may be intimate and substantial. It may have more impact on our environment than everything that has come before. This chapter is only a brief introduction to an exceptionally complex issue.

Molecular technologies may drastically affect the cost, complexity, availability, and recyclability of energy and materials. Here are examples of nanoproducts that are already entering markets and that will soon have an impact on ecology.

A glass-based *nanoglue* converts natural hemp into weatherproof particleboard for construction.[2] This may bring hemp, a generally eco-friendly product, back into widespread use. Nanocoatings are applied to car surfaces and sunglasses, making them scratch resistant, and reducing dispersal of coating residues throughout the environment by stopping material from rubbing off.[3] Nano-scale coatings are used as a replacement for chrome on ship and submarine components, extending the service life of machinery and potentially eliminating dispersal of chromium into the environment.[4] Biocide nanocoatings stop germs from sticking to sinks in homes and hospitals, replacing toxic cleaners, but raising the question of other biological interactions with bacteria.[5] Old sinks and bathtubs are renewed with nanoceramics, extending their useful lives.[6] For skin cream, *nanopigments* filter dangerous, cancer-causing UV rays.[7] For fire resistance, a nanoparticle gel forms a heat-impervious layer that's normally transparent, but expands to foam when exposed to fire. This reduces the required amounts of fire-resistant chemicals in doors and firewalls.[8] Another *nanogel* preserves books by neutralizing the sulphuric acid in paper.[9] A company is developing a toothpaste that uses particles of enamel-like material to fill tiny cavities.[10] This stops decay, and obviates the need for toxic mercury-based materials in fillings. Researchers from Unitika and the Toyota Technological Institute have developed new technology for making a biodegradable nanocomposite with elastic properties that make it suitable for disposable bottles and foam.[11] The New Century Nanometric Technology Research Institute in Changchun, China, uses nanometric technology to produce glass that decomposes dirt on its surface and sterilizes bacteria.[12]

These are among the first environmental indicators of a molecular era. Each has advantages and dangers. For the moment, let's look at the plus sides.

As early as 1985, Eric Drexler described potential benefits from trillions of solar cells embedded in the surface of every road, building, and structure.[13] Convergence of low manufacturing costs, high efficiency, and nanocomputing may make this possible. It might bring an end of the fossil fuel age, earlier than imagined.

First, it may solve the "energy deficit" that so-called sustainable energy sources have to overcome before they produce a net energy payback. For

Fig. 22: Energy from every conceivable surface. Flexible solar panels. UNI-SOLAR flexible panels can be mounted on curved surfaces such as hatch covers, decks, and vehicle roofs, or tied onto sail covers and awnings. Eventually, nano-scale solar cells will be sprayed invisibly onto surfaces, transforming roads and building exteriors into energy generators with potential output much greater than fossil fuels. (Courtesy of Bekaert ECD Solar Systems LLC; ovonic.com/unitedsolar/flex.html)

example, to make 2001-era solar cells, fossil fuel energy was required to mine the raw materials, transport them, manufacture the cells, then get them to a location, before they could "pay back" the energy used in their own manufacture. Yet, by using solar energy in the very process of manufacturing, nanotechnology will help to recalibrate this energy deficit by rendering solar power cheap and efficient. As the phrase goes, "It's payback time." Due to their smaller size and greater flexibility, solar cells might be painted invisibly on cars, buildings, and sidewalks, transmitting power to vehicles and appliances (see Fig. 22). In this new solar economy, transmission lines could disappear, along with other blights on the landscape: hydroelectric dams, offshore drilling, oil spills, and centralized power plants. Blackouts may also cease.

What about toxic industrial processes that have left their poisonous legacy everywhere? To reverse this, we may build nanobots to seek out specific chemicals in sediments. In the 1990s this type of remediation was destructive and expensive. Soil had to be dredged from the bottom, chemically treated in huge quantities, and then put back. This destroyed trillions of healthy microorganisms; stirred up other pollution, which flowed downstream; and disrupted the local ecology. In 2001 the U.S. Environmental Protection Agency forced a big electrical company to clean up such sediments in New York's Hudson River at a cost of hundreds of millions of dollars, using a controversial vacuuming and replacement process that some said was more destructive than leaving the poisons where they were.[14]

Nanobots may eliminate the need to suck sediments from the bottom to clean them. Instead, they'd burrow invisibly into the river bottom, breaking down dioxins, arsenic, heavy metals, and other by-products of "smokestack" industries, then deactivating themselves to become harmless pieces of sand.

Thus, we may be setting the stage for superior processes and products at lower prices with positive, rather than negative, impacts on the environment from manufacturing and disposal.

OVERABUNDANCE

But what happens if we do produce cheap materials with no toxic by-products, and cheap energy with no dangerous gaseous emissions? Some optimists point to zero pollution. Yet might we graduate to *overabundance* pollution? Rather than the scarcity so feared by the Club of Rome in their 1970s treatise *The Limits to Growth*,[15] we may see a contrary trend. Most of us know what happened with the paperless office: it didn't happen. Paper consumption rose alongside printer speed.[16] A national newsmagazine carried a composite of a software executive sitting atop a tree-high pile of paper, with a caption describing how many trees were saved by software. This prompted a colleague of mine to quip that the executive was sitting on the *Windows for Dummies* manual. In truth, the paperless dream turned out to be a tree-eating machine.

Might this overconsumption accelerate in the molecular economy? It depends. For example, overproduction doesn't have to threaten forests. We might make electronic paper from more plentiful feedstock such as silicon, or make cellulose products recyclable without having to send them to recycling factories, as we do today. Despite this, we still risk cluttering the world with trillions of tons of products. As material costs plummet and home manufacturing is commonplace, the temptation to satisfy our every material whim may not be limited by cost. The hyperconsumer society might quickly get out of hand. Moreover, these products may interact in complex chemical ways unless everything is inert. In the 1990s, for example, thousands of new chemical compounds entered the environment. Most were untested for how they reacted with each other. In the molecular economy, the variety of compounds may multiply. Drugs and chemicals designed for biological functions might have unknown toxic interactions. Therefore, the economy of plenty will require production principles to avoid this. (For examples of such principles, see chapter 14, along with appendix B.)

Then let's consider megaimpacts. Take the example of water: What happens when a billion water synthesizers produce H_2O, or Saudi Arabia decides to turn the desert into a forest? Whole regional climates may change. Clearly we'd require an international regime to regulate broad changes to the ecology, a practice sometimes referred to as *terraforming*.

Material abundance may generate enormous wealth that puts more pressure on land. When individuals grow wealthy, they spread out. They may have their apartment in compact Manhattan or telecommute from their farm in Bordeaux, but as they get more money they buy second homes, travel to them, and take more vacations. Every technology center from Seattle to Bangalore already experienced this in the 1990s.

Such migrations may accelerate when housing goes off the grid with self-sufficient energy and water, then commuting times drop as aerocars and supersonic tunnels arrive.

In a surprisingly contrary trend, the line between "natural habitat" and "human settlements" might get murky, as homes blend into ecosystems. Zoning definitions for "agricultural" and "residential" may become dysfunctional. Conventional arguments to save farms from urban sprawl may be rendered obsolete by urban superfarms and shrinking farmland requirements.

When these contrary trends collide, unusual questions arise. What happens if agricultural land use is cut by 90 percent, but suburbanization explodes? The land use relationship may be inverted. We may see this first in Europe, which for centuries has been stressed by population. Today, the Europeans are leaders in nanotechnology development. Tomorrow, if much of Europe's land is vacated due to a drop in farm area needs, other uses may take over. Imagine this scenario:

Molecular-era agriculture may eliminate land-intensive farms, first by slashing the amount of space per capita needed to produce crops, then, much later, by manufacturing food with molecular assemblers. This land may be quickly reoccupied when fast, affordable transportation empowers workers to commute hundreds of miles for work or play; the Internet lets most of them telecommute; and robotic machines let them inhabit hostile zones such as mountainsides or deserts. Furthermore, greater numbers of people may be able to afford second or third homes. We may also eventually have to deal with the space demands of intelligent Robo sapiens that we've manufactured. Like the robot in Asimov's *Bicentennial Man*, they may start to ask for their own living quarters and become robotic tourists, seeking their destiny.

To release the pressure valve on land demands, we might occupy some of the abandoned farms that agricultural shrinkage would give us. If we replaced half the liberated farmland with ecocommunities, this would still leave large tracts for reforestation and wildlife. Issues such as species extinction caused by habitat encroachment might disappear. Thus, one of the great goals of the environmental movement—habitat preservation—might be achieved. But this could only occur if we redefined our concepts of urbanization, agriculture, and wildlife habitat.

At the opposite end of the overabundance scale is the impact of trillions of nano-scale particles on cell ecology. For example, when we apply nano-scale particles as coatings, some particles escape into the atmosphere or water. For individual applications, these are inconsequential. Yet when multiplied by billions of potential products, they become significant. They may be emitted during manufacturing or normal wear and tear. What happens when such particles get past our natural defenses—such as skin, mucous membranes, and lung linings—by passing through them? Given that trillions of particles such as neutrinos pass through us each instant without any apparent harmful effect, we shouldn't be too paranoid about such a possibility. But still, we don't know much about the potential cumulative impact of man-made nanomaterials.

As the twenty-first century began, environmental pressure groups and government agencies appeared underequipped to cope with these challenges. Few were considering ideas such as nanoecology. There was no standard definition or curriculum for such a discipline, although some universities and environmental agencies were beginning to look at it. As this book was being written, the number of organizations that began looking into nanotechnology's ecological implications expanded significantly.[17]

Besides the awesome power it may give us to alter continental landscapes in a few weeks or months via self-assembling machines, nanotechnology may also give us the opposite ability—to minimize human impact by mimicking nature's own processes. If molecular assembly becomes real, it could help to eliminate pollution by rendering many types of globalized trade obsolete. If consumer goods are manufactured in the location where they're used, the need for polluting technologies that constitute many globalized supply chains may evaporate. For instance, a TV or computer manufactured in 2001 used hundreds of chemicals for thousands of parts. Many of these emit toxic gases into the atmosphere each time the set is turned on.[18] Those parts come from hundreds of places. They need mining, smelting, manufacturing, trucking, warehousing, handling, and assembly. Each has its own emissions, noise pollution, aesthetic pollution, energy consumption, and waste. Once the TV completes its useful lifetime, it becomes hazardous waste to transport somewhere for landfilling, incineration, or recycling.

Yet in the nanoeconomy, feedstock for most products may consist only of carbon, nitrogen, oxygen, and hydrogen.[19] These are already plentiful throughout our natural environment. They'd require no lengthy transport, although their extraction in large quantity would certainly require regulatory controls. Trace materials might have to be imported if they weren't found in local soils or rock, yet these would be in miniscule portions. Some materials transport may still be required, but manufacturing could be done in one operation, in the place where the set is going to be used, rather than in distant factories. There may be no toxic emissions, because the set will be constructed with atomic precision, one molecule at a time. It would have fewer components, because everything would be manufactured at once. The display screen would contain the whole unit and be only a few hundred thousand molecules thick, compared to the billions now required. Energy consumption or heat waste may be less than one-thousandth of today's liquid crystal display units, because molecular construction will make them virtually resistance free, with the distance traveled by power inside the unit a fraction of what it was before. (Some such efficiencies were already being achieved in 2001.)[20] When the unit reached the end of its life, nanobots would disassemble its components down to their original raw-material state, or incinerate them with zero emissions, as gases are recovered for reuse as raw materials. This whole process could occur in a car-sized or desktop-sized molecular assembler, located in your house or neighborhood.

If we need vast quantities of materials harder than diamonds yet more flexible than aluminum, we'll order them with software that commands a molecular assembler to pump out a carbon-based fiber that weaves into Kevlar-like sheets. These will be imbued with solar-power-generating qualities that make clean, cheap energy universally available.

Thus, it may be the beginning of a molecular renaissance.

Nevertheless, it's not without dangers. The "overabundance economy" may just be getting started, and the health implications of nano-sized materials are only now being explored. We'll discuss these further in part 4.

```
┌─────────────────────────────────────────┐
│                                           │
│                                           │
│            CHAPTER SIX                     │
│                                           │
│  REVIVING TROPICAL ISLANDS                 │
│                                           │
│                                           │
└─────────────────────────────────────────┘
```

CHAPTER SIX

REVIVING TROPICAL ISLANDS

Tropical-island tourist economies dot the globe. Many have problems that include urbanization, overdependence on tourism, and income inequities. How might they be transformed into knowledge-based powerhouses that restore the islands' ecological splendor? Here's a speculative peek at the molecular age sometime around 2070. As mentioned earlier, it's risky to predict dates, because these may be shunted forward or backward depending on many factors; but with that in mind, here's how it might look.

SAINT THOMAS, 2070

The American protectorate of Saint Thomas lies at the northeast corner of the Caribbean, where the Windward Islands meet the Atlantic. Like every Caribbean island, it used to depend on tourism for survival, which was a mixed blessing. Tourism enriched some, left many local residents poor, entertained millions of visitors annually, and made for an increasingly urbanized landscape, with concrete houses dug into volcanic hillsides. Coral reefs in the national park had suffered from cruise ship and charter boat effluent. Every decade or so, the economy took a beating when a hurricane put the whole place out of commission. In short, it was an economic roller coaster ride in Eden. Nor was Saint Thomas alone. Many islands faced similar challenges in the globalized economy.

Yet in 2070 Saint Thomas has morphed into a high-tech tropical paradise that's bolstered by an artistic renaissance and a stunning ecological rebirth. As a U.S. protectorate it was a natural first candidate for molecular technologies. Its trade wind climate, scenic mountains, and proximity to diverse cultures of nearby Caribbean islands were complemented by the advantages of a more broadly advanced technological infrastructure. These qualities had made it among the first magnets for American technologists who wanted to live in the tropics without the monotonous flatness of Florida. Dozens of other tropical islands were transforming themselves as well. But due to its Americanized status, Saint Thomas was among the first to adopt molecular technologies wholesale.

First to drive this revolution had been the special-effects artists from Los Angeles and New York who'd grown sick of the rat race. With molecular technologies at their fingertips they'd found a way to get out of Hollywood but still make money. How did they do it? Wireless broadband and real-time virtual reality.

Back in 2001 virtual representation had started getting real with the release of *Final Fantasy: The Spirits Within*,[1] a film based on a series of virtual-reality games. As a technical leap from earlier cartoonlike films, *Final Fantasy* was among the first with lifelike virtual characters created entirely on computer, using voices of well-known actors. The virtually real heroine, Dr. Aki Ross, made it on to magazine covers. The film and accompanying games spawned a culture of fan Web sites that testified to the dawn of online virtual reality.[2] Ross became, in some senses, a virtually real person, and so the trend continued.

As part of this trend, young movie producers began creating virtual personalities, or "nyms," for themselves that went to meetings and parties in cyberspace. This virtual life had opened the escape hatch for video artists. While their virtual nyms attended meetings on the Internet, their real-life counterparts migrated: many to Hawaii, some to Saint Thomas. These were among the first tropical islands to get megabroadband Internet access.

Some of their friends asked, If it's possible to experience everything in cyberspace, why bother moving to a place that has the real thing? Good question. Most didn't. They were content to stay where they were and let virtual reality do the rest. That's what saved places like Saint Thomas from overpopulation after millions got the freedom to move there.

Still, a few "naturalists" preferred the occasional feel of real sand, warm tropical seas, and fresh air. Those were the ones that escaped Los Angeles. In their spare time, these new residents had given Saint Thomas a marvelous makeover.

With invention of *chameleon coatings*–nano-scale layers with trillions of superstrong, color-coded, computerized components–buildings began blending into the background by taking on the appearance of their surroundings. The area was restored to its natural splendor. As hundreds of other buildings were painted, they started to get rave reviews from tourists.

Some visitors started to get lost because structures were so well camouflaged, but this was quickly resolved when everyone got a free global positioning system locator in a page of their e-guide. This *augmented reality* gave them their precise location, and video descriptions of products. Others worried that they might start bumping into camouflaged buildings, but it quickly became obvious that this was as likely as bumping into a tree. Tourists *did* bump into trees after too much sun and fun ... but most survived. The "trees" also had electronic collision-avoidance systems that communicated with approaching vehicles to prevent them from colliding. Thus, the dangers of camouflage were minimal.

The chameleon changeover also launched an energy revolution. The noise, smell, and cost of fossil-fuel generators that chained everyone to OPEC pricing disappeared, because the coatings contained solar cells that powered each building. Residents were happy because the coatings, which cost about a hundred dollars per house, eliminated electricity bills. Fuel cells were used at night, with hydrogen generated from the local desalinization plant. A bane of the electric age—transmission lines—disappeared. Chameleon coatings were tough enough to be painted onto road surfaces. The island converted to electric vehicles that drew power from road coatings. Thus, noise and pollution from internal-combustion engines became just bad memories.

Now, in 2070, hurricanes pass unnoticed, although years ago, a category 5 storm would've wiped them out—as had happened for centuries. Chameleon coatings double as nanotube-reinforced skins that protect buildings from high winds, and seal them from floods. The ten-millimeter-thick shell is virtually indestructible in a storm. Older structures keep their charming colonial designs: only the coatings are new. Furthermore, without a centralized power grid, electricity failures are a thing of the past.

Aerocars have also started appearing, based on prototypes first flown in the year 2000.[3] Many are shocked by how quickly the aerocar has taken over. Yet the seventy-year time span between the prototype and its widespread adoption wasn't unusual when compared to the time it took for cars to replace horses.[4] Aerocars now bring wintertime residents from the north, landing on nanotube-reinforced roofs. This has provoked a frenzy of air traffic controls to stop hundreds of airborne vehicles from swooping over private homes and resorts. Yet these vehicles aren't as invasive as their helicopter predecessors, which had thuk-a-thuk propellers. Instead, they sound more like a distant hum with their high-performance rotary engines. Visual pollution is minimized because chameleon coatings help them blend into the sky. Collision-avoidance systems help to minimize the risk of crashes, as described earlier. Still, aerocars must be controlled, and the world air traffic system has undergone a revolution to cope with them.[5] The guidance system industry has exploded to accommodate millions of pilotless craft on a worldwide grid. Many of those guidance systems companies are based in Saint Thomas. Hundreds of air traffic controllers were happy to move there from the frigid north. They're no longer tied to one location, because broadband lets them "be" in the control tower of every airport on earth. They work from whatever location they want, as long as it has broadband. Besides that, many traffic controllers are now Robo Sapiens.

For Saint Thomas residents, the world has changed dramatically. Freedom from hurricanes, along with zero air-conditioning costs and rapid mainland access, has brought a year-round tourism and business boom. Thousands of mainlanders have come with virtual offices in tow. The economy has transformed from tourism to a worldwide business hub, offering everything from video effects to genetic medicine to bridge engineering.

With this has come funding to educate every child, because thousands of professionals need skilled workers, and the government has insisted on local training to get away from the two-tiered economy that plagued the Caribbean for centuries. The tax base, once limited to a tourist trade that channeled funds into services for tourists, has blossomed into one that finances community centers, medical facilities, and civic services.

The flood of residents has caused local land prices to skyrocket, making millionaires of every homeowner, but forcing the government to provide affordable housing for renters. The yacht fleet has increased tenfold as land in the Caribbean becomes a hot commodity. Newcomers choose to live on boats or on the artificial reefs that are springing up throughout the islands.

The molecular assembler has just arrived. With desktop manufacturing, fabricators are producing most consumer products. For better or worse, many stores are disappearing: shopping centers, hardware stores, clothing outlets. Some are replaced by "concept shops," where visitors can socialize over a drink or designer drug and see the latest fabricator ideas. In the broadband economy, some people still want to be face-to-face, to escape from the virtual-reality world. It's a type of social backlash against artificial environments.

Meanwhile, buildings that at first were only disguised by chameleon coatings now merge with the ecology. Their walls are made of synthesized volcanic rock. Roofs support native vegetation that hasn't been seen on the island for three hundred years.

Thus, the tiny community of Saint Thomas has morphed into a virtual-world Mecca.

How might this molecular fantasy come true? Let's take a look at the real human inventors whose work may give us such a future. There are more contributors than we think. Many of us will be surprised to find that we already know them.

CHAPTER SEVEN

WHO'S DRIVING THE
MOLECULAR MACHINE?

Molecular technology today isn't just confined
to a few university think tanks. Nor is it confined to an elite among the superpowers, big business,
or big government. Its roots are embedded in the fabric
of our industries, research institutes, and military. It
extends to wealthy and poor nations. It's like a magma
dome rumbling beneath the surface of society, ready to
blast out from countless vents, burn away every obsolete
technology, and establish a new substrate on which to build.

It's hard to describe this group without starting an
encyclopedia, but here is a brief description of those who are
powering the molecular machine: who's doing what, where.
We'll focus on American and multinational players, but hundreds more are based exclusively in Europe and Asia, and
many of them are world leaders in their respective fields.

ALTERING WHAT WE ARE—
THE GENETIC MANIPULATORS

Most every pharmaceutical and agro-industrial company in the world
is investing in the life sciences. It's hard for consumers to tell what
these corporations are, due to a wave of mergers and divestitures. Companies are racing to divest themselves of old bulk-chemical and fertilizer
operations, while forming new partnerships to achieve the scientific and
financial muscle to get a lead into genetics. Beginning in the 1990s the
institute I managed, Hamburger Umweltinstitut in Germany, did a ranking

of the environmental performance of the world's top fifty chemicals and pharmaceuticals companies. Today the list of names we assembled is virtually unrecognizable, because mergers and divestitures have resulted in so many name changes. Names such as Zeneca and Novartis are blazing their way into consumer consciousness, as prescription drug advertising has been legalized in the United States. Many of these drugs result from genetic research. Moreover, start-up companies are playing a key role. In the human genome field, the company that forced the National Institutes of Health (NIH) to accelerate its own program was Celera Genomics, started by former NIH research scientist Craig Venter.[1] While government institutes of health in many nations are investing tens of billions in genetics, this is quickly jumping to the venture capital sector. There are hundreds of other biotech start-ups, and a look at one of the biotechnology stock market indices gives a good idea of who they are.[2] Then there are the household names that consumers might not identify as being in the biotech game: for instance, the consumer product division of Proctor and Gamble. They're interested in how enzymes might improve laundry detergents or accelerate the development of biodegradable plastics.[3]

Computer companies want to know how proteins fold and how DNA replicates itself, so they might use genetics to build molecular computers. The predictable ways that genetic materials replicate, or turn chemical reactions on and off, may help to drive computer calculations.

Finally, as we'll see with most of these technologies, the military is involved with research into genetic warfare and defenses against it, as well as investigations into genetics for building superfast molecular computers. Much of this is done via the Defense Advanced Research Projects Agency (DARPA), a name that keeps popping up across the range of molecular technologies.[4]

COMPANION MAKERS—
THE ROBOTICS MANUFACTURERS

With trillions of submicroscopic robots doing everything from cleaning the toilet to repairing DNA strands in our blood, robotics may in the future provide the eyes, ears, and legs of molecular technology. Robots have a long way to go—from hand-sized to molecular size; from dumb terminals to self-taught entities—but the investment levels are impressive. Pittsburgh is home to many of the world's top robotics experts. Many circulate around the Robotics Institute at Carnegie Mellon University.[5] Among their first applications was the Dante robot, designed to explore the insides of volcanoes. Similar such robotic crawlers were used by, for example, experts from the

University of South Florida's robotics institute to search wreckage of the World Trade Center, in pockets that were too small or dangerous for workers to enter.[6] Japan is still the world leader in commercial robotics, especially for automobile assembly but also for toys. Sony's robotic dog, AIBO,[7] has taken the toy world by storm although in 2002 it still cost upward of a thousand dollars (see Fig. 23). This "dog" arrives in the package as a "puppy," then its computer brain learns as it grows with your commands. It asks for attention but doesn't need to be walked or fed. AIBO also has companion "anibots," such as "Latte" and "Macaron." Their emotions and other abilities are determined by a removable "memory stick" that gives them special characteristics.[8] The customization of these capabilities was so successful that some programmers began to offer their own downloadable features on the Internet that let AIBO dance, speak, obey wireless commands, and share its video vision. AIBO owners were delighted with these new emotions for their "pets," but the programming sparked intellectual-property debates about who has the right to customize such emotions. The outcome of that discussion may have far-reaching implications for our relationship with robotic companions.[9] Likewise, Honda's four-foot-tall walking robot, ASIMO,[10] is a star attraction at commercial exhibitions. It's designed with a user-friendly appearance in mind, so that we mortals get used to the idea of having robots around (see Fig. 24). It walks eerily like a human. It can grasp things and move them. But to see fast robots, watch Germany. For several years running, the University of Freiburg has won RoboCup, the competition that pits robots against each other in soccer, as described earlier. On a more delicate scale, medical-research institutes are lessening the

Fig. 23: A friend for life? Like any human or animal, Sony's AIBO robotic dog goes through the developmental stages of an infant, child, teen, and adult. Daily communication and attention determine how AIBO matures. The more interaction AIBO has, the faster it grows up. (Courtesy of Sony Electronics Inc.; Entertainment Robot America; www. us.aibo.com/2d/electronics/aibo/index.html)

Fig. 24: Your robotic assistant. Honda Motor Company developed a lightweight humanoid robot named ASIMO that walks like a human being. Its people-friendly size lets it perform tasks within the realm of the human environment. Some experts say that the way robots look to us will determine whether we fear or embrace them throughout society. (Courtesy Sandia National Laboratories, MEMS and Novel Si Science and Technology Dept., SUMMiT Technologies, www.mems. sandia.gov)

invasive impact of surgery, or traveling into our bodies without surgery at all. Robotic surgery is being pioneered by Canadian hospitals. Biopsies, appendectomies, and cardiac bypasses have already been performed with help from robotic assistants. Known as *telementoring*, this lets surgeons view a case thousands of miles away from the patient via a robotically manipulated endoscope, and instruct on-site colleagues how to operate. Furthermore, robotic arms have been used to remove the gall bladder from a pig.[11] These arms perform delicate tasks, much in the same way that the robotic Canadarm does on the space shuttle and space station, except at a much finer scale. Companies in California, such as Intuitive Surgical Inc.[12] and Computer Motion,[13] along with European companies such as Karl Storz Endoskope in Tuttlingen, Germany,[14] do similar work.

THE NEXT GENERATION—
ARTIFICIAL INTELLIGENCE

Robotics, computing, and artificial intelligence are inextricably linked, so it's hard to identify a single group that leads this field. Military and security intelligence agencies throughout the world are involved because they're looking for ways to replace soldiers with more expendable machines, and also keep track of what potential enemies are doing with every technology, whether privately or publicly. The U.S. National Security Agency, for example, is a thirsty client. It sucks up every available bit of memory and processing power it can get its hands on. It's a major client for the biggest, baddest computers and software that can intercept, translate, and summarize information at an astonishing pace.[15] Yet the true brains in this are the military contractors and universities—the private and academic sectors—who are building the enabling tools. Companies such as IBM and Hewlett-Packard (HP), plus labs such as those at MIT in Cambridge, are working on genetic programming,[16] as discussed earlier. Stanford University does related work. Genetic programming may help the military to crack codes or design sophisticated weapons that are more efficient than their human designers make them now. The navy also has an artificial-intelligence unit, as does NASA's Jet Propulsion Lab in California.[17]

A MAINFRAME IN YOUR EYE

Big computer and chip companies such as IBM, HP, and Intel are racing to manufacture a nanochip that puts a thousand times the computing power of a desktop in a speck of dust so small that it would fit in your eye. By 2001, artificial retinas consisting of powerful silicon processors developed by Illinois-based Optobionics Corporation were implanted in blind human patients (see Fig. 21).[18] Yet these are orders of magnitude less powerful than what's being researched. The military wants to acquire such sensing and processing applications for weapons guidance and espionage. Imagine, for example, being able to send undetectable nano-scale eyes and ears into enemy headquarters, or to see every part of a battlefield in real-time 3-D and calculate the incoming trajectory of every missile without putting soldiers at risk. The military do such investigations through DARPA. A team headed by James M. Tour, a Rice University professor and cofounder of Molecular Electronics Corporation, is using DARPA funding to develop the first nanochip-type device. "I want to see us run up the tail of every chip maker around," Tour says. "This will change the landscape for some huge, global industries."[19] Meanwhile, researchers at Los Alamos Laboratory, where many components for the atom bomb were developed, are aggressively pursuing chemical self-assembly methods where microchips are manufactured in a beaker, thus bypassing about thirty lithographic steps and compressing the manufacturing time from days to hours.[20] Scientists at the University of Rochester, New York, have gone way past "micro" size by using light beams to perform computing operations, thus opening the door to computers that are thousands of times smaller and more powerful than those today.[21]

TURNING BYTES INTO ATOMS

In an economy where atomic structural formulas for building household furniture could be delivered over the Internet to personal fabricators, broadband Internet becomes the backbone. To transmit terabytes of data, we'll need big data pipes and superfast switches that let this data go everywhere it needs to go. Those networks will be everything from fiber optic to wireless. Cisco, Nokia, Ericsson, Nortel, and AOL/Time Warner are among the emerging giants of Internet broadband. Likewise, the software writers: Sun, Microsoft, and Oracle, among others, are competing to establish languages that allow us to build products in our kitchens by receiving data for fabrication via the Internet.[22] Thousands of smaller companies are also working on this. The rules by which these languages are developed are established by industry-wide groups such as the Internet Architecture Board (IAB),[23] Internet Software Consortium,[24] and Internet Engineering Task Force (IETF),[25] which try to make sure that Internet languages don't contradict each other or plug up the Web. The importance of these industry-wide consortia in establishing networks that everyone can use is not to be underestimated. This topic is further discussed in chapter 21.

DESIGNING INVISIBLE THINGS

Building blocks of molecular technologies are happening at a level we don't see; therefore, we need virtual-reality software that visualizes for us, in three dimensions, what we're building (see Fig. 7). The University of North Carolina codeveloped one of the first such systems. At another level, for years the imagination industry—Disney, AOL/Time Warner, DreamWorks, Miramax, and Sony—has been depicting how things might look in an era where imagination is the only limitation. Hollywood, with its special-effects expertise, may become one of the sources for translating ideas into software, software into bytes, bytes into fabricator instructions, and then those instructions into novel products. The filmmaker revolution that drives this is the digitalization of special effects—creating illusions from digital manipulation of images instead of using traditional physical effects. One of the undisputed leaders here is Lucasfilms's Industrial Light & Magic, which grew from the *Star Wars* films.

A CONVERSATION WITH YOUR CAR

We saw in chapter 3 that our cars are evolving into mobile homes and offices. Voice recognition and synthesis are essential for this. Companies such as IBM, AT&T Labs, BellSouth, Nortel, Bell Labs, Nokia, and Ericsson are vying for, or cooperating with, specialists such as Conversay, Nuance, Speechworks, and others for this trillion-dollar market. Lernout & Hauspie—once a leader in the field—took a dive from the top to bankruptcy in 2000, a stunning display of how companies can rise and fall dramatically. In 2001 its technology was acquired by Scansoft, another company on the rise.[26]

We already see examples of voice activation in banking and reservations systems, but so far, most of these talk in primitive ways that use prerecorded or precomposed synthesized voices. They require banks of energy-gobbling servers with huge memories. Nanochips may transform this market, because once memory and speed become supercheap and compact, it's possible to imbue everything from scalpels to notepads with voice recognition. Furthermore, imitation of your own voice will soon be possible. AT&T Labs, along with other companies, has developed a system that can reproduce a voice after absorbing a few hours of spoken segments. This has advantages and drawbacks. We might get our car to talk to us in our own voice, or we might imitate someone else's voice over the phone to misrepresent him or get privileged information.[27] Text-to-speech is especially important, and using something known as *concatenation algorithms*, we can program a computerized voice to read extended text. This can be used to voice-enable text-based Web sites. Lernout & Hauspie developed some such software, known as RealSpeak.[28] The barrier to using this for more than a few sentences is computing power. It needs a few gigabytes of random-access memory (RAM). This stretches the capacities of most servers, if they have to deliver the service to thousands of users. More RAM will be available in the next few years. This will let blind users voice-access most documents via the Internet. It's a race between voice synthesis and retinal implants, to see which comes first for vision-impaired persons: sight or sound. To minimize the risk of programmable voices being used for deception and fraud, encryption and voice confirmation technologies will be required. Hot tip: once the nanochip is invented, invest in the voice recognition industry and hedge your bets by investing in encryption, because each of us is going to want to protect or detect our own voice.

NUTS AND BOLTS MECHANICS– ENTREPRENEURS

Who will be the next Bill Gates? James Von Ehr started Zyvex, a Texas-based company that aims to build a molecular assembler. James Tour, a Texas-based nanocomputing pioneer, is after the nanochip. Naomi Halas of Rice University is part of a start-up that works on nanoshells.[29] Evelyn Hu is science codirector at the California NanoSystems Institute, and although she is an academic, she participates in programs that have led to patents. James Ellenbogen at the Mitre Corporation in Virginia is working on nanoelectronics. Joe Jacobson at MIT Media Lab is leading efforts on digital fabrication. Nobel prize recipient Richard Smalley founded Carbon Nanotechnologies Inc. in Houston. Thousands of researchers are pursuing a multitude of molecular money trails. Many of them are at the universities where nano research is underway. The Bayh-Dole Act lets federally funded researchers share in the ownership of their work, and has encouraged a flood of entrepreneurs. For information on start-up companies, contact the NanoBusiness Alliance, a new industry association started to promote the business of nanotechnology.[30]

BUILDING ARTIFICIAL ENVIRONMENTS

Constructed environments are prerequisites for us to work permanently on the ocean floor, study the bottoms of glaciers from crush-proof pods, or use the Moon and Mars as launchpads to populate the rest of the solar system. Molecular science is the foundation. The space agencies–NASA; the European Space Agency; and the space agencies of Japan, Canada, China, Brazil, and Russia–are pioneering such environments largely for the International Space Station, but also for the Moon and Mars. They're working on space medicine–the art of staying alive in extreme environments. We need robotics to build space-based settlements and space medicine in order to occupy them, as has already been shown with the space station. Big aerospace companies such as Lockheed Martin build some of the hardware, but smaller players, such as Canada's MacDonald Dettwiler and Associates, have perfected remote-sensing technology for tools such as the Canadarm.[31] A group of Mars enthusiasts–many who are members of the Mars Society cofounded by engineer Robert Zubrin and science-fiction writer Kim Stanley Robinson–plans to commercialize space colonization. In 2001 the Mars Society constructed a tiny artificial habitat in the Arctic as a precursor to a research station on Mars.[32] An argument rages over whether we'll need such artificial habitats for humans, or whether intelligent robots might do the work better and more cheaply for us in hostile places. In another field, the case for artificial environments may be indisputable: if our own earthly environment turns hostile due to volcanic eruption, asteroid hit, or climate flip. Space pioneers who plan artificial environments on other planets may be the first ones qualified to develop enclosed ecosystems here on Earth, so that large populations might survive a climate catastrophe. We'll discuss this further in the next parts of the book.

THE CYBERMONEY MANAGERS

Cybermoney and cyberbartering take on new meaning if cash takes a back seat to information as a means of exchange; materials are taken from the air or soil to build our consumer goods; and things are built from bytes. Corporations such as eBay, E*TRADE, and cybercash credit companies represent the seeds of what may be a globalized alternative currency system.

THE VENTURE CAPITALISTS

Although many got burned on the dot-com craze, venture capitalists are alive and well when it comes to molecular technologies. For biotechnology and robotics, seed capital companies number in the hundreds. For artificial intelligence and nanotechnology, the numbers are fewer. Big companies such as IBM, Hewlett-Packard, and Xerox are investing heavily. The *Nanotech Investor*,[33] an investment newsletter, exemplifies the explosion in venture capitalist interest. A California-based company called Technanogy[34] invests in nanometals with the aim of becoming the foremost nanotech metals company. Right now, the company works on rocket fuels. Texas-based Zyvex Corporation, as discussed previously, aims to build the first molecular assembler and is owned by a hands-on venture capitalist, James Von Ehr. Minnesota-based Molecular Manufacturing Enterprises[35] is an example of a nanotech seed capital company. Many more players will soon be around, and some of the ones mentioned here may soon disappear. The field moves as quickly as the research does. (For more on this, see chapter 18.)

SETTING THE STANDARDS—THE NIST

One of the best-kept open secrets of nanotechnology is the National Institute of Standards and Technology (NIST)—but many people might know it by its old name, the National Bureau of Standards. This is the place that sets the standards for how industry operates. It has special implications for nanotechnology, because its very nature is measurement. That's important to nanotechnologists. Split between Boulder, Colorado, and Gaithersburg, Maryland, it employed thirty-two hundred professionals with a budget of just under $700 million in 2001.[36] What does the NIST do and why is it so important? Basically, it's been doing what nanotechnologists have done for a long time—measure. It has three roles:

∞ *Standards.* The NIST sets standards for size, weight, speed, temperature, and density of just about everything that's sold. Without standards, manufacturers can't reproduce things exactly, and trading partners can't agree on precisely what it is they're trading.

∞ *Advanced measuring techniques.* The NIST invents machines and processes that make measurements, and often releases the devices to industry.

∞ *Materials characterization.* The NIST describes exactly what materials are.

For example, by studying how nanostructures behave, the NIST tests and catalogs their properties, then sets standards for measuring them. This gives the institute leverage for determining how nanotechnology is characterized in industry. Watch its work for clues about what's soon to arrive on the industrial scene.[37]

LAWYERS (NO, WE WON'T GET RID OF THEM)

As explained earlier, genetic programming—where computers may render patents outdated in hours or days—promises to create a legal quagmire that's just beginning. One example is the Napster case, where computer file sharing resulted in millions of music copies being distributed each day.[38] Fenwick & West is one law firm that handled the Napster case. They have also been involved in other high-profile intellectual-property cases, and represented many Silicon Valley companies. David Boies, an attorney with Boies, Schiller, and Flexner, represented the federal government in its antitrust case against Microsoft. He's known for his work on the 2000 presidential election ballot box controversy in Florida, where apparently the computer-human interface didn't work too well, and he also defended Napster.[39] In anticipation of the linkage between nanotechnology and intellectual property, at least one Washington, D.C., firm, Foley & Lardner, set up a unit to deal with nanotechnology, especially the intellectual-property and start-up sides.[40] No doubt, hundreds of law firms specializing in intellectual property will be drawn into the fray. If you're a student going into law, have a serious look at intellectual property.

A MATTER OF NATIONAL PRIORITY—ESTABLISHING THE COMMITMENT

The European Union, the United States, China, and Japan have each announced nanotechnology development programs that coordinate research across dozens of institutions, including Harvard, Columbia, Cornell, Rice, and Stanford universities, along with the University of Washington, the University of California, the National Science Foundation, the National Institutes of Health, the United Kingdom's Oxford University, Germany's Max Planck Institute, Minatec in France,[41] the University of Tokyo, China's Academy of Sciences, Zhejiang University, and the Chinese Society of Materials. In an unusual move that symbolized the vast new multinational networks that have emerged, the European Commission and U.S. National Science Foundation began collaboration on development of nanotechnology.[42] Apart from big national players, a subgroup of small-nation institutes is racing to use the technology to gain big-country clout. For example, in Israel, Tel Aviv University, Haifa's Technion Institute of Technology, and Jerusalem's Hebrew University started multimillion-dollar research centers in nanoscience.[43] Taiwan's Industrial Technology Research Institute is doing the same.[44] The University of Melbourne in Australia[45] has a program, as do several universities in Canada, including the University of Toronto and the University of Alberta,[46] as part of national commitments to nanotechnology.

THE ART OF MOLECULAR WARFARE

Military applications are among the most awesome drivers of molecular technologies. The players are referenced throughout this book. American efforts are led by the NSA and DARPA, along with a web of contractors and universities. The Russian, European, and Chinese governments are also investing. The Gulf War stunned Europe into realizing the superiority of U.S. military technologies. This has sparked a silent molecular arms race between Europe and the United States, although they are supposedly allies in NATO. The U.K. Defense Ministry is setting up its own academy to research such military applications.[47] Small countries and developing nations are investigating military applications of molecular technologies as a kind of "equalizer" that might give them a shortcut to military equality with the nuclear powers. One of the interesting early results of this research has been to render the United States's multibillion-dollar "stealth" aircraft visible to a new form of radar.[48]

THE INFORMATION BROKERS

Those who disperse news and views about new technologies are going to play an increasingly decisive role in a molecular economy, where information is everything. First there are the content providers: those who produce what we read, watch, and hear. General Electric, News Corp., AOL/Time Warner, Disney, Sony, Viacom, Seagram, Bertelsmann, and EMI are some of the controlling conglomerates. For general information about how the molecular age is unfolding, look to the *Dow Jones News Service*, the *New York Times*, the *San Francisco Chronicle*, the *Los Angeles Times*, the *Economist*, the *Financial Times*, *New Scientist*, *Scientific American*, *Science News*, *Discover* magazine, *Wired*, *Red Herring*, ZDNet, CNET, *Space Daily*, Space.com, and *MIT Technology Review*. For scientific details on the molecular era, look to journals such as *Science* and *Nature*, or trade magazines such as *Chemical & Engineering News*. They are a few among hundreds. In Germany, where molecular technologies are to be watched, a German-language science show, *Nano*, is about the only such daily show in Europe[49] or the world. *Frankfurter Allgemeine Zeitung*, a daily newspaper, took a leap ahead of many German newspapers by establishing a sometimes-controversial focus on the interplay between technology and society. It also has a good English-language section.[50] In Japan, try the technology section of *Japan Today*.[51] These are just small samplings. The field is unfolding so quickly that mergers and acquisitions may soon put this list out of date. For instance, upstarts such as Small Times, *Nanotechnology* magazine, Nanotech Planet, Nanotechnews.com, Nanoinvestor News, and nanomagazine.com,[52] and dozens of other nanotechnology Web sites may ascend quickly, languish, or be acquired.

Who among these masses of media is piquing our interest in molecular technology and telling us how it works? On television, cable networks are ahead of established broadcasters. The Discovery Channel runs numerous science series. As mentioned earlier, Germany's 3sat broadcasts the TV show *Nano*. PBS, and especially *The Charlie Rose Show*, has taken a lead in exploring the concepts and personalities that drive these new industries. Rose has done an admirable job of ferreting out the leaders, then giving them a chance to paint a picture in their own words. By contrast, the big networks such as ABC, NBC, and CBS seem stuck behind, aside from the occasional specials or news segments. In Asia, the *South China Morning Post* has a respectable technology Web site section for English speakers who want an idea of what's going on. India also has a variety of English-language technology Web sites, due to the rapid growth of molecular science in South Asia. For a look at the potential downsides, have a glance at the magazines and annual reports of environmental organizations. For both sides, read the *Futurist* magazine and ENN.com–the Environmental News Network.

DRIVING THE IMAGINATION

Though they are criticized for giving us unrealistic expectations, science fiction writers have also inspired many scientists. Among the older generation of science fiction writers, Arthur C. Clarke and the late Isaac Asimov–scientists in their own right–helped our imaginations run wild. Among the "young Turks," Kim Stanley Robinson shows how we might establish a new planetary civilization,[53] and Tad Williams gives us a dose of virtual reality in his *Otherland*[54] science-fiction series. In Hollywood, despite a gaggle of sci-fi films, only a few directors stand out. The late Stanley Kubrick was certainly one. Steven Spielberg is another, with *Jurassic Park* and *Close Encounters*, but more profoundly with his rendition of Kubrick's posthumous, *A.I.* Gene Roddenberry, creator of *Star Trek*, popularized molecular concepts, as did the team that followed his example after he died. Others are worth mentioning: Larry and Andy Wachowski directed and wrote *The Matrix*, about a world where human minds live in a computer-constructed virtual reality while their bodies are used as food by artificially intelligent creatures that took over the world after humanity nearly destroyed itself in a war. Saul David and Richard Fleischer, the producer and director of the Academy Award–winning 1966 film *Fantastic Voyage*, about a trip through the human body, created a masterpiece based on Harry Kleiner's screenplay that still airs today.

Painting Visions of What's Possible

Outside the realm of science fiction, a few visionaries have propelled us forward. Much early criticism was leveled at K. Eric Drexler and Chris Peterson for their allegedly unrealistic portrayals of what molecular nanotechnology might do. Yet it is they who fired the imaginations of so many researchers and gave us ideas of where these technologies might lead. James Von Ehr is a straight-talking Texan who isn't afraid to put his money where his mouth is when he says he's going to build a molecular assembler. There are many others. Among them, Douglas Hofstadter, John Koza, Ray Kurzweil, Hans Moravec, and Neil Gershenfeld are practitioners who give us visions of artificial-intelligence futures. Former *Fortune* magazine editor and renowned writer Alvin Toffler stands out for *Future Shock* and subsequent books in the 1970s and 1980s. Engineer Robert Zubrin prods NASA with a hot stick, and envisions how molecular technology might take us to Mars.

One of the drawbacks is that so few women have managed to make their mark at the visionary level, although the list is growing. On that list are Chris Peterson and Gayle Pergamit, who are discussed earlier, along with Naomi Halas of Rice University, Ottilia Saxl of the University of Edinburgh, and Evelyn Hu of UCSB. That's not to say many others don't exist, but rather that most of these technologies are still dominated by men. It's a disturbing situation that, while on its way to being remedied, still bodes poorly for a balanced gender approach to developing artificial intelligence.

One of the more significant scientific developments of this generation is that scientists and their wealthy technologist counterparts in the business world are starting to discuss the societal implications of technologies *before* they're invented, and the meaning of artificial intelligence *before* it exists. For example, a special issue of *Chemical & Engineering News,*[55] marking the 125th anniversary of the American Chemical Society (ACS), reveals that a new generation of chemists is pondering the moral dilemma posed by the vast power of molecular technology. Hundreds of conferences are also beginning to address such issues.

Yet only a few gatherings have brought together scientists and technologists who've written books aimed at a general readership and have also made a lot of money in technology. This is a rare breed.

On April 1, 2000, one of many conferences about the future of technology took place at Stanford University in Palo Alto, California.[56] The panelists who spoke at this conference were especially qualified to talk about the interplay between molecular technologies and humanity. First, many of them had published controversial best-selling books about the future impact of technology on society. Second, most of them had made a big impact in their respective fields and were recognized for their work with prizes ranging from the Pulitzer to the Nobel. Moreover, they were at this meeting to discuss an unusual topic: whether humans would be replaced by artificially intelligent machines in less than a hundred years—a subject normally reserved for *Star Trek* conventions.

Think about that. The scientific heads of some powerful companies and academic institutions got together to talk about replacing *Homo sapiens* in this century. Does that seem significant?

An outstanding feature of their presentations was that panel members were as much preoccupied by ethics as they were by the technologies themselves.

A glimpse at the panelists' biographies as summarized by conference moderators, reveals the breadth of their influence and experience, and also helps us appreciate why they warrant being listened to.

Frank Drake is chairman of the board of trustees of the Search for Extraterrestrial Intelligence (SETI) Institute, which, as the name suggests, specializes in scientific searches for signs of extraterrestrial life (they haven't found it yet). At the National Radio Astronomy Observatory in 1960, he conducted the first radio search for extraterrestrial intelligence. He is currently a professor of astronomy and astrophysics at the University of California at Santa Cruz, where he also served as dean of natural sciences.

Bill Joy, chief scientist and corporate executive officer of Sun Microsystems, is one of the most influential programming designers alive today. He was the principal designer of Berkeley UNIX, for which he received the ACM Grace Murray Hopper Award for outstanding work in computer science by someone under the age of thirty. He coauthored specifications for the Java programming language. Right now, Java is one of the Internet's universal languages. In 1997 Joy was appointed by then-President Clinton as cochairman of the Presidential Information Technology Advisory Committee.

Douglas Hofstadter is college professor of cognitive science and computer science at Indiana University. He is the author of numerous books and articles about artificial intelligence and has been identified by *Wired* magazine as one of the top thinkers in the field. His books include *Gödel, Escher, Bach: An Eternal Golden Braid,*[57] for which he was awarded both a Pulitzer Prize and an American Book Award in 1980.

Hans Moravec is director of the Mobile Robot Laboratory at Carnegie Mellon University. He is the author of *Robot: Mere Machine to Transcendent Mind* and *Mind Children: The Future of Robot and Human Intelligence.*[58]

Ray Kurzweil is a high-tech entrepreneur who pioneered voice recognition, which is now used on many personal computers. He founded, built, and sold four high-tech companies, all using artificial-intelligence and pattern recognition technologies. He is the author of *The Age of Intelligent Machines,* named Most Outstanding Computer Science Book of 1990 by the Association of American Publishers, and *The Age of Spiritual Machines: When Computers Exceed Human Intelligence.*[59] Kurzweil is the principal developer of the first omnifont optical character recognition, the first print-to-speech reading machine for the blind, the first charge coupled device (CCD) flatbed scanner, the first text-to-speech synthesizer, the first music synthesizer capable of recreating the grand piano and other orchestral instruments, and the first commercially marketed large-vocabulary speech recognition. He also received the 1999 National Medal of Technology, the nation's highest honor in technology.

John Holland is professor of computer science and psychology at the University of Michigan. In the early to mid-1960s he laid the foundation for the theory behind *genetic algorithms*—using evolution and natural selection to solve computational problems.

Kevin Kelly is editor-at-large of *Wired*. From 1984 to 1990 he was publisher and editor of the *Whole Earth Review*. He is also the author of *Out of Control: The New Biology of Machines, Social Systems and the Economic World*.

Ralph Merkle is a principal fellow at Zyvex Corporation and a pioneer in the field of nanotechnology. He is the recipient of the 1998 Feynman Prize in Nanotechnology for Theoretical Work.

John Koza is consulting professor of medical informatics at the School of Medicine and consulting professor at the School of Engineering, Stanford University. He is the inventor of genetic programming, the way in which computers approach artificial intelligence by learning to solve design problems for themselves.

A notable feature of their panel discussion was the lack of consensus on how safe or dangerous it might be to let corporations control self-replicating, intelligent machines. Some said they are too powerful to be left just in the hands of commercial interests. Capitalism-optimists countered that companies aren't about to kill off their customers and destroy their own markets. Ethics-skeptics said that regardless of who makes the rules, they will act in their own interests, and so we shouldn't speak so piously of ethical behavior. Finally, a few panelists remarked that the arguments still had to be carried through to a more complete understanding of their implications.

This lack of consensus wasn't due to poor communication skills, because these are some of the best communicators in the business. Rather, it was due to a lack of consensus among them about how to cope with the new power of the molecular future.

Moreover, never throughout this discussion did panel members mention that a natural catastrophe might play a role. In discussing whether we'd be replaced in a hundred years, no one considered that the equation might be altered by natural events, independent of our own actions. It just wasn't part of the discussion.

The absence of such recognition may signify our Achilles' heel.

PART 2

NATURE'S TIME BOMBS

We humans are like two-year-olds, just beginning to understand the immensity, wonder, and hazards of the wide world.

–Peter D. Ward, Donald Brownlee, *Rare Earth*[1]

CHAPTER EIGHT

ARE WE GETTING MORE OR LESS VULNERABLE?

On Saturday, June 23 2001, emergency managers in Hawaii and on the West Coast mobilized to deal with the threat of a tsunami from that morning's earthquake in Peru. The earthquake, initially reported as magnitude 7.9 but rapidly upgraded to 8.1, killed more than a hundred people in Peru and generated a tsunami which hit the local shoreline with waves 5 meters [15 feet] high. Would the tsunami be large enough to threaten shorelines of the north Pacific? The U.S. Tsunami Warning Centers in Alaska and Hawaii monitored tide gauges on the coast of South America and at Easter Island, but the data were ambiguous. Waves were expanding across the Pacific, but because of complicating effects such as harbor resonances, it was impossible to tell how large the open-ocean waves really were.

That placed emergency managers in a quandary. If Hawaii were to enter a Watch, then without more information, a Tsunami Warning, implying full shoreline evacuation, would automatically be triggered at 9 p.m., three hours before wave arrival time. The potential hazard of what was clearly going to be only a small tsunami had to be weighed against the costs and hazards of a nighttime evacuation. Based on the available data Civil Defense heads and the Pacific Tsunami Warning Center together decided not to put Hawaii into a Tsunami Watch.

At midnight, the waves reached Hilo, Hawaii with a peak amplitude of 40cm [two feet], far larger than anticipated, and only a little below the level where harbor facilities and pipelines begin to suffer damage. Not until a week later, when seismologists had a chance to analyze more data, was it clear that the earthquake was actually the largest earthquake worldwide in 25 years, with a magnitude of 8.4. If the true magnitude had been known June 23, a Tsunami Warning would unquestionably have been called in Hawaii. The shoreline would

143

have been evacuated and many potential shoreline sources of pollution removed, shut down, or protected. It was a very, very close call.²

At the dawn of the third millennium, we still don't have dependable networks to tell us whether our coastal infrastructure might be wiped out in a few hours.

So much has been written about the dangers of, skepticism about, and defenses against natural disasters that it's justified to ask how this book could add significantly to the many volumes already published. What's missing from current literature that might make these chapters worth reading?

An odd thing is happening as we discover more about nature's time bombs. On one hand, our survival chances are improving, because we're finding out more about them. This is good. Yet on the other hand, our confidence may be badly misplaced, because we're also learning that we're still defenseless against truly big disasters. Furthermore, we're building into more zones that are prone to disruption. As we grow more confident in our technologies, our exposure is increasing. This is bad.

It's the contrast between what scientific investigation tells us, and what we may be lulling ourselves into, that bears investigation.

Here are some symptoms of the disease.

In May of 2001 the whole electricity-generating grid of Brazil was crippled by a drought that dried up the nation's hydroelectric dam reservoirs. Normally, these dams provided 90 percent of the electricity supply. The presidential palace went on half power. Streetlights were disconnected. Rio de Janeiro considered shutting down its famous hilltop statue of Christ. Within weeks, supplies of candles and generators for a nation of 170 million were sold out. The crisis put an abrupt halt to Brazil's economic expansion and led to increased unemployment. As a true indication of how serious the crisis was, night games of Brazil's national favorite pastime—soccer—were canceled.³

It was a stark display of how fragile we are. Despite two decades and hundreds of billions of dollars in expenditures on "renewable" water power, nature was able to undo much of it with a few years of drought. Moreover, dryness also hit many U.S. reservoirs in 2001.

Imagine if such a drought were to continue for five or seven years, as has happened in many regions of the world. In America one such drought—exacerbated by bad soil management—led to the "dust bowl" years of the Great Depression, where thousands died, millions were impoverished, and political instability emerged.⁴ Parallels are chilling.

Politically, the Brazilian time bomb ignited quickly. Even before power rationing came into effect, leaders of big industrial conglomerates were quick to criticize the government's "incompetence." The public soon joined in.[5]

Brazil's crisis shows what happens when an industrialized economy comes to depend on a centralized source of energy. What's the solution? One answer is to build twice as many dams. Yet besides the environmental costs, this might also worsen the economic situation later. For instance, once those dams' energy supplies were absorbed by new demand, if another drought struck, the cycle would repeat itself on a much more destructive scale.

Brazil isn't the only place where this risk exists. In Miami a similar drought threatened the region's drinking water, forcing the state government to declare a disaster and consider pumping sewage back into aquifers.[6] In New York City, overheated wires and an excessive demand for air conditioning during the summer of 1999 strained the electric power grid, blacking out much of Manhattan for a day[7] and causing millions of dollars in losses and damage.[8]

Despite these problems, the conventional wisdom among disaster experts, scientists, and citizens is that our modern society is far better equipped to deal with big natural disasters than it was before.

Jesse Ausubel, director of the Program for the Human Environment at the Rockefeller University, gives a classic interpretation of this widely held view:

> A range of social and technological developments have lessened human vulnerability to the natural environment, including climate. The lessening trend is widely repeated throughout the world, explained by industrialization, better-built structures, telecommunications, and institutional innovations, in short, development.
>
> Compare yourself to your grandparents and great grandparents. Climate surely mattered more for our ancestors who crossed perilous seas in windblown boats, struggled with horses and wagons through the mud when it rained, prayed for a shining harvest moon, and dried fruits and canned vegetables to tide them over the long winter.
>
> Numerous facts confirm the lessening. For example, the tornado death rate has decreased sharply in the United States in this century. With indoor malls for shopping and domed stadiums for athletic events, climate matters less.[9]

At one level, this view is accurate. Yet at another, it doesn't account for less frequent but more devastating anomalies. Furthermore, these may not be anomalies at all, but rather phenomena that march to the beat of a different drummer whose tune we haven't yet learned.

In a paper entitled "Existential Risks"[10] Nick Bostrom, an expert in probability theory and the philosophy of science at Yale University, gives a summary of really bad things that might throw such a rosy outlook onto the junk heap: threats that could lead to our extinction, or destroy the potential of Earth-based intelligent life.[11]

Bostrom argues that some of the most profound risks to humanity have gone largely unnoticed. Several of these possibilities seem extremely remote. Yet the advantage of having a probability theorist rather than a disaster expert compile such a list is that it puts our conventional view of things into a different perspective. This is vital, because one thing we've learned about the future is that surprises tend to come from places where the status quo isn't looking. September 11 was a small-scale example.

Humans cause some of the risks we'll be discussing, so it may seem strange to lump these together with "natural" risks, such as a storm or earthquake. Yet many technologists argue that human technology is only another part of nature, and in many ways mirrors it. In his book *The Age of Spiritual Machines*, Ray Kurzweil describes the parallel:[12]

> A key requirement for an evolutionary process is a "written record" of achievement, for otherwise the process would be doomed to repeat finding solutions to problems already solved. For the earliest organisms the record was written (embodied) in their bodies, coded directly into the chemistry of their primitive cellular structures. With the invention of DNA-based genetics, evolution had designed a digital computer to record its handiwork.[13]

Kurzweil goes on to explain how the pace of learning accelerated, with cells organizing themselves into societies and those organizing themselves into plants, then animals. As primates emerged, the pace quickened from hundreds of millions to tens of millions of years. Then, fifteen million years ago, humanoids emerged. After that, *Homo sapiens* came along only half a million years ago. Today the pace of natural genetic change is down to hundreds of thousands of years.

Kurzweil completes the link by arguing that evolution has since been taken over by a human-sponsored variant that we call *technology*. This mimics natural processes at a pace that's been gathering speed since the beginning of known time. He says that such technology, through its very existence, leads to other evolutionary paths. Thus, it's part of evolution (see Fig. 1).

Looking at it that way, we see how the risks created by technology may be an outgrowth of natural events, just as we're a product of evolution. The link between technology and evolution is central to our understanding of where we sit in the evolutionary chain.

Patrick Moore, an ecologist who cofounded Greenpeace but now disputes its methods, also advocates the technology-evolution continuum in his book *Pacific Spirit*:

> Perhaps the central myth that has been created in the war of words over the environment is that human activity is somehow "unnatural," that we are

somehow not really part of nature but apart from it. . . . [Yet] the central teaching of ecology is that we are part of nature and interdependent with it.[14]

In the context of what Kurzweil and Moore argue, it makes sense to evaluate risks we create alongside the ones nature creates, because we're part of natural processes.

Nick Bostrom of Yale divides such risks into four categories:

Bangs. Life goes extinct in a relatively sudden disaster.

Crunches. The potential of humankind to develop is permanently damaged, although in some form it continues.

Shrieks. Some form of advancement is attained, but on a very limited basis.

Whimpers. The things we value as a human species disappear gradually, or things are achieved only to a fraction of what could have been done.

Here's a summary of what he says those families of disaster might encompass. Each category is arranged in descending order of probability.[15] For example, in the "Bangs" category, "Runaway global warming" is considered the least probable.

Bangs

Deliberate or accidental misuse of nanotechnology. Rogue nanoreplicators eat up the biosphere or destroy it by poisoning, burning, or blocking out sunlight.

Nuclear holocaust. Even if some survive the short-term effects of a nuclear war, it could lead to the collapse of civilization. Humans living under stone-age conditions may be just as vulnerable to extinction as other animal species.

We're living in a simulation and it gets shut down. As in the film *The Matrix*, it turns out we're living in a computer simulation. If we are, the risk is that the simulation may be shut down. A decision to terminate our simulation may be prompted by our actions or other factors.

Badly programmed superintelligence. We mistakenly give a superintelligent entity goals that cause it to annihilate humankind. For example, we tell it to solve a math problem and it complies by turning all the matter in the solar system into a giant calculating device.

Genetically engineered biological agent. A tyrant, terrorist, or lunatic creates a doomsday virus with high virulence and mortality. For example, Australian researchers recently created a modified mousepox virus with 100 percent mortality while trying to design a contraceptive virus for mice for use in pest control. The technology to do this is already widely available.

Physics disasters. The Manhattan Project bomb builders worried about an A-bomb-derived chain reaction that would consume the atmosphere. There is other speculation that high-energy particle accelerator experiments may result in an expanding bubble of total destruction that would sweep through the galaxy and beyond at the speed of light, tearing all matter apart as it proceeds.

Naturally occurring disease. What if a disease such as AIDS, for example, mutates to be as contagious as the common cold?

Asteroid or comet impact. The real but small risk that we're wiped out by an asteroid or comet.

Runaway global warming. Release of greenhouse gases into the atmosphere, it turns out, creates a self-reinforcing feedback process and we cook.

Crunches

Resource depletion or ecological destruction. Natural resources needed to sustain a high-tech civilization may be used up.

Misguided world government decides to stop technological progress. As with Prohibition, some fundamentalist movement dominates the world and permanently puts a lid on humanity's potential to develop to a posthuman level.

"Dysgenic" pressures. Currently it seems that there is a negative correlation in some places between intellectual achievement and fertility. If such selection were to operate over a long period of time, we might evolve into a less brainy but more fertile species.

Technological arrest. The sheer technological difficulties in making the transition to the posthuman world might turn out to be so great that we never get there.

Shrieks

Takeover by a transcending upload. An *upload* is a mind that has been transferred from biological brain to a computer that preserves the original mind's memories, skills, values, and consciousness. If this process is sudden, it could result in one upload reaching superhuman levels of intelligence while everybody else remains at a roughly human level. The upload could invent technologies that prevent others from getting the opportunity to upload.

Flawed superintelligence. A badly programmed superintelligence takes over and implements the mistaken goals it has been given.

Repressive totalitarian global regime. An intolerant world government, based perhaps on mistaken religious or ethical convictions, is formed, is stable, and decides to realize only a very small part of all the good things a posthuman world could contain.

Whimpers

Our potential or core values are eroded by evolutionary development. In essence, we become something else that resembles nothing like humanity. Love, hate, altruism, pleasure, happiness, and sadness are replaced by consumption and efficiency in a repeating loop.

Our "cosmic commons" are burnt up in a colonization race. Selection ends up favoring forms of artificial life that spend *all* their resources sending out colonization probes. Resources are burned up as everything is occupied, consumed then converted to something else. Humanity disappears.

Killed by an extraterrestrial civilization. We run into a superior race that clears us away for a galactic superhighway.

Which, if any, of these might get us first? Bostrom concludes that our own technologies pose some of the greatest near-term (i.e., the next few centuries') risks, more than so-called naturally occurring phenomena. But, he adds, "we shouldn't be too quick to dismiss the . . . risks that aren't human-generated as insignificant."[16]

Bostrom's list of risks focuses on high-level threats to civilization. That's helpful, because it gives us the lay of the land.

In this book, we focus more on what might delay our progress; what might stop us just as we're on the cusp of technologies that may help us adapt to some of these onslaughts. There's a close relationship between the threats described by Bostrom and those described in the next chapters. A smaller blow may send us into a vicious cycle that opens us to risks catalogued by Bostrom. Just as a wounded animal is more apt to meet with death, so a society is more apt to die if it's disabled. On the other hand, a harsh but sufficiently superficial blow might expose our vulnerability, and inspire us to prepare for the worst. Either way, such blows may influence us.

With that in mind, let's have a look at absolutes: those things that we know have happened before and are virtually certain to recur. These are the glancing blows that may throw us off balance and expose us to worse assaults, or wake us up to a new imperative.

CHAPTER NINE

SHAKING UP TOKYO AND THE GLOBALIZED ECONOMY

C an a natural catastrophe contribute to causing a world war?

The 1923 Great Kanto earthquake flattened Tokyo, causing more devastation in some ways than the nuclear bombs that were to later hit Hiroshima and Nagasaki (see Fig. 25). The quake happened over a wider area and crippled underground superstructures as well. In ninety seconds, Japan's centralized economy was annihilated. Widespread poverty followed. It was so severe that, several years later, the country was in no shape to weather the Great Depression. Such instability helped to open the door for an expansionist military government. After the military took over, war in Southeast Asia and then the Pacific ensued. Thus, Earth changed our history. Yet virtually none of our history books draw this link between the sudden impoverishment of a national economy and the onset of war in the Pacific.[1]

Many of us are only vaguely aware of how much the globalized economy depends on Japan, and how much Japan depends on Tokyo. In 2001, despite years of recession, Japan had the second largest national economy and was the largest lender internationally. Its trade underpins the Asian region and many Western economies. Most of this trade is based in the Tokyo region.

Tokyo also sits atop the intersection of three of the planet's most active tectonic plates. Therefore, it makes sense for its trading and financial partners to ask, How well did the city learn from the 1923 quake? How well prepared is it for the next one? The reply, unfortunately, is not well enough. A 1995 earthquake in Kobe shocked Japanese experts into

廊遊原吉   前失燈

廊遊原吉   後失燈

Fig. 25. Think it couldn't happen again? Tokyo before and after a big quake. On September 1, 1923, an earthquake flattened the densely populated industrial cities of Tokyo and Yokohama, Japan. Today, millions of homes in the same area could collapse and burn just as catastrophically. Houses are still made from flammable materials. Residents still cook with oil that spills and catches fire in a quake. One difference is that Tokyo is home to some of the biggest industries and financial institutions in the world. Thus, the globalized financial system may be at risk. (Courtesy of Saint Louis University Archives; www.eas.slu.edu/ Earthquake_Center/ 1923EQ/)

admitting that, despite billions spent on prevention, cities such as Tokyo may be just as exposed in some ways as in 1923, due to new development based on vulnerable structures.

Geologist-turned-journalist Peter Hadfield, who lived in Japan for many years and is consulted by the media when earthquakes strike, observed that the maximum acceleration of the ground at Kobe was estimated by Japanese experts as double that of the 1923 Great Kanto earthquake, which killed about 140,000 people and wiped out most of Tokyo. The degree of lateral shaking at Kobe caught many seismologists by surprise. Experts at Tokyo University found that parts of Port Island and Rokko Island, near Kobe, moved laterally by up to twelve feet instead of four to eight inches–the expected norm.[2] Writing in *New Scientist*, Hadfield observed:

> The unexpected intensity of the quake may prove a problem for Japanese engineers. Most highways in Japan are built to resist a ground acceleration of . . . less than half that which shook Kobe. The Ministry of Construction has now set up a panel to decide if building standards should be tightened.[3]

Furthermore, someday we might hear that an earthquake hit Tokyo but it wasn't that bad. We might share a sigh of relief. Unfortunately this reprieve may be short-lived. Such *chokkagata* earthquakes–those that occur directly beneath the city–are damaging but not catastrophic. The one to watch for is the more broadly destructive regional quake near the industrial belt, between Tokyo and Nagoya.

Banks are not prone to cry wolf. Japanese banks are especially loath to admit liabilities, as seen in the past years of hangover from a 1980s lending bubble. It's therefore sobering to read a 1989 Tokai bank study[4] warning that the next Tokai earthquake, as it has already been officially named, could cripple an interwoven globalized economy. A *Risk Management Systems* study also explains effects of such a quake.[5]

The studies estimate losses of up to 70 percent of Japan's GDP. More than eight hundred thousand houses and buildings may be destroyed and 2.5 million burned, leaving millions homeless. The damage: over $1 trillion. The capital of Japan would be partially functional, but with telecommunications and transport disrupted, and gas, power, and water lines broken. The region's international trade and banking would be interrupted. Toxic pollution would arise from fires and ruptured tanks.

In *Sixty Seconds that Will Change the World*,[6] Peter Hadfield repeats what many Japanese geologists have explained: that such an earthquake is a near certainty. The murderous shaking of this regional event is historically cyclical–and overdue. Detection of crustal tensions by seismic sensors forebodes a tragedy.[7]

An economic time bomb ticks under the Tokyo-Nagoya region, menacing global security. Yet the world seems unprepared. In the face of incontrovertible evidence, nations seem disbelieving that one of the most advanced economies, despite a history of earthquakes, could be so vulnerable. G8 countries, international financial institutions, and environmental agencies appear to have no publicly acknowledged plan to avoid an international panic, or to introduce improved technologies to the reconstruction of Tokyo if it is crippled.

Complacency in North America and Europe may be reinforced by a lower occurrence rate of large earthquakes in Western economies during the 1980s and 1990s.[8] It may also seem that because Japan spent billions on earthquake prevention, it is ready. Yet a look at what happened in the 1995 quake in Kobe dispels this false sense of security. The globalized economy is exposed if something similar occurs in Tokyo and this international lending capital goes down. Western markets rely on Japanese investment and trade. In 2001 upward of half a trillion dollars in U.S. debt and securities were held by Japan, in the form of Treasury bills, shares, bonds, and bank loans.[9] More critically, by 2001 about one-fifth of Japanese stocks were owned by foreign investors—up substantially from decades ago.[10] What happens if the Japanese stock market crashes after an earthquake, and Japan liquidates its foreign holdings to pay for reconstruction of Tokyo? What happens to just-in-time global supply lines for American and European products when so much Japanese capacity closes down?

The 1994 Northridge earthquake in California had a similar sobering effect on U.S. experts, who were caught by surprise when an undetected fault slipped. The impact of the quake on the publicly financed structures of Los Angeles are only now being fully felt. For example, Pulitzer prize–winning medical journalist Laurie Garrett, in her book *Betrayal of Trust,* chronicles how the breakdown in LA's public health-care system was caused largely by financial distress from the Northridge quake.[11] The city, its resources already stretched, could not cope with the cost of rebuilding its infrastructure.

The disruptive societal impacts of the 1994 Northridge and 1995 Kobe earthquakes plus Hurricanes Andrew, Floyd, and Mitch, are far from over— *they are still going on.*[12] They take years to make their way through national economies. The damage they wreak in terms of weakening of public infrastructures sometimes remains hidden for a decade, and becomes apparent only when we discover, for example, that we're having to close hospitals because the capital expenditures required to keep them in shape were redirected to earthquake emergencies many years ago. Thus, such events aren't onetime shots where we dust ourselves off and walk away. They endure for years, sometimes reaching far into the future.

Other centers of economic activity, such as New York and St. Louis, are unprepared for tremors as big as those that have hit their regions in the past.[13] One of the largest recorded earthquakes in America's history occurred outside St. Louis in the 1800s, causing the Mississippi River to reverse its flow. If it were to recur, one of America's busiest continental transportation hubs may be flattened (see Fig. 26). New York City might also be hit by a serious quake. There are fault lines close enough to New York to cause one. Furthermore, many Manhattan buildings are not built to earthquake standards. A tremor could temporarily shut the city, with multiplier effects on world financial systems. Most New York State residents would be shocked to find that their state is near the top of the list for potential annual damage to buildings from earthquakes.[14]

Fig. 26. Think you're safe from earthquakes if you're not on the West Coast?
Although earthquakes in the central and eastern United States are less frequent than in the west, they affect much larger areas. This is shown here with examples of two areas affected by earthquakes of similar magnitude: the 1895 Charleston, Missouri, earthquake and the 1994 Northridge, California, earthquake. Inner zone indicates minor to major damage to buildings and their contents. Outer zone indicates shaking felt, but little or no damage. Infrastructures in the Midwest are far less prepared for such a shock than on the West Coast. Moreover, the quake depicted above was far less violent than one that struck New Madrid, Missouri, in 1811. (Image from U.S. Geological Survey Fact Sheet 198-95 by Schweig, Gomberg, and Hendley; quake.wr.usgs.gov/prepare/fact sheets/New Madrid/)

Alarmed by the potential for such disruptions, the National Science Foundation and the Japanese government have been cooperating to find solutions in a multiyear project.[15] Yet, as Peter Hadfield has observed, regardless of their conclusions it's politically impossible for Japanese politicians or officials to launch the type of preventative construction programs that might protect 40 million persons in the Tokyo region from a big trembler. Nor is it reasonable to expect that St. Louis, New York, or other quake-prone centers are going to be allocated funds for similar work. New York, for example, is focused on recovering from the September 11 attacks, perhaps for many years. Public awareness about earthquakes isn't sufficient to generate the political will for added expenditures that may range into the billions. And how many residents of Missouri are aware that if an earthquake of the same magnitude that struck in the early 1800s hit again, they might see their state infrastructure destroyed? Because of this lack of aware-ness, there is a broad reluctance to spend funds to prevent catastrophic loss. Yet as we saw from the terrorist attacks in New York, the cost of ignoring such seemingly improbable threats is far greater.

That's not the worst of it, either. Scientists tell us that we have a few more items to put on our worry list . . .

CHAPTER TEN

SO, YOU'RE BORED
BY DOOMSDAY?

Galileo gazed at Jupiter through his telescope and saw its moons circling, and he
understood that Earth is not the heart of the universe.
The Catholic Church saw it and understood that it was
a threat to their own universe, so they forced him to
recant the reality he saw.

Four hundred years later, the *Galileo* spacecraft
gazed at Jupiter through telescopic eyes while a comet
tore Earth-sized holes in its atmosphere.[1] From Earth we
too, saw this. But did we understand it?

Hollywood understood it and made a lot of money from
films about the end of the world. Astronomers understood it and
urged politicians to guard against a similar event here on Earth.
But did we as a civilization understand what it meant for us?

The history of the Shoemaker-Levy 9 discovery shows how
far we've come–or not–since the days of Galileo. On May 22,
1993, the scientific world sat back in astonishment when
astronomer Brian Marsden published calculations showing that frag-
ments of a comet discovered by Carolyn Shoemaker, the late Eugene
Shoemaker, and David Levy were going to smash into Jupiter. Some
said that its parts might blow holes as wide as Earth in the planet's
atmosphere.[2] Many scientists were skeptical about the potential effects.
Planet-altering collisions were supposed to occur only every million
years or so. The chances of us being around to see one, much less having
a satellite staring straight at the impact zone, were infinitesimal.

But on July 16, 1994, it hit. And we saw it, through scores of earth-
bound telescopes and a newly repaired Hubble Space Telescope (see Fig.

Fig. 27. Dress rehearsal for Earth? Image of Jupiter's cloud tops after the impact of the first fragment of Comet Shoemaker-Levy 9, on July 16, 1994. The impact site is visible as a dark streak and crescent-shaped feature, several thousand kilometers in size, in the lower left of the image, magnified in the box. Many of us saw this event from Earth, but did we understand its implications for our future? (Courtesy of NSSDC/NASA; nssdc.gsfc.nasa.gov/planetary/sl9/html/hubble_images.html)

27). Furthermore, the *Galileo* spacecraft, from its miraculously advantageous vantage point en route to Jupiter, gave us the only direct view of fragment collisions on the far side of the giant planet. *Galileo* stored the data in its onboard recorder, then transmitted them to Earth. Surprisingly, the upper-scale predictions came true. A planetary blowout scarred the atmosphere half a solar system away.

Suddenly, after centuries of ridicule, catastrophe got respect. On Earth, this smoking galactic gun sent tremors through much of the scientific community, galvanizing the United States and Europe into funding the first official search for rogue comets and asteroids, commonly known as *near-Earth objects*. The Spaceguard Foundation was born: the first agency designed especially to seek threats to Earth that come from outer space. Offices were established in Rome, London, and Tokyo. Finally, a serious search was being undertaken. Since then, satellite and ground searches have produced a vivid new map of Earth—one pockmarked with impact craters that confirm how our planet has been irregularly bombarded.

Comet Shoemaker-Levy 9 and subsequent near-Earth object discoveries shook the traditional segments of the Darwinist world to their roots. They tore a gash in the comfortable concept of evolution as just a gradual process, where life develops smoothly over eons. Suddenly, we had eyewitness proof that catastrophic leaps occur, and may be the catalysts for evolutionary changes. The significance of this event was to establish that violent, rapidly occurring phenomena can play at least as important a role in evolution as gradual developments.

This also reignited interest in smaller disasters that had been ignored or attributed to other causes. For instance, it's now generally accepted by scientists that a near-Earth object caused the Tunguska explosion of 1908 in Siberia.[3] Investigations into hundreds of other known or potential impact sites around the world have since been initiated.

Comet Shoemaker-Levy 9 demonstrated one other point: civilization may be at risk.

DISASTER PORN

Despite, or perhaps because of, the legitimacy accorded a potential apocalypse by the Shoemaker-Levy 9 event, catastrophe became the media rage. We were, and still are, subjected to a flood of Hollywood films where Armageddon happens daily. Television networks cover every "extreme" phenomenon: avalanche, tornado, flood, hurricane, and tsunami. By 2001, despite conjecture that doomsday excitement had played itself out with the end of the old millennium, some networks still carried programming that claimed Atlantis was in Cuba, and that a comet collision twelve thousand years ago had first stripped the continent of life, then submerged it, by causing glaciers to melt and seas to rise.[4]

Some producers say that catastrophe overkill is causing calamity fatigue among viewers. Others say we're only at the beginning of an era that will imbue our culture with as much violence as the gladiator era in Rome. Screenwriters have a name for it: *disaster porn.*

Despite its perverse side, such attention is far from completely bad. It opens our minds to the real risks and impact of catastrophic occurrences. The media sensationalize such extremes, but they also shine a light on researchers who are cracking open mysteries about natural catastrophes.

These investigators range from geomorphologists traipsing across Australia in search of flood evidence, to scientific journalists warning of an earthquake that may annihilate communities and rock the world economy, to astronomers seeking wayward asteroids. Their tools include space telescopes, DNA analysis, enhanced dating methods, sediment core sampling, and computational engineering. Much evidence comes from new discoveries, but more comes from reinterpretation of old data.

This diverse work is bringing us a consistent, compelling message. Earth is not what it seems. Evolution may be a much more irregular process than formerly believed.

UPSETTING THE GREEN APPLE CART

Comet Shoemaker-Levy 9 also punched a hole in the core of environmentalism.

Conventional wisdom holds that Earth is a *closed system* in relation to matter: Little of consequence comes in, and nothing leaves. Schoolchildren are taught that, besides energy from the Sun, nothing affects our world from the outside. Big, gradual changes are caused only by the workings of this planet.[5] Moreover, we're told that human beings are the ones most likely to destabilize the biosphere. Such teachings promote a view of Earth as being isolated from the rest of space, and a view of humans as being most responsible for the stability of its biosphere.

Yet it now seems that there are exceptions to these rules. Such exceptions may call into question the rules themselves. For example, asteroids, comets, and meteors are more than just pebbles that pester Earth. Although they bring only a few tons of materials to Earth each day,[6] they also disrupt ecosystems through rare but violent collisions.

Moreover, the Sun's energy may not interact as unwaveringly with Earth as we'd like to think. Earth's protective magnetic field ebbs, flows, and flips, causing variations in the amount of solar radiation reaching the surface. This affects, for example, the rate of mutation. Ionized particles that are carried by the solar wind also disrupt our satellite and electricity grids.[7] As we learn more about interplay between Earth's magnetic field and the solar wind, it's becoming apparent that the radiation cycle is a force to be reckoned with.[8]

These factors suggest than Earth is more of an *open ecosystem* than we may perceive. Matter may not leave the planet in appreciable amounts, but it undoubtedly interacts with energy and materials from outer space. This has fundamental repercussions for environmental doctrine.

Such profound realizations, once they take hold in our society, may be on par with discoveries that Earth is neither flat nor the center of the universe. They may force a revision of the principles of environmentalism. With the preponderance of evidence one might ask, Why hasn't this reassessment happened already?

As an environmental manager, I remember where I was when Comet Shoemaker-Levy 9 hit Jupiter, as clearly as when a man landed on the moon or when Kennedy was shot. Standing in a Brazilian hotel lobby on the edge of the Atlantic forest, watching bright flashes and dark holes play across the TV screen in grainy time delay, I was struck by the realization that every preconception we had about the environment's natural stability had just changed in an instant. When I discussed this with my colleagues afterward, they thought it was interesting but unremarkable. Today, most of them still seem unconcerned. Just as it took the church—and hence, the world—a generation or more to be reconciled with the impact of Galileo's discovery, so the diverse philosophies of the environmental movement have yet to be reconciled with this gift of knowledge that the *Galileo* satellite brought from hundreds of millions of miles away.

Comet Shoemaker-Levy 9 being broken up and then swallowed by Jupiter also supports a theory that the orbit of Earth in our solar system is a special one. On one side it's protected by Jupiter's huge mass, which acts as a gravitational magnet, keeping planet-killing projectiles out of our orbit.[9] On the other, Jupiter also throws these objects off course into our path, as may have occurred, for example, in the 1800s with Biela's Comet.[10] We still haven't figured out if our relationship with Jupiter is benevolent or dangerous, or a combination thereof, but in either case it seems we're tied to each other. Our better known, yet often forgotten special relationship is with the moon, whose gravity moves our oceans every minute of every day.

Other revelations by astronomers and astrobiologists show how the conditions that grant us life are influenced by more than our own planet or ourselves. We've theorized for some time that the oceans were formed partly from comet bombardment. New discoveries in 2001 lend credence to that.[11] We now also know that, every instant, trillions of neutrinos pass through each of us, then back out into space. They seem to do us no harm, but their presence is everywhere.

Together, these show that Earth is not a closed system. It's an integral part of a solar system constantly in flux. Its environment can be dramatically altered by external conditions.

Thus, we start this millennium with the realization that our environment is open to the galaxy rather than insulated from it, and that interplanetary blips disrupt our home from time to unexpected time.

Besides the ecological significance of such a shift, the *perceptual* significance is high. The view by some policy makers of outer space as something separate from us, therefore perhaps not worth the cost of exploring may have been reinforced by the perception of Earth as a closed system. Yet Earth is not closed. Rather, it is open to energy *and* matter.

This marks the beginning of a transition from inward-looking environmentalism to a broader view based on new disciplines such as astrobiology and nanoecology, which tell us we're at risk from the same systems that give us life. Far from being a peaceful place for gradual evolution, the planet's surface sometimes turns on us with or without external intervention.

We're learning, for example, that climate changes aren't only attributable to human interference. The occasional volcanic explosion may alter our climate more than we've been able to in our total history as a species. Atmospheric temperatures may sink or rise suddenly, due to factors that are explained later in this chapter.

Part, though by no means all, of this risk is described in the controversial field of *catastrophism*, which, according to the *Anthropology Glossary*, is defined as "The belief that the fossil forms represented in each layer of the earth were destroyed by a catastrophic event and that the next set of plants and animals represented a new creation event and were organisms that survived the catastrophe."[12]

In plain language, that means most things were wiped out and then started again.

Catastrophism is ignored or ridiculed in the scientific community for many reasons. It attracts nuts. It lacks a factual foundation for its theories. It's often misused to justify a "God created the earth in seven days" dogma that enrages scientists.

Thus, it's essential to differentiate between catastrophism and the argument that catastrophe is a defining part of evolution. They are not the same. Elements of catastrophism, for example, are used often to forward religious theories that have no scientific foundation.

Deep and often justifiable scientific skepticism about catastrophism has overshadowed the search for the truth about catastrophic influences. For example, when evidence such as the Comet Shoemaker-Levy 9 collision with Jupiter surfaces, it's sometimes rationalized away as a fluke. Some scientists explain that just because it happened to Jupiter, with its much bigger gravitational pull, it doesn't necessarily mean that Earth is at risk.[13] Efforts to reinforce the "highly infrequent" theory of catastrophic near-Earth-object collisions are embodied in an article that appeared in the *Skeptical Inquirer* in 1997, analyzing a spate of asteroid collision books that hit the shelves in the years after Shoemaker-Levy 9 hit Jupiter:

> Fortunately for us, impacts large enough to produce mass extinctions are
> rare, taking place at average intervals of tens of millions of years. However,
> there is a spectrum of comet and asteroid sizes, with many more small
> impacts than large ones. Based on what we know today, impacts much
> larger than the Cretaceous-Tertiary (K-T) event are possible in the future
> (although very improbable). And impacts smaller than the K-T event – say
> by objects one kilometer or a few kilometers in diameter – occur much
> more frequently. The planet is struck by a one-kilometer asteroid or comet
> at average intervals of about 100,000 years.[14]

In this article, author David Morrison is careful to qualify his statement
by saying that such events are also randomly distributed. One may occur as
soon as tomorrow, or a million years from now. Yet the interesting thing
about this analysis is that it presents the one-hundred-thousand-year frequency data[15] for less catastrophic impacts as being more or less established
(see Fig. 28). It would be more accurate to say that we are only beginning to
find how and when they strike. For example, there is a gap in our knowledge
about clusters of space objects that exploded in the atmosphere. These may
have left widespread devastation, but few traceable marks on Earth's crust.
The Tunguska event is one such example. Other projectiles may have hit ice-
age glaciers, driving miles into their icy depths and releasing vast amounts
of climate-altering gases without leaving craters on Earth's surface.

The point is not to deny Morrison's analysis, but rather to say that it's
too early to pin the tail on the donkey in terms of frequency of smaller near-
Earth-object collisions.

Why is this so important?

For centuries, an acrimonious argument has raged between established
scientists and upstarts who suggest that the catastrophic version of evolution
deserves a serious look.

> In the early days of science, geologists were divided into two schools, those
> who believed that Nature worked by occasional catastrophes alternating
> with periods of inactivity, and those who thought that natural agents
> worked smoothly and ceaselessly throughout the ages. . . . The Unifor-
> matarians prevailed over the Catastrophists and until recently appeared
> completely victorious. However there are indications that the doctrine of
> Uniformity has been carried too far.[16]

As explained earlier, catastrophism tends to attract macabre fascination,
but for many years it was discounted as serious study because established
scientists often mocked it, and most reasonable persons correctly perceived
that there was nothing to prevent it, so why worry?

Since the 1970s a new breed of scientist has been working to break through such attitudes. Research by physicist and Nobel laureate Luis Alvarez and his son Walter uncovered the iridium anomaly: a layer of iridium found around the planet near the surface, and thought to come from an asteroid impact. This was instrumental in demonstrating that a near-Earth-object collision led to extinction of the dinosaurs.[17] This was followed by discoveries of several large-impact craters that dated to the time of the iridium anomaly.[18]

What did this have to do with our present civilization? Although it may have led to the rise of our own species by annihilating the dinosaurs and making room for mammals, sixty-five million years is an irrelevant time span to most voters, and to politicians who decide what projects to pay for that may identify catastrophic threats to society. Yet the folly of this "wait till we see the smoking gun" approach was tragically demonstrated when another kind of disaster thought to be unimaginable–despite warning signals–resulted in the loss of thousands of lives at the World Trade Center, along with a blow to the U.S. economy. Has this taught us about preparing for other types of seemingly improbable catastrophes? Not yet. The infrequency of near-Earth-object disasters has plagued efforts to get money for more research. Until the Shoemaker-Levy event of 1994, there was no broad awareness that such events occurred more than every few million years, or might be a threat to us. Since 1994, U.S. and European governments have allotted only some funding.

When it comes to discussing near-term catastrophic threats, analysts still inhabit a twilight zone shared by serious science and the paranormal. They face the task of separating fact from the hysteria of religious sects whose chosen dates for the end of the world are consistently wrong. Researchers have to weave their way past UFO cults, pyramid worshippers, and conspiracy theorists to track down the real stuff.

The dominant view still seems to be that the deep past is disconnected from present-day potential threats. Those who suggest that modern-day planetary conflagrations may be probable are regarded with deep cynicism. This self-reinforcing loop has hampered financing for investigations, and it may prove to be our undoing.

NOT JUST ONE GREAT FLOOD

In 1994, when an earthquake off the Russian coast triggered a tsunami alert in Hawaii, Coast Guard helicopters hovered over the seashore with bullhorns, urging surfers to leave the water and seek safe ground. The surfers refused. They wanted to "catch the big one." Meanwhile, some bewildered tourists thought it was a "salami alert" and swore off sandwiches for fear of food poi-

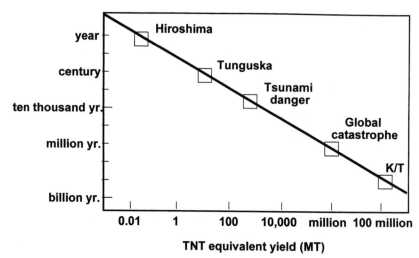

Terrestrial Impact Frequency

Fig. 28. How often are we hit? Frequency of impacts by near-Earth objects, ranging from a nuclear-bomb-sized explosion to the globally catastrophic event that wiped out the dinosaurs. The level of certainty for this graph is still unknown. The problem is that such collisions could occur next week or next century. We're still not good at forecasting because we don't have the tools to see what's out there. Nanotechnology could help solve that, with millions of tiny sensing satellites that would scour the galaxy for rogue objects. (Courtesy of David Morrison, NASA)

soning.[19] One expert who saw this noted, in a wry understatement, that most people have no idea of what these waves are or of the power behind them.

We're sea-loving animals. More than a third of the world population and economic capacity is less than two hundred miles from a coastline. As long as our love affair with the ocean persists, we face the constant threat of mass destruction by tsunami, hurricane, and sea rise or fall. Still, many of us seem blissfully unaware of the implications.

Sediment samples from around the Pacific strongly suggest that terrific waves suddenly inundated areas from Sydney to Seattle three hundred years ago.[20] Examination of other such samples by geomorphologists, such as the Smithsonian Institution's Mary Bourke, suggests that extensive coastal and inland flooding occurred within the last five thousand years. Creation and destruction of inland seas is a regular planetary event. When Hurricane Floyd inundated New Jersey in 1999, the *Wall Street Journal* quoted Bourke's conclusions that "these events can be bigger and more frequent than we have ever known in the earth's current climate."[21]

In the year 2000 the discovery of ancient structures hundreds of feet beneath the Black Sea surface appeared to support other evidence that, seventy-five hundred years ago, it was transformed from a lake to a sea by a gigantic flood, when a dam that had held back the Mediterranean collapsed.[22]

If such collapses seem too fantastic or remote to believe, one need only look back to Alaska in 1958, when a giant piece of a peninsula collapsed, carrying fishing boats toward land on a titanic wave, and then out to sea again. Several eyewitness accounts, photographic evidence, and records gathered by established scientists confirmed this event.[23] An interview from an October 2000 BBC *Horizon* documentary reveals the potential of such landslide-induced megatsunamis. Here's an eyewitness account by survivor Howard Ulrich, who was fishing in the bay.

> . . . there was a large rumbling noise from up at the head of the bay. . . . There was a slight pause. I thought that everything was over with, but some movement up there caught my attention . . . and . . . what I observed was a, like an atomic explosion. After this big flash came a huge wave. . . . I had 40 fathoms of anchor chain and it started running out off the boat. Came to the end of the 40 fathoms just snapped it like a string and then we were free and, but we were still on the front of the wave. We were swept up over the land and up above the trees. That's where I assumed that we were going to end up. I had never heard or seen of anything like this. It was unbelievable. I couldn't imagine what could have caused anything. I kept wondering just what mechanism could cause something like that.[24]

This was the highest wave ever recorded—over fifty times the height of an ordinary tsunami, and as high at landfall as any building in the world.[25]

There are larger and more devastating threats, some not known until the past few years. In the 1960s Jim Moore, a geologist with the U.S. Geological Survey in Menlo Park, California, was studying early bathymetric maps of the seafloor around Hawaii and found what appeared to be fields of volcanic rock strewn across the seabed. He theorized that they were debris from titanic landslides. Yet it took until the 1990s for his suggestion to be taken seriously, when higher-resolution maps of the seafloor showed evidence of many landslides in the area.[26] Now, scientists at the U.S. Geological Survey and other agencies are considering landslides as a cause of tsunamis.[27]

Hawaii and the West Coast of North America have long been known to be at risk of tsunamis, but most residents of North America's *East Coast* would be shocked to learn that they're also in the potential path of such a monster. *New Scientist* writer Tristan Marshall puts it concisely:

Any day now, a gargantuan wave could sweep westwards across the
Atlantic towards the coast of North America. A mighty wall of water 50
meters [150 feet] high would hit the Caribbean islands, Florida and the rest
of the eastern seaboard, surging up to 20 kilometers inland and engulfing
everything in its path. . . . The Atlantic wave . . . will start its journey 6000
kilometers away, when half an island crashes into the sea.[28]

A team led by Bill McGuire of the Benfield Greig Hazard Research
Center at University College, London, discovered that part of La Palma–a
volcanically active island in the tourism-intensive Canary Islands, just off
the northwest coast of Africa–has destabilized. Similar surveys near Réunion in the Indian Ocean, the Marquesas in the western Pacific, Tristan da
Cunha in the South Atlantic and El Hierro in the North Atlantic have shown
that oceanic volcano collapses occur worldwide.[29]

If part of the La Palma volcano collapsed into the sea, it would drive a
tsunami similar in type to the 1958 wave in Alaska–except on a far more powerful scale. Like most tsunamis, this wouldn't be just one wave. It would be a
series of waves that hit in elongated intervals, over about an hour. Also like
most tsunamis, these would travel at upward of *350 miles per hour*–until they hit
shorelines such as the Caribbean islands and the East Coast of North America.

Most of the U.S. eastern seaboard is flat. There are few barriers to stop
a flood or sea surge. This was convincingly demonstrated during Hurricane
Floyd in 1999, when many parts of the coast were inundated (see Fig. 29).

With that in mind, consider the scenario presented by McGuire, who is a
Benfield Greig Professor of Geohazards and director of the Benfield Greig
Hazard Research Center. His group is leading a study of the La Palma col-
lapse risk, which has received serious attention from the government of the
United Kingdom. In a chapter of his book *Apocalypse*,[30] McGuire describes La
Palma's violent geological history, along with how its next eruption and col-
lapse might affect North America. Here is my summary of his description,
with an emphasis on the qualitative rather than the quantitative:[31]

∞ A tsunami, such as one generated by the collapse of La Palma,
would travel virtually unnoticed across the Atlantic at the speed of
a jumbo jet as a series of mostly undersea waves, then pile up in the
shallows of the continental shelf to a height of 150 feet, still going as
fast as a locomotive.

∞ The waves may stretch the length of the eastern seaboard, from
Maine to the Bahamas, with the power to penetrate miles inland.

∞ America's space launch platforms would be swept away at Cape
Canaveral in Florida.

∞ Key Largo and the city of Miami Beach may cease to exist, as would waterfront communities in Fort Lauderdale, West Palm Beach, Cape Canaveral, and Jacksonville, Florida; Charleston, West Virginia; Norfolk, Virginia; Annapolis, Maryland; Atlantic City and the rest of the New Jersey shore; Long Island; New Haven, Connecticut; Providence, Rhode Island; and Boston Harbor. Nor would Philadelphia; Washington, D.C.; or Baltimore be spared, because such waves would move along inland waterways, although with less power.

∞ Ships in the naval yards at Norfolk would be picked up and thrown ashore.

∞ Wall Street might physically collapse, along with structures such as the United Nations building.

∞ Such smashed structures wouldn't just settle in the muck. They and their debris would lumber along in a huge mass, knocking down other structures. This might occur along the entire coastline of North America.

∞ Thousands of fuel and chemical storage tanks would burst on impact. Sewage treatment facilities, many located close to the sea, would be shattered, spilling their toxic mix into the onrushing water. Oil refineries would explode.

∞ McGuire guesses that roughly 12 million people would die.

∞ The economy of the United States would immediately be thrown into depression.

Why might so many die? Wouldn't they have warning?

If they were lucky, they'd get a few days' alert about the imminent collapse of La Palma. Yet if we look at the history, for example, of the Mount St. Helens eruption, we see that the exact moment of such events is extremely difficult to predict. So authorities are left with a dilemma: when to evacuate.

Furthermore, as hurricane evacuations show, each day that the East Coast of North America is left empty costs tens of millions of dollars. Throngs of evacuees put stress on the rest of the nation as they look for places to stay, are separated from their jobs, and run out of patience. Moreover, history shows that a good portion of the population sometimes doesn't believe the threat, or figure they'll just ride it out. Thus, millions might stay behind.

If La Palma were to collapse, the warning would be about seven hours from the time of that event in the Canaries to the onset of the first wave.[32] Seven hours would give enough time to make it inland—that is, if everyone got the message immediately, if they left on time, if they left everything behind, if they didn't sit in traffic waiting for it to clear up, and if they headed in the right direction. In New York City, Cambridge, or Cape Cod, for example, millions of evacuees would have to head miles south or north, or along exposed riverside routes, before heading inland.

Afterward, what might be left to come back to? Farmland would be saturated with a toxic chemical soup that would also have contaminated groundwater supplies. Rivers and coastal waterways would be oxygen-deprived dead zones.

Further south, the offshore banking offices on every Caribbean island would have disappeared. Not a single structure within a hundred vertical feet of the shore would be undamaged. Havana, Nassau, Freeport, Charlotte Amalie, San Juan, and Basseterre: all might be gone.

It's important to emphasize that the estimates by McGuire's team of how big such a wave might be are not subscribed to by everyone in that professional community, and that the modeling calculations still have to undergo more peer scrutiny. Yet the fact that such a serious group could make such a serious estimation is alone cause for concern. It suggests that at a very minimum, we'd be well advised to allocate greater resources to analyzing such a potential.

Furthermore, what are the contingency plans for such an occurrence? None. While we've started looking for wayward asteroids, no major disaster preparedness agency in the United States seems to be working on plans in case of a volcanic collapse and subsequent giant tsunami, despite the geological proof of past inundations, and new scientific investigations that show La Palma to be unstable.[33]

Going back to the pronouncements of experts who say we're better equipped to deal with big disasters than we were, say, a hundred years ago, we don't need to look far to see how misplaced this sense of security may be. For example, almost two hundred years ago, in 1811, when America's largest earthquake hit the New Madrid fault near St. Louis, Missouri,[34] there was no sophisticated, intertwining infrastructure, and not much toxic soup in storage tanks. Contrasts now include: population size, infrastructures along waterfronts, millions living where swamps and wetlands once existed, an interdependent globalized banking system, and a mix of toxic and/or radioactive chemicals. In a huge earthquake or tsunami, these aren't strengths—they're liabilities.

Yet catastrophe doesn't always come at the instant when such events occur. It also arrives in the prolonged aftermath. Scientists are just beginning to uncover how this happens.

Fig. 29. Here's what makes fossil fuels and sewage systems unsustainable in a flood. Franklin, Virginia, after Hurricane Floyd, September 25, 1999. Aerial view of flooded downtown shows oil slick from flooded tanks and cars. Such contamination could be reduced by replacing fossil fuels with solar energy, and retrofitting storage tanks with nanotube skins that self-seal. (Courtesy of the Federal Emergency Management Agency [FEMA])

WHAT THE TREES TELL US

The study of tree rings is known as *dendrochronology*. Some scientists spend their lives in this pursuit. Their experience shows that it's a good thing we didn't cut down every old tree in the world, because we'd have lost irreplaceable evidence about our climatic past.

For some time now, scientists have sought such clues from trees that grow for thousands of years. The Irish oak is one such tree. The distances between its annual growth rings give us accurate records of climate conditions that occurred throughout the life of the tree. If whole trees are examined, their rings can be compared to those found in beams of old buildings, because the spaces between their growth rings match. It's possible to see what part of a beam comes from what century by matching its rings with that of a known control sample. This helps build a statistically significant database about local climate conditions.

After studying hundreds of such specimens, then comparing them with other tree ring reports from around the world, Mike Baillie, professor of palaeoecology at Queen's University in Belfast, Northern Ireland, found that around 540 C.E., trees virtually stopped growing in the Northern Hemisphere. He explained it this way:

> These were the narrowest rings in the lifetime of the tree. We began to realize that for that to happen you're actually almost talking about damage. It's more than just cold, it's something where the tree is actually physically suffering in some way. And we've recently found some trees which are actually physically damaged; they have damage scars and show all the evidence of having been brutalized in some way. . . . From chronologies built by colleagues in Germany, Finland, North America . . . we were able to see the same event, a growth downturn in trees all the way from Siberia across Scandinavia, across Northern Europe, across North America down into South America, it became obvious that this was a global event and that the confines of it are between 536 and 545 A.D.[35]

Debate rages over what exactly caused the hostile conditions, but whatever it was, a persistent dust cloud accompanied it. As early as 1984, Richard B. Stothers of the NASA Goddard Institute of Space Studies catalogued the scientific evidence and written historical observations of this cloud. He remarked in the scientific journal *Nature*:

> The densest and most persistent dry fog on record was observed in Europe and the Middle East during AD 536 and 537. Despite the earliness of the date, there is sufficient detailed information to estimate the optical depth and mass of this remarkable stratospheric dust cloud. The importance of this cloud resides in the fact that its mass and its climatic consequences appear to exceed those of any other volcanic cloud observed during the past three millennia.[36]

Whether it was a volcanic cloud, comet impact plume, or other phenomenon is still debated. Some scientists, such as Baillie, theorize that it could have been caused by an encounter with one or several near-Earth objects, while other scientists favor a volcano, due to the apparent discovery of sulphur compounds in ice core samples, accompanied by an alleged absence of iridium that might have signaled a comet or asteroid encounter impact.[37] As of yet, there is no consensus. For our purposes, it's not important to know what was the cause; only to know that it likely wasn't human.

In *The Years Without Summer*,[38] various academic authors trace the possible relationship between this climatic event and drastic subsequent downturns among civilizations around the world. The book's editor, Joel D. Gunn, a researcher at the Department of Anthropology, University of North Carolina at Chapel Hill, summarizes its conclusion this way: "During the fifteen years after 536 . . . the world was fraught with famine. Among other things, the population of Europe fell to its nadir between Classical and modern times."[39]

In his book *Catastrophe*,[40] archaeological journalist David Keys also concluded that the climate event devastated world agriculture and caused the collapse of civilizations.[41]

From 535 to 555, there is evidence that the worst weather ever recorded hit the Middle East, causing heavy snowfall and famine, followed by flooding. China experienced drought and famine, with parts of the country getting snow in August. South America had a two-year drought. Trees in North America stopped growing for several years. These disruptions continued not for a year or so, but for a generation.[42]

Furthermore, studies of glacial ice cores suggest that other big climate changes occurred in only a few years, rather than across centuries. Professor Richard Alley, a professor of geosciences at Pennsylvania State University, describes this in his book *The Two-Mile Time Machine*.[43] Alley undertook five expeditions to Greenland and two to Antarctica to gather ice core samples that hold records of climate changes. These are similar to the stories that tree rings tell, except that other clues such as dust are included and help to broaden the picture. Alley won several awards for his work and published his results in journals such as *Science* and *Nature*.[44] Among the more interesting findings are that until about ten thousand years ago, the earth's climate fluctuated far more violently than it does now. The period of calm climate that *Homo sapiens* have enjoyed in their later development seems to be a relatively new phenomenon, as Alley explains:

> Large, rapid and widespread climate changes were common on earth for most of the time for which we have good records, but were absent during the few critical millennia when humans developed agriculture and industry.[45]

Furthermore, and most disturbingly for our contemporary civilization, the shift from one climate pattern to another can occur with remarkable speed:

> . . . the ice cores also show that the ice age came and went in a drunken stagger, punctuated by dozens of abrupt warmings and coolings. The best known of the abrupt climate changes, the younger Dryas event . . . ended about 11,500 years ago when Greenland warmed about 15 degrees Fahrenheit in a decade or less.[46]

It's not hard to imagine the impact on present-day agriculture, homes, heating, cooling, and related infrastructures if the Northern Hemisphere experienced a similarly dramatic shift in average temperatures now. For example, in northwestern North America, a four-year absence of cold winter temperatures, which normally kill insect larvae, led to an explosion of mountain pine beetle populations by the fall of 2001. This normally minor tree infestation affected millions of acres of forests, threatening viability of the region's forestry industry.[47] It was caused by an apparently cyclical temperature fluctuation that was tiny compared to what happened millennia ago, but that nevertheless caused economic havoc. Most of us don't seem to give credence to such natural impacts, perhaps because they seem too distant, or perhaps because the popular media, along with many environmentalists, tell us these phenomena are more likely to be caused by human rather than natural factors. Yet, if the evidence of such past climate upheavals is accurate—as tree ring and ice core studies from around the Northern Hemisphere suggests it is—then we face a risk. The same thing could strike at us, perhaps crippling our high-tech society by decimating our agricultural base and shifting disease vectors via temperature changes.

Furthermore, such climate switches appear to have multiple causes.

OTHER POTENTIAL CULPRITS

Besides asteroid hits and giant volcanic explosions, there is evidence that global warming was earlier caused by underground methane hydrate deposits that escaped to the atmosphere in quantities far larger than the climate change gases we put into the skies today.[48] Some scientists now view methane hydrate deposits as one of the great potential fuel sources for the twenty-first century. A lot of research is going into finding them.[49] Known deposits contain up to three thousand times the amount of methane in the atmosphere right now.[50] Methane is twenty times more potent than carbon dioxide as a global warming gas. At sixteen hundred feet below sea level, it's stable at 41°F, but escapes if heated only two degrees more.[51] These deposits have the potential to release more than 350 billion tons of methane into the atmosphere.[52]

This may alter the whol global-warming discussion. William Dillon, a scientist with the U.S. Geological Survey, succinctly summarizes it this way:

> Methane trapped in marine sediments as a hydrate represents such an immense carbon reservoir that it must be considered a dominant factor in estimating unconventional energy resources; the role of methane as a "greenhouse" gas also must be carefully assessed.[53]

He further postulates that methane hydrate may cause landslides on the continental slope:

> Seafloor slopes of 5 degrees and less should be stable on the Atlantic continental margin, yet many landslide scars are present. The depth of the top of these scars is near the top of the hydrate zone, and seismic profiles indicate less hydrate in the sediment beneath slide scars. Evidence available suggests a link between hydrate instability and occurrence of landslides on the continental margin. A likely mechanism for initiation of landsliding involves a breakdown of hydrates at the base of the hydrate layer. The effect would be a change from a semi-cemented zone to one that is gas-charged and has little strength, thus facilitating sliding.[54]

If so, some of the larger landslides might make the Atlantic Rim a candidate for tsunami.

Other research is aimed at uncovering past cataclysms that range from the Toba volcano in Sumatra about 74,000 years ago, to a magma dome in Yellowstone Park that last erupted 640,000 years ago but is still considered active today.[55] Each eruption seems to have changed the world climate.[56] To see how these might affect us, we need only look back to the much smaller 1815 explosion of a volcano at Krakatau, Indonesia, which seems to have altered global temperatures when its veil of dust spread around the globe, dimming the sun's rays.

Thus, while we fret—perhaps rightfully—about what we may be doing to the climate, the climate may be getting ready to bite us, as part of its own natural cycle.

What are we doing to get ready for this?

Not much.

CHAPTER ELEVEN

AN ELEPHANT IN THE ROOM OF ENVIRONMENTALISM

Nature's hiccups can undo our great advances. A rash of studies has exposed our disintegrating ability to cope with powerful phenomena. Worse yet, our own actions are exacerbating the situation.

Property loss is increasing worldwide, due to unrestricted development in risky hurricane and earthquake zones. Perversely, this can actually improve economic conditions for some sectors in the short term, by fueling construction booms after disasters. Such short-term rebounds are often generated by insurance settlements. Yet underneath, a cancer grows. This foundation for economic stability—insurance—is collapsing. Reinsurers are withdrawing from large economic sectors, such as storm protection in Florida and earthquake coverage in California and Japan. It's already hard to get insurance in hurricane zones such as the Caribbean. Coverage is often unaffordable, or unavailable in sufficient amounts, to residents of high-risk areas. Moreover, taxpayers may get hit with the cost of bailing out insurance companies that can't cover their onerous obligations. A 1998 study by the Arkwright Mutual Insurance Company estimates the eastern United States could incur hurricane losses of $50 to 100 billion, driving insurers into insolvency. The insurance giant Munich Re has estimated that the number of natural catastrophes that affect insurers has tripled since the 1960s, with costs to the globalized economy increasing by a factor of nine, and costs to insurers rising by a factor of fifteen. They attribute such increases to climate change, as well as greater development in high-risk zones.[1]

Developing nations are worse off. Their citizens often can't pay for insurance. In China, flooding of the Yangtze River in the 1990s displaced 223 million people–nearly the equivalent of the U.S. population.[2] Moreover, social disintegration is a real risk. Here is sobering reading from the *World Disasters Report* of the International Red Cross:

> The explosive combination of human-driven climate change and rapidly changing socio-economic conditions will set off chain reactions of devastation leading to super-disasters. . . .
>
> In 1998, natural disasters created more "refugees" than wars and conflict. The report indicates that declining soil fertility, drought, flooding, and deforestation drove 25 million "environmental refugees" from their land and into the already vulnerable squatter communities of fast-growing cities. They represented 58 percent of total refugee population worldwide. . . .
>
> This insidious combination is throwing millions more into the path of potential disaster. Already, 96 per cent of all deaths from natural disasters occur in developing countries. One billion people are living in the world's unplanned shanty towns and 40 of the 50 fastest growing cities are located in earthquake zones.[3]

BOTCHED DISASTER RECOVERIES

Furthermore, Didier Cherpitel, secretary general of the International Federation of the Red Cross, introduced the organization's *2000 World Disasters Report* by summarizing: "Too often, efforts at reconstruction after a major disaster don't lead to recovery. Instead they end up rebuilding the risk of danger in future disasters by ignoring economic realities."[4] Thus, most disaster relief is actually contributing to future disasters.

Much of the responsibility for disaster prevention falls to insurance companies and government agencies, although multinational corporations are beginning to take this on themselves as insurance proves too expensive or unavailable. Government risk-reduction tools include trend forecasting, construction codes, and discouraging development in risky zones. Sometimes these work, but increasingly they fall victim to growth and sheer wealth, or sheer poverty. The richer we get, the more we build our most expensive real estate in high-risk zones such as coastlines. As the numbers of poor people without land increase in developing nations, they, too, migrate to risky flood zones. The classic case of this is Bangladesh, where millions are flooded out periodically.

A sad commonality links disasters such as the Kobe earthquake with the destruction of Homestead, Florida, by Hurricane Andrew; the annihilation of Nicaragua's economy by Hurricane Mitch; and the hobbling of the Los

Angeles infrastructure by the Northridge earthquake. In many cases, after emergency relief crews have left, victims are exploited.

For example, the California earthquake insurance system suffered a blow to its credibility when its insurance commissioner, whose role was to protect consumers against questionable insurance industry practices, was found by the California state auditor to have "abused his discretionary authority in the settlement of enforcement actions"[5] against insurance companies that handled claims from the 1994 Northridge earthquake. The scandal made newspaper headlines for more than a year.[6] It was especially pertinent because it occurred in one of the wealthiest and most closely regulated economies in the world, rather than, for example, in a developing nation. Thus, it appears that no one is immune to exploitation after a natural disaster.

Reconstruction is also often botched. In the rush to rebuild, the same mistakes are repeated. Often we fail to take advantage of the opportunity offered by such disasters to modify the underlying infrastructures that led to the problem in the first place.

A good example of this is the region around Mount Vesuvius in Italy. Despite vividly preserved archaeological records and media footage of recent eruptions, the city of Naples has grown up around one of the planet's deadliest volcanoes. Millions have been spent researching the caldera, and many scientists are concerned that an eruption of the type that destroyed Pompeii wouldn't give enough warning for evacuating the city's residents.[7] Such an event could kill thousands and destroy the region's industrial base. As we've seen, the same can be said of Tokyo, where a large earthquake is practically guaranteed to maim a nation. Much of Florida has been built or rebuilt in the path of hurricanes, with structures that are incapable of withstanding the battering of a big storm. Thus, despite expenditures on emergency relief and reconstruction, our systems are still fragile in the face of natural disasters.

A DEEP CRACK IN ENVIRONMENTALISM

This breakdown is also an elephant in the room of environmentalism.

Sustainable development has many differing definitions that together form a basis for contemporary environmentalist philosophy. In 1987 the World Commission on Environment and Development defined sustainable development as "satisfying the needs of the present without compromising the ability of future generations to meet their own needs."[8] This is one of the most frequently quoted definitions. Yet environmental groups and industry often disagree on the meaning of terms such as "compromising the ability of future generations."

Some international agreements show how disaster preparedness is supposed to fit into this definition. Disaster preparedness was written into a sustainable-development code, adopted by world leaders at the Rio Earth Summit in 1992. This code, known as Agenda 21,[9] is a book on its own. As with many parts of that code, the implementation of its disaster preparedness recommendations has been sketchy. For example, the U.S. National Science and Technology Council (NSTC) report *Bridge to a Sustainable Future: National Environmental Technology Strategy*,[10] which was written years after Agenda 21, omits natural extremes such as earthquakes and hurricanes from its priority list. Thus, natural disasters and sustainable development are often still split among different branches of governments, though some links seem to be improving.

The Internet Web sites of many environmental organizations virtually ignore natural disasters, except where human intervention seems to be making them worse. Many environmental groups maintain that their job is to focus on the problems humans cause, and that natural disasters are something for someone else to deal with. Human rights organizations seem more willing to recognize nature's nastier side, perhaps because they view it from a different perspective: as something that restricts human liberty. This may offer hope for innovative solutions.

Nevertheless, if a megatsunami of the magnitude described earlier were to hit the East Coast of North America right now, no "sustainable" technology advocated by environmental groups or agencies would prevent or mitigate the impact (see Fig. 30).

Some awakening to this problem is evident. Maurice Strong, one of the chief organizers of the 1992 Earth Summit, signifies this in his autobiographical book, *Where on Earth Are We Going?* when he concludes that we need to prepare for natural disasters and extraterrestrial threats such as asteroids.[11] Yet Strong is an exception. The evidence suggests that if we continue to rely on existing ideas of "living in harmony with nature" we may be thrown backwards centuries when disaster strikes. Survivors wouldn't be "the fittest," they'd just be lucky. An age-old struggle would begin again, as we tried to crawl back toward the stages of development that we've attained today.

Right now our chances of avoiding this fate are remote. A war is going on between environmentalism and "technologism." This war may be distracting us from the true environmental challenge.

"VIGILANCE AND COURAGE"

As proof accumulates that our defenses against smaller disasters are failing, some scientists are calling on policy makers to put more resources into investigating what once was regarded as a paranormal sideline.

Fig. 30. Are these structures environmentally sustainable? The definition of "sustainable development" gets clouded when a truly big disaster hits. Top: Tornado damage to a house. Bottom: Tsunami damage at Kodiak, Alaska, following 1964 Good Friday earthquake. New technologies are required to protect our communities from the devastation of natural forces such as these. (House: courtesy of Federal Emergency Management Agency [FEMA]; tsunami: courtesy of National Oceanic and Atmospheric Administration/Department of Commerce)

In his book *What Remains to be Discovered*, Sir John Maddox concludes that the biggest threats to humanity are resurgent diseases, climate changes, near-Earth objects, and the possibility that the human genome is defective.[12] While he says the first two are more immediate concerns, he gives credence to the remainder by emphasizing that we need to start working on them. Maddox is no fringe scientist. He's an internationally respected Fellow of the American Academy of Sciences and was editor of the scientific journal *Nature* for twenty-three years.

A short time ago, many scientists would have dismissed such statements about asteroids and climate catastrophes as hogwash. Furthermore, the thought that humanity's genes might be defective wasn't on anyone's radar. That Maddox could be awarded high marks for his concerns by scientific magazines and newspapers of record, such as the *New York Times,* shows how much the landscape has shifted. Part of this is due to a shift in scientific knowledge. For example, the Human Genome Project made it possible to track evolution of chromosomes and genes by examining DNA sequences. Some scientists have shown the history of sex chromosomes to be marked by a disruption to the Y chromosome and the adaptation of the X chromosome to compensate.[13] Thus the human genome is in flux, as Maddox postulated. He therefore lays this challenge on environmentalists:

> The possibility that the genome may be unstable is a stringent test of what people mean when they insist that Homo sapiens should find its proper place in Nature. Would enthusiasm for what is called "sustainable development" persist if it entailed the extinction of . . . the human species.[14]

Yet, far from harping about the shortcomings of green philosophy, Maddox is guardedly optimistic about our chances for finding solutions:

> In reality there is no reason why any of the potential calamities now foreseen, even the most scary among them cannot be avoided. But avoidance requires vigilance and courage. Will we have enough?[15]

CHAPTER TWELVE

LOST MESSAGES FROM ANCIENT TIMES

No book about natural threats is complete without describing the exploits of a group that doggedly chases the fate of lost civilizations.

The word "Atlantis" is inflammatory. Skeptics rightly point to thousands of pages of scientific literature that conclude there's still no proof of it. They regard many Atlantean promoters as menaces to science. For example, the publishers of *Skeptical Inquirer*,[1] who have close ties to the publisher of this book, support that view. On the other side of the fence, believers rail at the skeptics, saying they haven't looked hard enough, and that the scientific status quo is prejudiced against the field. It's a loud, public argument.

Yet for this book, such discussions are secondary. When describing nature's time bombs, it's not necessary to prove that such civilizations did or didn't exist. Proof of their existence and violent demise certainly might reinforce the legitimacy of natural threats to contemporary society, but such proof is not a prerequisite for validating those threats. Rather, the challenge is to determine if these investigators have made scientifically valid contributions to knowledge about natural extremes that visited the earth thousands of years ago.

Among the most famous and disputed ancient civilization advocates are Graham Hancock, author of a series of investigations documented in *Fingerprints of the Gods*,[2] and Robert Bauval, who probed ancient knowledge of the heavens in *The Orion Mystery*.[3] They are associated or have crossed paths in one way or another with Atlantis sleuths

Rand and Rose Flem-Ath, authors of *When the Sky Fell*;[4] ancients engineering expert Christopher Dunn, author of *The Giza Power Plant*;[5] the late Charles Hapgood, who was the first to argue in *Maps of the Ancient Sea-Kings*[6] that Antarctica was discovered and mapped long before the history books say it was; and Jane Sellers, who linked the tracking of earth's multithousand-year precessional wobble with Egyptian astronomical observations in *Death of the Gods in Ancient Egypt*.[7]

Each of these authors produced books about ancient civilizations. They don't necessarily share each others' views. Many of them have their own conflicts with skeptics.

The controversy is summarized by Peter James and Nick Thorpe in *Ancient Mysteries*.[8] Although this book tends to lump many different works together as a group, which—as we'll see—is problematic, it still gives a general summary of what the argument is about. It goes something like this: Some researchers claim that somewhere on earth, advanced civilizations existed prior to written history and died out due to unknown cataclysms. They left us messages based on astronomy, encoded in ancient monuments such as the pyramids of Egypt, South America, and China. We haven't figured out what the messages are yet, but we're getting there. The message may save us from a similar unhappy fate.

James and Thorpe, along with scientists writing in publications such as *Skeptical Inquirer* and high-level journalists from programs such as the BBC's flagship *Horizon* program, have dissected such theories. Some characterize the postulations as being close to quackery. Others suggest that their authors promote fantasy for monetary gain.

In response, Hancock and Bauval especially have mounted a counterattack by admitting fault with some of their early suppositions, but steadfastly sticking to the concept and presenting arguments as support. The fight went as far as the British Broadcasting Standards Commission, which in 2000 heard a complaint by Hancock and Bauval against alleged editing prejudice of a BBC *Horizon* program that challenged their theories. The commission granted that the BBC had erred in one section of its program, but rejected other complaints. The documentary was subsequently reedited and rebroadcast.[9]

Hancock and Bauval continue to publish new works on related matters, while research on the downfall of ancient civilizations remains embroiled in controversy.

As explained earlier, researchers who investigate ancient catastrophes tend to be thrown into one basket by critics, due to inherent skepticism about catastrophism. This isn't improved when research by serious authors is lumped together by the media with outlandish claims by less accomplished theorists. The end result: much of the scientific community tends to dismiss the whole group as Atlantis fanatics, and attempts to banish them to the hinterland of pseudoscience.

Yet certain characteristics of this group warrant further attention. As indicated earlier, they are by no means uniform. Some are engineers or mathematicians rather than archaeologists or historians. By attempting to link legends with scientific evidence, some of them approach catastrophe as much from a technical perspective as a biblical or mythical one.

Graham Hancock may be controversial, but he makes this compelling case for aggressive investigation into prehistory:

> In 1997 a chain of mountains almost 2000 kilometers long and more than 3000 meters high was discovered in the South Pacific. Nobody ever knew the mountains were there before because they are under water–as, in fact, is 70 per cent of the earth's surface. Marine archaeologists–who are looking for targets much smaller than mountain-ranges under the sea–can therefore be forgiven for finding just 100 submerged sites more than 3000 years old in the past half century. Even at the crude mapping level, it is one of the absurdities of scientific priorities that we now have a better map of the surface of Venus than we do of the 225 million square kilometers of our own planet's sea-floor.
>
> . . . let's remember as well that along continental margins and around islands across the world an area bigger than the United States of America was inundated at the end of the Ice Age: 3 million square kilometers (an area the size of India) was submerged around Greater Australia alone; another 3 million square kilometers went under around South-East Asia; the Florida, Yucatan and Grand Bahama Banks were fully-exposed off the Gulf of Mexico; huge areas of land were swallowed up in the Mediterranean, the Black Sea, the North Sea and the Atlantic, etc, etc, etc–the list really does go on and on.
>
> In my view the possibility of a serious "black hole" in scientific knowledge about recent prehistory is plausible, reasonable and worthy of consideration. I therefore propose that the conclusions of modern archaeology regarding the origins and early evolution of human civilization should be treated as provisional until a comprehensive, global, marine-archaeological survey of continental shelves down to depths of at least 120 meters has been undertaken.[10]

Such a position deserves further serious attention. It was bolstered in late 2001 by the discovery of what some researchers claimed was a "lost city" off the coast of Cuba, an astonishing twenty-one hundred feet below sea level. Scientists from Canadian-based Advanced Digital Communications, along with the Cuban Academy of Sciences, began mapping the site with a deep-sea submersible, a year after side-scan radar tests pinpointed circular and triangular structures on the ocean floor. Initial estimates put the age of the seven-square-mile area at up to six thousand years—more than a millennium older than the pyramids at Giza in Egypt. The team emphasized that it was far too early to tell what had created the formations, but that didn't stop the media from speculating about Atlantis.[11] Archaeological experts, on the other hand, discounted the discovery, explaining that the area probably contains natural rock formations that occur throughout the world and are commonly mistaken for man-made structures.[12] The point of mentioning this controversy is to show that our own civilization has yet to accurately understand large areas of the ocean floor. Until that is done, we'll probably continue to see these conflicting viewpoints.

In each of these controversies, no one disagrees that ancient civilizations such as the Egyptians and Mayans existed. No one disagrees that they possessed great expertise that was lost as their civilizations crumbled. Ice core samples prove that those civilizations existed under environmental conditions, such as climate, that were different from what they are today.

The questions remain: What altered those conditions, and might similar factors cause our own decline?

In *A Green History of the World,*[13] politics professor Clive Ponting of the University of Wales at Swansea chronicles the breakdown of ecosystems that supported Egypt, Rome, and the pre-Columbian Americas. He lists the impact of natural events on political history, and overexploitation by civilizations of the resources on which they depended. From wildlife to soil, he describes how they bankrupted their natural wealth and doomed their societies to gradual decline. From Vesuvius to Lisbon and Tokyo, he chronicles seismic disasters that changed the direction of civilizations, destroyed one power base, and gave rise to another. Moreover, when Ponting wrote his superbly researched book he was not privy to a mass of newer data, as explained in previous chapters of this book, that suggests nature had as much to do with altering local conditions as humans did. Climate change, volcanic eruptions, and earthquakes played pivotal roles in bringing these civilizations down.

Taken together with more recent investigations, such as those described in earlier chapters, it makes a convincing case for more research into the implications of natural upheavals for our own society. Yet scientific skepticism about the ancients, combined with historians' resistance to an engineering view of ancient civilizations, have hindered the financing of serious investigation into these phenomena.

The controversy goes further than the pyramids. It extends to the foundations of the human species. In *The Hidden History of the Human Race,*[14]

Michael Cremo and Richard Thompson chronicle evidence from archaeological digs worldwide that call into question conventional wisdom about
where and when the first *Homo sapiens* appeared, and how much technological capability they had in prehistory. They ask how many times our species
may have risen to some level of technological advancement, only to be
thrown backward. This challenge to conventional theories has received a
frigid reception from the established archaeological and anthropological
community. Noted anthropologist Richard Leakey savaged the book, calling
it "pure humbug."[15]

The reputation for scientific accuracy of investigators of the ancients
hasn't been helped much by an earlier and broadly popular book, *Chariots
of the Gods*,[16] which claimed that aliens from space built ancient monuments.
This spawned a series of 1972 television specials worldwide, and was read
or seen by more than 50 million persons. Yet it was severely criticized in scientific circles for factual misrepresentation.[17] Researchers such as Hancock
have distanced themselves from such works, but the job isn't easy, partly
because Hancock's book *Fingerprints of the Gods* is easily confused with the
earlier title by Erich Von Däniken. The books are totally unrelated.

This range of controversies shows the high level of opposition that catastrophe researchers face. Persistent and sometimes justified skepticism,
international politics, professional jealousy, and media hysteria challenge
their research.

Completely separate from those factors, a big roadblock to research has
been thrown up by national security. For example, high-resolution satellite
remote-sensing research was blocked until recently by outdated Cold War
mentalities. Take the example of Space Imaging Inc., the first company to
launch a nonmilitary high-definition Earth observation telescope. Efforts to
put civilian spies in the sky were impossible for half a century due to Cold
War military pressure not to launch. Civilian high-resolution imagery was
deemed too disruptive to national security. Oddly enough, Russia and
France had been selling such imagery on the open market for several years
before the American government gave the go-ahead. When Space
Imaging's *Ikonos 2* satellite was finally launched in 2000, after numerous
delays, some U.S. legislators tried to limit what it could look at by declaring
some installations off-limits.[18] The effect of this resistance was to frustrate
efforts at getting a detailed look at the earth's surface from viewpoints other
than missile-launching sites or weapons of mass destruction. Nonetheless,
pioneering efforts such as Space Imaging's are still bearing fruit as more scientists start to use the service, and the company gains new permission for
higher-resolution photographs.[19]

Cold War leftovers have also hindered the search for ancient settlements off the coast of Cuba, because American companies have been prohibited for decades from working with Cuban organizations on research.[20] Then again, politics and archaeology have always been entwined. The search for evidence from ancient realms has been held back elsewhere, when scientific explorers underestimated resentment by developing nations against foreigners plundering their treasures. Pharmaceutical companies reinforced this fear by originally undercompensating countries for drugs found in their rain forests, although some of these transgressions have since been rectified through revenue-sharing agreements.

Fortunately, new technology is empowering investigators to get around such roadblocks. The backyard sleuthing of geologist Eugene Shoemaker, Carolyn Shoemaker, and David Levy with high-powered, yet increasingly affordable, telescopes led to the eyewitnessing of how big objects can and do collide with planets, thus exposing the risk that such objects pose to us. In their case, they had the great advantage of a comet proving them right when Comet Shoemaker-Levy 9 hit Jupiter shortly after their discovery. Most researchers of ancient evidence don't get confirmation as quickly.

∞   ∞   ∞

Let's now look at some of the molecular tools that are being developed for detection of, and protection against, a potential catastrophe.

PART 3

BLUEPRINTS FOR A
MOLECULAR DEFENSE

"An organism that cannot adjust itself to its changed environment is bound to perish."

–John Fiske, American historian,
philosopher, and Darwinist, 1842–1901[1]

CHAPTER THIRTEEN

WHY GO THERE?

To inoculate ourselves against nature's occasional tantrums, we may have to construct powerful molecular defenses. Yet, as we'll see later, these defenses themselves may threaten our existence, due to their potential for abuse. Some say that these risks outweigh the potential gains.

So, if it's such a risk, why go there?

Evidence suggests there may be no alternative. Right now we have no credible defense against catastrophes such as asteroid collisions; only meager defenses against high-end earthquakes, tsunamis, or climate changes; and marginal defenses against hurricanes or floods. Our ecological strategy, according to environmental agencies and groups, should be to maintain nature's status quo by adopting "sustainable" lifestyles that don't throw the ecology out of whack, while repairing damage we've done so far, then hoping the system rights itself.

By doing only that, we may paradoxically open ourselves to calamity. Without an adjunct, the strategy of "sustainability" may condemn us to waiting for one of nature's galactic clearance projects to discard us from an ever-changing universe. It wouldn't be the first time that a species has gone extinct.

Some argue that such risks are remote. Yet research summarized in this book makes it clear that phenomena such as titanic volcanic eruptions and natural climate changes have occurred in the past, with relatively brief warning. By ignoring these lessons, we're playing a game of Russian roulette. In this respect, environmental theory is deficient, and warrants revision.

On the other hand, this doesn't justify ravaging the earth with new terraforming technologies that might be capable of altering the ecology at a whim. Nor does it justify "business as usual," supporting outdated, polluting technologies that are maintained only by virtue of wasteful subsidies.[2] There is great value in not hastening the demise of natural systems that have sustained societies for centuries and underpinned our technological advancements. But we also have to reevaluate our confidence in those balances, because sometimes they go out of equilibrium by themselves. Our natural history shows that such conditions shift drastically.

Therefore it may be time to reassess our definition of "sustainability." Otherwise, we may find ourselves entrenched in our Maginot Line of sustainable development, while natural forces bypass our defenses and strike us from the rear. We may be, as they say, sitting ducks.

To solve this, we have three parallel tasks ahead of us:

1. Build defenses against natural risks that we already know about.
2. Begin to ask the right questions about the things we don't know.
3. Establish principles to guide the development of risky technologies that may help protect us.

Right now, our society is unprepared for these challenges. We're slow to develop our defenses, we often don't ask the right questions, and we are only beginning to struggle with principles that might see us into this new era without annihilating ourselves.

The rest of this book explores how we might take on each challenge.

We start with the easiest part—the known risks.

CHAPTER FOURTEEN

TOOLS FOR DEFUSING TIME BOMBS

For millions of years, species have either adapted to their environments, or perished. Adaptation comes instinctively and genetically. Yet the technostructure on which we've come to depend has evolved so fast that it's like a powerful but gangly adolescent whose defenses are inept.

We've seen how governments, the insurance industry, and disaster agencies have spent billions studying climate change, earthquakes, and volcanoes, only to reach a sobering conclusion: our systems are vulnerable to large-scale natural attacks. In some ways, such disasters are worse than nuclear enemies, because they don't negotiate. There is no mutually assured destruction to deter one party from attacking another. Asteroids, for example, are indifferent to the size of our intellect, missile stores, or bank account. Moreover, we're at that awkward stage where small events can cause cascading failures in our adolescent digital society. Failure of a lone satellite can knock out millions of pagers. One hacker can close down a multinational corporation's Web site. A milliliter difference in the amount of oil used on a bearing causes thousands of hip implants to fail. A small variation in reinforcing clips for roof joists leads to whole neighborhoods disintegrating in a hurricane.

Furthermore, rich and poor are in this together. No amount of personal wealth is sufficient to protect someone against big natural catastrophes and subsequent societal disintegration. The Hamptons and Palm Beach stand square in the path of an Atlantic tsunami, for example. Malibu, Santa Barbara, and Monterey on the West Coast are exposed to similar risks from the Pacific. Wealthy communities are just as exposed–perhaps even more so–as poorer communities to truly big risks.

So how do we cope?

Let's look at how experts prepare us to avoid, endure, and clean up after natural disasters, then compare this to where we need to go, and how molecular science might take us there.

THE UN DISEASE

Due to its pervasiveness and the investments made in studying it, climate is a good place to start when we look at how experts approach adaptation.

A central player in the climate controversy is the United Nations. Most of us yawn when we hear the term "UN." Many Americans believe it's a left-wing plot for world domination that aims to control local governments and take away the right to bear arms.[1] The United States often refuses to pay its UN dues because there's no domestic pressure to do so and much sentiment not to. Elsewhere, the UN is perceived by many as a nice idea in theory, but rife with pork barreling and corruption, enveloped in a shield of useless publications and even more useless administrators, who get caught up trying to negotiate an end to endless wars, then send in ill-equipped soldiers to do policemen's work. On the scientific side, UN science is sometimes perceived in the United States as a politically motivated plot against capitalism. This is despite the U.S. National Academy of Sciences having supported findings of the United Nations Environment Program (UNEP) Intergovernmental Panel on Climate Change.[2] Furthermore, a deadlock has developed, where rich nations say that poor nations need to cut back on "smokestack" technologies, while poor nations tell the rich to lead by example and clean up their act. Finally, the acronym-laden text of UN documents tends to put readers to sleep. Not the stuff of best-sellers.

In this atmosphere, quoting UN materials is often a death sentence for authors. When potential readers such as corporate managers or those who consider themselves red-blooded Republicans see the term "UN," their eyes sometimes glaze over with distain or boredom. Everyone has a predetermined opinion. Yet in the area of natural disasters, UN organizations play a big role, so we have to contend with what they do. Surprisingly enough, it's not all boring or muddled. Sometimes, thanks to innovative outside help, they produce good work.

This chapter focuses on a small group that normally works outside the UN, but has produced some of the best UN materials on adapting to natural hazards.

After most nations signed the United Nations Framework Convention on Climate Change in 1992, the UNEP produced a climate change handbook in cooperation with the Institute for Environmental Studies, Vrije Universiteit, Amsterdam. The *Handbook on Methods for Climate Change Impact Assessment and Adaptation Strategies*[3] was written by top experts in the field, with financing from Netherlands and Denmark. It describes tools used today and being considered for tomorrow for adapting to extreme weather. These tools could also be applied for earthquake, tsunami, and other natural-disaster preparedness, because of similar methodologies. Many UN documents are infamous for convoluted language, but this handbook gives a concise summary.

The first thing the authors do is point out differences between *mitigation* and *adaptation*. Such terms are often confused in disaster-related literature. The confusion gets worse when different disciplines apply different meanings to the same terms. Clarifying these definitions is crucial, because much climate change work focuses on preventing climate changes, rather than adapting to them. Mitigation measures such as greenhouse gas reduction get most of the available resources, while efforts to open our adaptive umbrellas receive few. This is happening despite conclusions by many experts that climate changes are already underway and we need to adapt to them.

The editors of the climate change handbook started by saying, in the simplest terms: "mitigation" means trying to prevent climate change by stopping the things that cause it; "adaptation" means getting ready to survive it.[4] There may be crossover, because some adaptation might also help prevent climate change, but this is by sheer chance because we still don't know how to manipulate a climate that's constantly in flux.

Adaptation and mitigation are sometimes distinguished from each other by using the concept of *variety*. This was introduced into industrial-systems theory fifty years ago by British cyberneticist and researcher W. Ross Ashby,[5] then developed by steel industry research manager Stafford Beer.[6] "Variety" describes the sophistication of a system. The greater the variety, the more complex the management job. The Law of Requisite Variety[7] states that variety in a given environment has to be matched by an equivalent amount of variety to manage it.

The main ways to manage environmental variety are:

∞ *Reduce the variety of the environment*; that is, knock it down to our level of understanding.

∞ *Increase the variety of responses*; that is, diversify the abilities of those who manage it.

Brad Bass, a scientist with Environment Canada's Adaptation and Impacts Research Group, is among those who've applied the concept of variety to ecology.[8] He explains that, to cope with variety in the natural ecology, we've traditionally used mitigation. This reduces environmental variety to match our limited responses. That is, it tries to knock nature down to our level of understanding. Historically, big engineering projects such as dams have reflected this. The other approach is adaptation, where we try to increase the sophistication of our responses to match nature's sophistication.[9] Scientists such as biologists, biochemists, ecologists, physicists, and meteorologists work on investigating such sophistication.

The editors of the UNEP handbook take a good stab at describing how we adapt to natural changes. They use well-established categories[10] adopted by the Intergovernmental Panel on Climate Change (IPCC). With apologies to UNEP and the handbook authors, here is a further simplification of those terms–a layperson's glossary of adaptation strategies–and a brief outline of where we are with them today.

Take Your Lumps—Bear Losses

Many communities accept a certain level of destruction because there's nothing else to do. They don't have resources to get away from or protect their communities, so they watch damage happen, try to stay out of the way, then pick up where they left off. In most of the world this is still the most commonly used adaptive tool. Families see their homes crumble in an earthquake, burn in a brush fire, or wash away in a flood. Then they rebuild. Unfortunately, this sucks energy from the economy, tends to kill victims who have no protection, and affords no insurance against recurrence. Yet for billions of people it's the harsh reality. It's also the baseline that scientists use to determine what would happen if we did nothing to prevent a disaster: What's the worst case for those who are affected? What would they have to absorb, and are they capable of doing that?

Doing nothing is an example of the Law of Requisite Variety in action. It may seem like the worst of all worlds, but when we look at some of the disastrous "fixes" that governments have tried, as explained later in this chapter, we see that doing nothing sometimes is preferable because it doesn't make things worse. This may also apply, for example, to threats as big as asteroids, where doing nothing and taking one hit may be preferable to blowing up a projectile and taking numerous scattered hits.

Spread the Load—Share Losses

Since the seventeenth century, insurance has been a widely practiced loss-sharing strategy. This is done to protect local populations, but also to support the role those populations play in the broader economy by working in high-risk zones. For instance, flood plains are fertile agricultural areas and we need people to farm them, so we subsidize some of their losses.

Often, this loss sharing works well, but sometimes it's overwhelmed by the magnitude of the disaster, as was the case, for example, when Nicaragua was hit by Hurricane Mitch. Other disasters of recent years, including Hurricane Andrew and the Northridge earthquake, sent smaller insurers into bankruptcy, or forced them out of the regions they were insuring. In these cases where commercial methods fail, an impressive network of families, villages, municipalities, state governments, federal agencies, and disaster relief organizations have shared the cost of destruction and reconstruction. One thing we've learned, though, is how quickly those mechanisms break down during sustained or widespread environmental changes. Our collective shock absorbers have their limit, and it's often breached. Their variety can't cope with the variety of the environment.

Counterattack—Modify the Threat

Altering the environment to eliminate, reduce, or delay natural threats is a drastic, yet frequently used strategy. Such big counterattacks are known as terraforming: redesigning parts of the earth. This has only been possible since the nineteenth century, when engineering agencies such as the U.S. Army Corps of Engineers began molding watersheds to hold back floods. Much of America has been terraformed in one way or another, from the Mississippi Basin that covers most of the central United States, to the Columbia River Basin on the West Coast, to the state of Florida in the south, and the Gulf Coast states of Texas, Louisiana, and Alabama. The impact of the work done by the U.S. Army Corps of Engineers has often been understated. Taken together, it represents a stunning feat of human ingenuity and invasiveness, much of it designed to ward off nature's temper tantrums. This practice has since extended worldwide and been adopted by developing nations, with the enthusiastic approval of disaster relief agencies.

Yet terraforming is a controversial defense. Its proponents and detractors are typified by their positions on the largest terraforming project in the world: the Three Gorges Dam.

For centuries, China has been blessed and plagued by enormous river systems. On one hand, they replenish soil fertility; while on the other, they kill millions with flooding. The greatest natural disasters of the modern world, in terms of loss of life, occurred in China, when China's second longest river, the Huang He–commonly known as the Yellow River–flooded. In 1887 about nine hundred thousand died. In the 1930s perhaps more than 3 *million* lost their lives.[11] With such murderous natural rampages on their minds, and driven by a thirst for power sources to fuel their post-industrial revolution, China began damming its rivers in the same way North America and Europe did a century ago. China did so with strong encouragement from Western funding agencies and Fortune 500 companies.[12] Despite international furor over ecological effects of the Three Gorges Dam on the Yangtze and its displacement of more than a million inhabitants, the Chinese government is committed to completing the largest project of its kind in the world. The dimensions are staggering; sixty stories high over a length of 2.3 kilometers, costing $27 billion, and scheduled to generate 18,200 megawatts–equal to the output of eighteen nuclear power plants. It will create a lake more than three hundred miles long that sub-merges hundreds of villages.[13]

Half a world away, most of the 13 million Florida residents live where they do today because of a project that in its time was of similar magnitude: the channeling of the Florida Everglades. Today, government agencies desperate to gain new water sources are undertaking what they call the biggest environmental restoration project on the planet: replumbing those same Everglades. About two-thirds of funding for this "restoration" will go toward underground reservoirs for water storage.[14]

Parallels between the Three Gorges and Everglades megaprojects are striking. Both arise from a need for fresh water and flood control. Both affect huge populations. Both involve terraforming parts of the earth. Opponents to the projects claim that each government is hoodwinking the public by overestimating benefits and underplaying the downsides.

Such terraforming is sometimes considered an adaptive strategy, but in its present form it's really mitigative. It tries to eliminate variety in the local environment so we can use the resources without increasing our own variety to adapt. That is to say, we bring the ecology down to our own relatively simple level of understanding or tolerance by manipulating it.

The future of such terraforming in the so-called developed world is questionable. Efforts at replumbing the ecology have met with limited success. After half a century of attempting to control the Mississippi River, for example, the Army Corps of Engineers has admitted that flooding still jumps the river's artificial banks. It's been argued that their work worsened such flooding by creating narrow channels that back up heavy flows. Dams on other American rivers face competing use from fisheries. The fishing industry and other allies have successfully petitioned to have dams removed to accommodate spawning, despite the fact that this removal increases dependency on fossil fuel energy.

Conversely, in some developing economies, dams are seen as a defense against both floods and dependence on foreign oil. Thus, disaster agencies are caught in a vice. Coalitions of environmental and social justice groups, along with affected residents, fight terraforming. Developed nations are suffering the effects of earlier terraforming failures. On the other hand, international and domestic agencies still advocate large-scale flood control as a disaster preparedness strategy for developing nations. This paradox is bound to exacerbate an already intense struggle.

Resist the Onslaught—Prevent Effects

When we see disaster coming but can't stop it, we resort to changing our own human environment to adapt. When a drought threatens to parch crops, bring new pests, or destroy the livelihood of a region, farmers alter crop management to survive. They press governments to pipe water across greater distances or dig more wells, while subsidizing more fertilizer and pesticides.

This can be broadly defined as a strategy of *resistance.*

For example, one defense against earthquakes, tsunamis, and ice storms has been to build resistant structures. Communities in storm or earthquake zones shore up structures to withstand a once-in-a-lifetime hit. Transmission towers and bridges are built to withstand high winds. A moderately successful example of quake-resistant design is the city of San Francisco, whose building standards enabled it to withstand serious shaking without significant loss of life in the last big quake. A bad example is the former Soviet Union, which imposed cookie-cutter designs from geologically stable regions, such as Moscow, onto seismically active regions such as Azerbaijan. The result: thousands died in a quake of the same intensity that killed less than a hundred in San Francisco. Nevertheless, a capitalist society does not guarantee a resistant one. Tokyo and New York are each vulnerable to quake damage, due to inadequate designs of low-rise residential and office buildings, along with water and chemical storage tanks.

In 1998, 3 million people lost power in the northeastern United States and southeastern Canada when an ice storm toppled ten thousand power poles. Design standards were inadequate to cope with what most engineers saw as a highly improbable event: a super ice storm. In such cases, engineers and other authorities estimated that it was too expensive to design for these infrequent scenarios. In so doing, they unilaterally decided that millions of residents would "bear the loss" in such freak conditions. Unfortunately, the costs of those losses and the business interruptions they cause were sometimes underestimated, with devastating results.[15]

Moreover, resistant buildings and adaptive agricultural practices don't prevent losses from other types of destruction that accompany such disasters. Thousands of square miles of crops can be destroyed by flooding from hurricanes, or pests that thrive during a drought. In these cases, resistance on its own doesn't solve the problem of economic loss, because it fails to increase agricultural variety to an adequate level.

Make Your Living Another Way—Change Use

Where climate change makes an economic activity impossible or extremely risky to continue, we change land use. A classic example of land-use changes occurred at the beginning of the twentieth century, when farmers brought new drought-tolerant crops to the dry prairies. Unfortunately, they neglected soil management and this partial solution led to the dust bowl of the Great Depression. Thus, changing use also requires a comprehensive approach, to match the environment's complexity. This is another example of the Law of Requisite Variety.

Avoid the Threat—Move

Getting out of the way includes multiple tactics, such as building a fast escape route, staying out of risky zones, or moving altogether. In some ways we're getting better at this, but in others we're getting worse. A fast escape route is typified by the United States's interstate highway system. This lets millions rapidly escape hurricanes, but is useless when floodwaters submerge short but crucial links. Japan, France, and Germany each have fast, superefficient rail systems, but in the case of the Tokyo region, for example, this does no good in a disaster, since the trains stop automatically in a quake, and besides that, the area's 30 million residents would overwhelm trains in case of a mass exodus to avoid, for example, a resulting tsunami. Residents of other megacities such as Shanghai, Mexico City, and Jakarta have no place to go on short notice. Cities such as Naples have virtually no chance to escape a sudden eruption. Yet there are no plans to "move" such settlements from harm's way.

Staying out of the way has always been a rough road. Human beings have a propensity to build in the worst disaster zones: from hurricane-hit Florida, to cyclone-prone Bangladesh; from the quake-prone Ring of Fire that circles the Pacific, to the countdown to catastrophe in Naples. Our boundless optimism convinces us that we can overcome these threats. We often think we have, but then the Law of Requisite Variety bites back. After wasting tens of billions of dollars on megaterraforming projects that are subsequently destroyed at the flick of nature's hand, we're beginning to assess the economic folly of going further. This strategic reappraisal took on new life in the 1990s when the Federal Emergency Management Agency (FEMA) launched Project Impact[16] to prevent disasters before they happen through innovative community designs, rather than only trying to terraform. This includes, for example, zoning regulations that discourage building in some flood-prone areas of the Mississippi River Valley.

Nevertheless, such policies are subject to the whims of powerful developers and politicians who, for financial and other reasons, often disregard the threat and plow ahead as before. The Mississippi Valley has seen a struggle over construction in flood-prone regions,[17] involving FEMA, developers, insurers, farmers, and environmentalists. In 2001 the Bush Administration proposed cutting Project Impact from the federal budget, but the cuts were reversed by the Senate after pressure was exerted by beneficiaries of the program.[18] In Florida builders have tried to water down building standards despite a history of hurricane damage being exacerbated by inadequate construction codes.[19]

These battles demonstrate that staying out of the way is not our strong suit. We prefer to put ourselves in harm's way, then figure out how to deal with it. This is part of our nature. We don't like to back off. Neither do ants or beavers, so we're not alone. The urge seems embedded in our genetic code.

In the global-warming discussion, it's been suggested that moving agricultural production from parched or overheated to cooler, wetter zones could be an option.[20] There is some precedent for this. For example, much of the agricultural production in southern China has been shifted to western and northern regions to accommodate the urbanization of the Shenzhen-Guangzhou corridor. This used to be one of the richest agricultural regions in the world. Now much of it is concrete. Taking such land out of agricultural production and replacing it with less fertile land in the north and west has also meant increasing fertilizer and energy consumption. If climate changes force similar strategies elsewhere, farmers are going to face the same types of choices. It's not just a question of picking up, moving somewhere else, and carrying on as before. Different regions mean different soil, climate, and pest conditions. Again, the Law of Requisite Variety applies.

Research New Adaptive Strategies

By themselves or together, none of these strategies seem to work well. We still build in the worst conceivable locations. The whole of North America's Pacific and Atlantic coastlines are severely at risk from—and underprepared for–tsunamis. Florida is a disaster waiting to happen. Tokyo is a global economic meltdown with a fuse attached. Italy is waiting blissfully for a replay of Pompeii. Researchers are looking for answers.

There is no shortage of areas in which to invest research funds. More resilient crops, stronger structures, less vulnerable energy transmission, safer water systems, and early warning systems are just a few. One way to start is improving our ability to see what's coming at us.

We're getting better at knowing when, where, and how disaster is going to hit. For climate-related phenomena we rely on weather prediction that's enhanced by computer modeling. Satellites play a big role in tracking hurricanes. Sea-based tsunami detectors are placed offshore to warn us before the waves hit shore. Dangerous volcanoes near heavily populated areas are dotted with sensors that detect tremors and groundswell. Deep-sea submersibles are opening our eyes to underwater trenches where continents collide. At the farthest reaches of space, we're starting to look for rogue asteroids that may transect Earth's path.

On the other hand, tornado detectors are only marginally effective due to the extremely fast development of such weather systems (see Fig. 30). We've tried our hand at predicting earthquakes, but this has so far been a failure, to the extent that the United States has halted major research on it.[21] Thus, while detection plays a central role in our ability to adapt, it still has far to go.

Compared to world population distribution, relatively few disaster researchers are based in developing nations, where the vast majority of disaster fatalities occur. This is reflected, for example, in the composition of the lead author group for the UNEP climate change handbook. The handbook covers developing nations, and contributors from those regions are mentioned in the handbook's acknowledgments. The UNEP's Nairobi office also coordinated some work. Yet the absence of input by lead authors from developing economies reflects a general tendency in the disaster prevention and environmental communities. I've witnessed this often during my years in the environmental remediation field. This missing link skews research, and leaves developing nations with the impression that they're left out of decision making in areas that affect their welfare. Such exclusion isn't a reflection on the editors or authors of the UNEP handbook, it's just the way things are. Much of this occurs because, compared to the world population, Europe, America, and Japan have a disproportionate share of the scientific resources. Yet therein lies a flaw: Although science is regarded as an international language, and many scientists from developed nations work in developing regions, they still don't *come from* those regions. They lack exposure to the realities that have hindered the transfer of solutions from research to application. A classic example of this is in the water industry. In wealthy northern countries, fresh water is usually regarded as a valuable resource. In most cases, people don't fear it; they seek it. Conversely, in wet, tropical regions, water is often regarded as an enemy. It washes homes down hillsides, stops traffic, and destroys human health by spreading many of the world's deadliest diseases. Men and women who shave might notice the virulence of bacteria when they shave with tap water and their skin gets infected. Water is also more dangerous because continuous parasitic cycles in the tropics confound health solutions developed in temperate climates, where winter kills more parasites. Thus, people in the tropics spend much of their time battling against water. Psychologically, their approach to water differs from that of temperate countries.[22] This is hard for visitors from temperate climates to fathom, especially when they see local residents working and playing in resorts at the beach. Many temperate-climate-based scientists who develop solutions to water-based problems are only marginally aware of the depth of this perceptual problem, unless they run into it. Many do. Their SUVs get submerged in flooded parking lots. Telephones fail in the rain. Electronics are destroyed by mold. Moreover, visitors lack immunity to local water-borne illnesses, so they get sick. Yet for those who spend only short periods in the tropics, these are usually just inconveniences. For residents, they're a debilitating daily struggle.

Thus, solutions proposed by scientists from temperate, industrialized nations often don't work in tropical and subtropical economies. This often-unintentional "ecoimperialism" has had negative consequences for countries located between the tropics of Capricorn and Cancer, where most of our world population lives. Tropical rains in the south destroy sewer systems that work well in the north. This happens when systems are undermined in poor soils, badly built by corrupt officials who get kickbacks from multilateral bank projects,[23] or overwhelmed by occasional torrents. The litany of such disasters is long. For the purposes of this book, it's necessary to say only that disaster mitigation and adaptation research has been hobbled by this perceptual flaw that renders many solutions unworkable.

Spread the Word—Educate, Inform, and Encourage Behavioral Change

There's hope that if we're given the right information, we'll change the way we do things. If we know our houses will fall apart in a hurricane, we'll fix them. If we know a flood plain is unmanageable, we won't build there. It's been suggested that we put more resources into public-awareness campaigns to achieve this. Yet the track record does not bode well for a change in human behavior. First, our thirst for material wealth usually overcomes our other instincts, for example, by inspiring developers and politicians to make money from building on high-risk land.

As a result, politics that are dominated by special interests play a negative role in setting standards. In Florida, for example, pressure from builders plus incentives from cheap insurance premiums led to poor construction standards in vulnerable coastal zones, causing tens of billions of dollars in damage from events such as Hurricane Andrew.[24] The seeds of such insurance catastrophes were sown during a flood of retirees to the Sun Belt, abetted by all-too-willing insurance companies who were eager to do business. The same has occurred in flood-prone river deltas. That's not to say there haven't been successes. On the tiny island state of Mauritius, in the middle of the Indian Ocean, the same experience led the government to require that buildings be constructed with concrete roofs. This law has prevented huge economic losses from cyclones.[25] Nonetheless, reconstructing Florida to the same standards would cost billions of dollars and is politically unachievable. So, every hurricane season, residents just hope they're not in the path.

The other problem with getting large populations to change their behavior is that we're still not able to predict, for example, what climate change might do to some regions. Will the prairies become a lake? Will the Gulf Stream move south and plunge Europe into a deep freeze? Will South China become a desert? Climate change experts conclude that one of the first symptoms we'll see is not a change in constants but rather a change in extremes. So New York might be much colder in some years, and much hotter in others. Storms might increase in severity. How do we tell people to modify their behavior when the climate itself is (mis)behaving unpredictably?

∞   ∞   ∞

Those are the main methods suggested by the climate change community for adapting to environmental upheavals. They each have their own strengths and drawbacks.

A host of other cultural roadblocks also frustrate the search for solutions. In 1999 an Adaptation Learning Experiment[26] was held by some of the same experts who produced the UNEP climate adaptation handbook. This group examined a series of natural disasters such as ice storms, unexpected snow, drought, and flood that had hit various regions of Canada over the previous decade. Canada is considered one of the world's best-equipped nations to handle big disasters. The country participates in disaster relief operations worldwide, and has a sophisticated disaster response infrastructure at home. Canada is also a good example to examine, because it has a small population living in a large area, so it depends on sophisticated technologies to overcome physical obstacles. It has, for example, one of the most advanced communications and energy industries in the world. That also makes the findings of this experiment more disturbing, because they show that even the most technologically sophisticated countries are prone to culturally based pitfalls when it comes to disaster preparedness.

Using examples from across the country, participants in this learning experiment came up with a good summary of the barriers that bedevil our attempts to prepare for big disasters. (Note: topic headings were developed by the workshop. Many of the examples shown here are mine.)

We forget lessons we've learned. As the cliché goes, those who ignore history are condemned to repeating it. Nowhere is this more prevalent that in the field of natural disasters. As discussed previously, we build in the same places that have been wiped out dozens of times before. Our institutional memories are often erased by time. For example, there hasn't been a level 8 earthquake along the New Madrid fault in Missouri for a few hundred years, so we've constructed an infrastructure that may well disintegrate when it recurs. (See chap. 19 for more on our cultural amnesia.)

Disaster response neglects long-term solutions. Whenever we clean up a mess, usually there's so much pressure to put things right that we rush ahead, making the same mistakes. After the earthquake in Kobe, Japan, the local government flipped from a state of paralysis in responding to the immediate disaster to a state of hasty reconstruction to put the local economy back on its feet. An excellent opportunity to do things differently was lost. In Bangladesh, Florida, and other storm-prone zones, we see the same story repeated ad nauseam. In the 2001 Gujarat quake in India, the government personified this approach by doing nothing.[27] Months after the disaster hundreds of thousands of people remained homeless; the economy of a region was obliterated.

Public support for adaptation is missing. Our governments pour billions of dollars annually into disaster relief, because when catastrophe strikes, demands from the media and the public for an immediate solution are overwhelming. Yet calls for preventative actions usually die as soon as the TV cameras have left the scene. Moreover, due to the increasingly litigious nature of our society, much effort is spent trying to cover up the cause of a disaster, or arbitrarily selecting someone to take the fall for everyone else. The classic technological example of this was the *Challenger* space shuttle disaster. After it struck, the entire space program underwent microscopic scrutiny. Despite blatant evidence that safety had been ignored in favor of meeting deadlines, efforts were launched by contractors and NASA insiders to block information about the real cause of the problem: faulty design. It was simply too big to accept, and too many people had too much to lose. Only due to the brilliant efforts of individuals such as Richard Feynman, as discussed in chapter 2, was the true cause exposed and a program launched to clean up manned spaceflight program design processes.[28]

This trend of societal resistance isn't just limited to high-technology insiders. Numerous studies have shown that disaster prevention is often slow to gain public acceptance, or is overcome by societal pressures, despite convincing evidence of the benefits associated with adopting new approaches.[29] This goes a long way toward explaining the other disaster-adaptation shortfalls identified in the learning experiment, namely:

∞ we have options, but we're not using them;
∞ solutions are not effective at the local level;
∞ information sharing is inadequate; and
∞ there are too few incentives to change our approach.

MAKE WHOLE SYSTEMS ADAPTABLE

The most profound finding of this review is that it is difficult to prepare local adaptation measures in advance. Rather, it makes far more sense to build adaptive capacities into our infrastructures. The reason is plain: although some disasters can be accurately foreseen, many cannot. Therefore, the team concluded that *whole systems* have to be made adaptive, rather than just building one set of defenses for one set of circumstances. An example of this would be a storm-prone earthquake zone that's also at risk from giant waves, drought, and flood. Too silly to contemplate? That's the entire West Coast of North America, plus Indonesia and Japan.

In summary, the Canadian team says that preparing for individual disasters in such regions is too expensive to be realistic. Instead, the entire infrastructure needs to be adaptable. The good news is that some scientists have started considering how to do this.

RESILIENCE

Hawaiians learned a bitter lesson from giant tsunamis that occasionally rip through coastal cities: we can't yet affordably build seaside dwellings to withstand hits from such waves. So instead, they began designing the lower floors of buildings to accommodate the waves, rather than resist them. This also works in areas with tidal surges from storms. Many buildings have been constructed with flow-though ground floors. When moderately high waves tear through, upper floors are left intact. Cars in the parking lot don't do so well, but at least the occupants of the building are left alive.

This is the difference between *resistance* and *resilience*. One method tries to stand up to the threat, while the other opens the door to let it pass through unhindered, thereby saving the larger infrastructure. A variation on this concept was illustrated by researcher C. S. Holling, who demonstrated that trying to resist forest fires and pests through control measures led to the buildup of fuel, so when one measure failed, a domino effect was unleashed where the infestation or fire became far worse.[30] In such cases, it may make more sense to let fires run their course, and instead adapt human settlements to such risks via, for example, landscaping strategies.

David Etkin, a researcher who contributed to Canada's Adaptation and Impacts Research Group, says that resistance is, in many cases, futile. Terraforming hasn't worked, for example, in a major fifty-year experiment with the Mississippi River. Instead, such scientists argue for another way: resilience. Etkin explains it this way:

> It is worth distinguishing between two types of mitigation; "resistance," and "resilience." Resistance refers to the capacity of society to withstand external forces, while resilience refers to its capacity to "bounce back" to its pre-disaster state.
>
> Perhaps best illustrated by example, a house resistant to floods is one with a dike around it, while a house resilient to floods might be a house that floats. Another example of a resistant system would be a power line built to resist high ice loads. A resilient power system, however, might have redundancy built into it, or perhaps break-away arms designed to fail just below the design value of the rest of the tower, so that repair is inexpensive and rapid.
>
> Designing a society resistant to disasters only postpones them, and may contribute to worse disasters in the future. Designing a society that is resilient as well as resistant will reduce the future toll of human loss and suffering.[31]

This is akin to bending like a tree in the wind, then snapping back afterward. It's the least developed strategy, but the one that Etkin and his colleagues say gives us the best odds. Such a method is being applied by Brad Bass with vegetation-covered roofs and "vertical" gardens in urban areas. Governmental agencies, construction companies, and environmental groups are developing these for resilience to floods, droughts, air pollution, and water pollution.[32]

One of the foundations of resilience is *redundancy*. To make up for our vulnerability, we build redundancy into systems so that enough is left to support the rest in case of widespread failures. It was redundancy, for example, that saved the *Apollo 13* astronauts from death; when four backup systems failed, there was enough left in remaining systems to get them home. Redundancy was also one early rationale for the Internet. If half of the United States is wiped out in a nuclear attack, the Internet may still function because it depends on no central source.

Redundancy works well when parts of a system go down. But when the whole infrastructure comes under pressure, cascading failures can multiply. If half of North America's computer capacity is wiped out in an attack, or a solar-flare anomaly disables electronic systems, cascading failures may shut down nearly every server, with the exception of some that are shielded and run on generators. Thus, the whole system might be temporarily knocked out.

That's what happens locally in an earthquake or hurricane. Fortunately, these are only local, so unaffected systems from outside the damaged areas are able to send in help. Yet if a whole continent is disrupted by, for example, climate changes from an eruption or comet collision, the stress is too great. No "outside" resources are left to be flown in to fix the problem. Instead, systems start to collapse and social disorder follows quickly. Thus, redundancy is only effective if we have enough backups to deal with crop, water supply, and other shortfalls.

Unfortunately, much of this discussion about resilience and redundancy is moot, in view of political and technological realities. Roger Pielke, of the National Center for Atmospheric Research, and Daniel Sarewitz, of Columbia University's Center for Science, Policy and Outcomes, describe the problem in relation to climate changes:

> Intellectual and financial resources are also poorly allocated in the realm of science, with research focused disproportionately on understanding and predicting basic climatic processes. Such research has yielded much interesting information about the global climate system. But little priority is given to generating and disseminating knowledge that people and communities can use to reduce their vulnerability to climate and extreme weather events.
>
> For example, researchers have made impressive strides in anticipating the impacts of some relatively short-term climatic phenomena, notably El Niño and La Niña. If these advances were accompanied by progress in monitoring weather, identifying vulnerable regions and populations, and communicating useful information, we would begin to reduce the toll exacted by weather and climate all over the world.[33]

RELINQUISHMENT

After years of debates, it's now more or less accepted by most scientists that the world needs to try to prevent climate change by reducing the amount of

CO_2 we put into the atmosphere, although such a strategy is still vociferously challenged in some quarters. A precursor to the CO_2 reduction approach was the Montreal Protocol on reducing other types of chemical emissions that destroy ozone. This strategy has mostly involved cutting out old "smokestack" technologies and replacing them with improved ones.

Another strategy has also emerged: blocking new technologies until they're proven not to be harmful. This idea of suppressing new technologies for what we *don't* know about them has the scientific community sweating profusely. Known as the *precautionary principle,* it's characterized this way by the United Nations' Environment Program:

> The precautionary approach is innovative in that it changes the role of scientific evidence. It requires that once environmental damage is threatened, action should be taken to control or abate possible environmental interference even though there may still be scientific uncertainty as to the effects of the activities.[34]

Entrenchment into European legislation of this "don't do it till we know if it's harmful" approach also has American legislators looking nervously over their shoulders. Industry and many academics oppose such legislation because they claim it will prevent scientific progress. This controversial and complex issue is covered later, in chapter 20.

IMPLICATIONS OF THE MOLECULAR ASSEMBLER

We've seen how adaptation is a powerful mechanism to protect us from planetary extremes, but the present situation does not bode well for putting it into practice. That's where molecules come in. Once the earlier-described molecular assembler becomes real, our adaptive technologies will have the potential to explosively develop. We'll face new choices for defending ourselves from natural calamities.

Here are examples of adaptive technologies that we might develop, and where they may take root. Many are described earlier, but this section depicts their disaster preparedness applications. It must be emphasized once again that, because of reasons explained earlier,[35] time lines for the introduction of such technologies are uncertain. We may see some applications arrive ahead of others. Some may be set back by unforeseen circumstances. These are only a few of many possible futures.

Preempt the Enemy

A 2001 earthquake in Seattle spurred the California Institute of Technology and the U.S. Geological Survey to enter the final deployment phase of an Internet-enabled earthquake-warning system for southern California. This is the first of its kind in the United States, and one of the most advanced in the world. It's *not* a prediction system. Rather, it's designed to give cities such as Los Angeles a few precious seconds warning, from the time an earthquake occurs up to 130 miles away, until the shock hits. The system wouldn't help in a quake that happens right under the city, but it may give residents time to prepare when they hear an alarm that warns of more distant tremors.[36]

Many regions of the world need improved earthquake sensor systems. Hundreds of cities in quake zones have none. The main barrier to implementation is the perception of cost versus benefit. The capital and maintenance costs of installing such systems run into the millions of dollars. That's not a lot of money for a major U.S. city. Yet, according to California's state geologist, James Davis, a survey done in 1991 about cost of the southern California system found that many residents didn't want to pay for it.[37] It has the same ring to it as resistance by passengers, airports, and airlines to the potential cost and inconvenience of more security, until the far greater costs of lax security became apparent with the attacks on the World Trade Center and the Pentagon. A similarly complacent mentality applies to earthquake and tsunami detection systems. The political will seems to be missing for governments to pay the preventative ante.

The same goes for near-Earth-object sensors. For example, the Spaceguard Foundation, dedicated to protect Earth against rogue asteroids, has been hampered in detecting near-Earth objects by the relatively small cost of computerized telescopic detection, along with the greater costs of operating detection satellites and interpreting their data. Thus, detecting a hundred-meter-wide incoming asteroid that could annihilate humanity is improbable in the time frame required to do something about it.

Yet with the help of nanotechnology, we might be able to install such detection systems in space, and across Earth, for a fraction of the cost of today's technology. Nanosensors will be molecular extensions of the human senses. They might perform similar functions as the satellites we send to other planets now, except they will be smaller by a factor of a thousand or so. On Earth, they might have the capacity to sense, for example, changes in pressure, magnetic field, temperature, light, and movement, then transmit the data to a remote computer for analysis. Self-directed nanosensors may operate in the inhospitable conditions of space, ocean depths, and volcanic vents. With their microscopic size and weight, they might be attachable to animals, microbes, and dust particles. We'll have trillions of eyes, ears, and noses–at rock-bottom prices. Warnings about volcanic eruptions, crustal slippage and near-Earth objects may be easy to get. The high costs of maintaining such systems could plummet: Nanosensors might be self-repairing, or so cheap that they'll be expendable. At the receiving end, nanocomputing may let us analyze data from each location and share it instantly across globalized networks. Thus, a geologist on a mountaintop in Chile might have in her wristwatch real-time access to already-interpreted data on every seismic zone everywhere on earth, plus the capacity to enter new data, then have it cross-checked and instantly added to the international database.

This could give every seismologist and meteorologist an accurate picture of how the surface of the earth moves: weatherwise, seismically, magnetically, gravitationally, and tidally. With such an omnipotent view, our capacities to forecast and prepare for environmental extremes would multiply logarithmically. We'd be able to track many of the land- and sea-based organisms on earth. Although this raises a new set of privacy concerns for humans, it's a step toward managing environmental resources and adapting to nature's extremes.

The late humorist and science fiction writer Douglas Adams postulated in his *Hitchhikers Guide to the Galaxy* series that the real aliens are the mice, and while we spend our time searching for flying saucers, the mice are busy running society. In the molecular era, he may be right, at least about the size. The truly omnipotent devices will be the smallest ones.

Absorb the Shock

We could be protected from injury in earthquakes, tsunamis, and tornadoes by *utility fog*. Computer systems architect John Storrs Hall, a research fellow with the Institute for Molecular Manufacturing (IMM), describes utility fog this way:

> Nanotechnology is based on the concept of tiny, self-replicating robots. The Utility Fog is a very simple extension of the idea: Suppose, instead of building the object you want atom by atom, the tiny robots linked their arms together to form a solid mass in the shape of the object you wanted? Then, when you got tired of that avant-garde coffee table, the robots could simply shift around a little and you'd have an elegant Queen Anne piece instead.
> . . . Another major advantage for space-filling Fog is safety. In a car (or its nanotech descendant) Fog forms a dynamic form-fitting cushion that protects better than any seatbelt of nylon fibers. An appropriately built house filled with Fog could even protect its inhabitants from the (physical) effects of a nuclear weapon within 95% or so of its lethal blast area.[38]

Utility fog is one of those remarkable concepts that we may find unbelievable, because we haven't yet comprehended the profound capabilities of nanotechnology. Yet, if invented, it may radically alter the way we deal with disaster risks. For instance, if a utility fog were to be developed with the properties described by Hall, victims caught in disasters such as the World Trade Center attack might conceivably survive a building collapse of such magnitude in shock-absorbing cocoons. As described earlier, such collapses could also be avoided altogether by nanotube structural reinforcing, but nonetheless we can also see the potential for being protected from other types of explosive shock by utility fog.

Manufacture Water

Conventional water systems are expensive to build and fix. In a disaster, damaged fuel tanks, sewage lines, and agroindustrial effluent contaminate water. Anyone who's lived through a hurricane or earthquake knows it can be weeks before drinking water systems are restored. Disease is the biggest threat, but long-term toxic contamination is also a by-product, because everything becomes saturated with industrial waste. Molecular science may solve this in graduated steps. In the early years, nanomaterials could reinforce our piping systems so they're robust and watertight enough to withstand big hits. Such strength may also improve the resilience of our water and sewer systems against terrorist attacks, then cut the cost of reconstruction. Thus, taxpayers might avoid the heavy recovery burdens such as those incurred in New York after September 11. Next, portable *water synthesizers, purifiers,* and *waste recyclers* may treat drinking and wastewater at the source. Vast water pipe networks would be unnecessary. Initially, water synthesis might occur by drawing vapor from the atmosphere, but as things progressed, we'd have the means to combine two hydrogen atoms from a cheap source with one oxygen atom—to build our own water molecules.

Fabricate Food

Farms and food distribution systems are always susceptible to climate changes. Years after Hurricane Mitch struck Central America in October 1998, much of the region's agriculture and distribution remained decimated.[39] Then, in 1999, the fragility of eastern North America was exposed when Hurricane Floyd wiped out the Carolinas' hog industry, causing nutrient pollution from manure ponds that washed across the states into homes, industries, and fishing grounds.[40]

Molecular biosynthesis and *robotic replenishment* may allow quick replacement of production, so we wouldn't have to depend on centralized systems to grow and deliver our food. In the first, primitive stages of molecular assembly, we'd build packaged greenhouses, radically different from those today, that would allow local or individualized production by millions who know nothing about farming.[41] The greenhouses would have transparent enclosures made from carbon nanotubes that prevent glare. They'd be self-maintaining in ways that avoid other environmentally negative effects of conventional greenhouses, such as effluent pollution.[42] Agricultural software could be downloaded from the Internet and fed into nanobot hardware that plants seeds, waters them, controls pests, and harvests crops.

At the next stage of molecular manufacturing, food synthesis could occur directly, without growing crops or livestock. This infinitely more complex task might take decades to achieve, but may transform the landscape by eliminating many farms, livestock operations, and slaughterhouses. Such technologies would also be essential for warding off famine and disease, in the drastic event that one of our hemispheres becomes a twilight zone due to fallout from an asteroid hit or volcanic eruption.

Make Healthy Products

Burst pipes, chemical storage tanks, garbage, sewage, and cars leave a toxic mess in every catastrophe. There's only one way to finally solve this: by designing safer products that don't require toxic ingredients and don't leave contaminated residue when they're used.

For decades, industry has complained that this was impossible because of the cost, but in the molecular economy we'd make waste healthy rather than toxic. Imagine, for example, a house that emits fresh air instead of toxic gases from its materials, or roads that slough off fresh water in the rain, instead of polluted runoff from traffic.

Nanotechnology experts such as rocket scientist Gayle Pergamit, coauthor of the nanopioneering book *Unbounding the Future*, emphasize that regardless of how small nanotechnology gets, it still adds up to the macro scale.[43] Thus, we'd still have to look at the real amounts of contaminants being put into our environment, including their impact on human health. For example, as nanoparticles become ubiquitous, the threat of toxic nanomaterials entering the human body through lung, mucous, or stomach linings rises substantially. Nano-sized particles infiltrate such linings more easily than larger dust particles.

So how might we build nanoproducts that are a boon rather than a bust for health? One way is to redefine the way that we look at product components. In the early 1990s I was part of a team that developed the *intelligent product system* (IPS). This was conceptualized and applied by Michael Braungart, a chemist who heads numerous private institutes and consults to companies and governments on the award-winning methodology, along with Justus Engelfried, now a professor in Merseburg, Germany.[44] In the early 1990s, IPS was among the first methods to describe how molecular chemistry could be used to make every product compatible with environmental cycles, instead of wrecking them. Since then, the concept has been featured in a range of media,[45] while being adopted by companies in North America, Europe, and other parts of the world.[46] Here are some highlights from the methodology:

∞ Consumable goods such as textiles, fuel, or packaging that end up in the environment as they are used, must *biodegrade* along defined, traceable pathways. An example is Climatex, a furniture textile manufactured by Rohner Textil AG in Switzerland.[47] The product, introduced in the late 1990s, is designed to biodegrade in a composting facility.

∞ Products that provide an ongoing service and are meant to be more durable are designed for *disassembly*[48] and recycling after being returned to the manufacturer or otherwise discarded. These include goods such as cars, computers, and washing machines.

∞ Manufacturers use IPS principles to recycle processing chemicals and other reusable substances in *closed-loop systems*.

These principles may seem obvious now, but in the early 1990s they were as revolutionary in industry as they were suspect. Now, though, such systems are used by automobile, footwear, and floor-covering companies to make their products more compatible with natural environmental processes.[49]

Other examples of IPS are found around the world in the form of integrated farming systems (IFS).[50] Such systems, used for recycling farm and urban waste, have penetrated agricultural and wastewater treatment markets under the auspices of numerous organizations.[51] The variant that most closely resembles IPS was codeveloped by Hamburger Umweltinstitut in Germany[52] and O Instituto Ambiental in Brazil. Such systems combine wastewater recycling with flood control and agriculture, to leave water cleaner than when it enters an agricultural production facility. IFS characterizes how intelligent product systems are supposed to work at the molecular scale. Each component of the nutrient chain is assimilated and disassembled via mechanical, chemical, or biological means. Nutrients and their components are recovered. Pathogens are eliminated. Safe products are produced. The whole cycle is self-reinforcing.

With digital fabrication and molecular assembly, that self-reinforcing cycle may serve as a guideline for manufacturing products. The way materials are extracted from, and returned to, the environment will be more, rather than less, important. This is due to the variety of materials that might be manufactured, along with the potential for so many new, small particles to get past the body's defenses.

According to IPS, digitally fabricated products would be disassemblable or biodegradable. This would avoid flooding the world with materials that couldn't be either absorbed by the natural environment or taken apart again. The technical possibility for molecular machines to disassemble themselves has been discussed for years, but there is still a dearth of discussion among digital-fabrication researchers about how to do it. Incorporating methodologies such as the intelligent product system may help speed that discussion.

In turn, digital fabrication researchers may also be able to perform a reciprocal favor for "sustainable" methodologies such as the intelligent product system. Many "sustainable" methods are hampered by a built-in weakness. The recycling chain that is sometimes necessary to achieve sustainability often requires cooperation along a line of raw-materials producers, manufacturers, consumers, and waste managers throughout the life cycle of the product. This is difficult in industries that aren't fully integrated or that get their supplies from globalized networks. Quality-assurance programs can go some way toward facilitating this, but in many cases the increasing complexity of product components makes it a struggle. Furthermore, industries are reluctant to tie themselves into recycling or other closed-loop cycle contracts that restrict their ability to get bids for the lowest price from diverse suppliers. Thus, "sustainable" manufacturing faces serious roadblocks.

Happily, molecular technologies may offer an elegant solution to such shortcomings. Years before the IPS and other such methods came along, K. Eric Drexler described how nanotechnology products could *self-disassemble* into reusable raw materials.[53] Everything we make could be discarded and disassembled, and the component materials could be reused for other applications. With desktop manufacturing this could be done on-site, without sending materials back to the factory or a recycling plant. Furthermore, molecular manufacturing might help to reduce dependence on worldwide material flows, because many products would be made locally and disassembled locally after use.

Another weakness of environmental methods, as described earlier, is susceptibility to natural disasters. In such catastrophes, "sustainable" networks fall apart and are expensive to reconstruct. For example, if a tsunami were to hit a coastline, sustainable agriculture might be wiped out. Manufacturing facilities—such as the thousands built on waterfronts—might have their closed-loop systems smashed, with attendant toxic pollution. Sustainably manufactured vehicles might be destroyed or rendered useless on roads that are blocked by wash-outs. Most sustainable methodologies do not account for these types of horrific natural assaults.

Rendering some products biodegradable may help to mitigate the damage.

Yet that alone would be insufficient to make the overall infrastructure of the "recycling economy" resilient against such onslaughts. Similarly, if a catastrophic volcanic eruption changed the climate suddenly, most sustainable methodologies wouldn't provide the rapidly adaptive mechanisms we'd need to stay on our feet.

Once again, molecular assembly and digital manufacturing may help to solve this by eliminating the need for closed loops and complex recycling chains. Products would be manufactured in one place without complex globalized delivery systems. Molecular codes could make them self-disassembling or biodegradable under specified conditions. In a natural disaster, destroyed products could degrade or disassemble themselves for later robotic recovery and subsequent reuse as raw materials. If products or their factories blew apart or were submerged, they'd pose no toxic threat. Thus, after catastrophe struck, the task of reoccupying disaster zones would be safer.

Such concepts for assembly and disassembly point to a safer future for nanotechnology. The intelligent product system and Drexler's disassembly concept are good examples to start with. Other examples include the Foresight Guidelines for Molecular Nanotechnology (see appendix B). So far, though, the molecular research community hasn't paid much attention to linking nano-scale manufacturing with macro-scale impacts in such ways.

Respond To, and Clean Up, the Mess

In the aftermath of disasters, rescuers often have to stand by helplessly, or risk their lives while parts of structures continue to fall and victims die. This was poignantly illustrated in the minutes and hours after the World Trade Center collapse, where many rescuers died from falling debris, others were injured by toxic smoke or dust, and still others had to hold back to avoid later collapses. There are more such lessons from the World Trade Center calamity: a newly built emergency response center for the city was one of the first things to be knocked out, because it was housed in what was thought to be a secure building that subsequently collapsed. Similar situations are often replayed, tragically, after earthquakes. From these, we see a pressing requirement to redesign our crisis management systems so they respond in more flexible ways. Tools such as utility fog might do this; they may protect individual victims from the initial onslaught. Yet many could still be trapped afterwards. In such situations, *rescue bots* could perform emergency searches and extractions. Such autonomous bots aren't so far off. In 2001, organizers of RoboCup, the robotic event described earlier, started a new event–RoboRescue–to test such technologies for situations where it's too dangerous for emergency crews to go. It was inspired by the devastating loss of life in the Kobe quake. One of its goals is to establish a robotic rescue brigade by the year 2020.[54]

To recover from big hits, we must also cut the cost of rebuilding. Molecular robotics may let us produce millions of *construction bots* and *repair bots* that perform remotely managed work in hostile environments. Moreover, with cheap molecular construction, our appliances could become replaceable commodities instead of expensive durable goods. If our homes were destroyed, we could replace our appliances quickly and cheaply. In today's economy, one of the greatest barriers to disaster recovery is energy. Networks of oil drilling, pipelines, refining, transportation, and delivery are susceptible to catastrophic interruption. The molecular economy, on the other hand, could improve *energy resilience.* We could produce *solar coatings* that are painted onto surfaces, generating energy that replaces power from vulnerable centralized energy sources. These coatings may wrap our economy in a resilient blanket by diversifying our energy generation base. Moreover, the same technology could improve *systems redundancy.* Instant data backups and retrievals via the Internet could save businesses, in the event unexpected disasters wiped them out. These systems will become ubiquitous as they get smaller. In space, self-launched nano repair bots might help satellites damaged by solar flares to fix themselves. Nanotubes may make it realistic to build structures that withstand sustained hurricane-force winds and upper-end seismic shock. This would also make them resistant to conventional terrorist attacks. Moreover, such defenses could be complemented by expendable self-disassembling materials that are designed to fall apart organically or benignly. Together, these materials would give us the option to stand and fight a disaster, or cut and run and return to rebuild, knowing we were safe from toxic waste. For example, had they been invented at the time of the World Trade Center and Pentagon attacks, nanotube-reinforced construction could have saved those structures from collapse when jets crashed into them. Self-disassembling materials could have speeded the recovery of victims, obviated the need for a massive cleanup, and eliminated the risks from toxic materials.

Together, these advances hold the potential to free us from the ruinous costs of disaster recovery, not only in the advanced industrial nations, but also in nations that are running to catch up—especially in the disaster-prone tropics, where most of the world population lives.

Escape from the Danger Zone

What good is an environmentally sustainable, energy-efficient vehicle if the roads it runs on are under water or buried by volcanic ash? Despite the advantages of utility fog and structural nanomaterials, we'd still have to evacuate if, for example, volcanic eruptions or megatsunamis couldn't be stopped in time. One escape route is to go *up.* Advances in materials, guidance systems, and fuel efficiency are about to make the pilotless personal aerocar feasible. In 2001 prototypes were already being flown.[55] Today's roads can't support a mass exodus from coastal zones in a giant tsunami, but with short warning, aerocars will be able to: by going up. How do we prevent them from running into each other? In 2000 NASA began work on guidance system corridors—invisible superhighways in the sky.

Moreover, waterproof underground *electromagnetic transport tunnels* may also serve as evacuation routes. Dozens of times faster and more numerous than today's subways, they could evacuate millions from coastal zones in minutes. The first commercial-scale electromagnetic train was being built in China in 2001, although this was an aboveground version.[56]

As enabling technologies for aerocars and supersonic tunnels grow more affordable, we might see debates intensifying over their relative merits, including noise, aesthetics, and personal freedom. We shouldn't underestimate the importance of such debates. They may determine whether our next big transportation move is underground, overhead, or both.

LIBERATING THE INDIVIDUAL

Right now, when disaster hits, we're dependent on insurance companies, disaster agencies, and banks to pull us out of a jam. Yet time after time, we see that whole regions are impoverished by such disasters, and often don't ever quite get back on their feet. Prime examples of this are Homestead, Florida, after Hurricane Andrew; Nicaragua after Hurricane Mitch; and western India after the 2001 Gujarat quake.

The molecular age may free us from this umbilical cord of undependable insurance and inadequate disaster relief. Today, for instance, if we live in a place where we need earthquake insurance, we can't get it. Tomorrow, our homes may be more resilient at lower cost, owing to new structural materials. Our consumer appliances, such as computers, stoves, and refrigerators, will be harder to break and cheaper to replace. Later in the molecular revolution, desktop manufacturing may give us clean water, manufacture drugs to treat injuries and illness, and let us reconstruct everything from scratch with locally available raw materials.

By empowering individuals to protect themselves from disaster and clean up afterward, we may enter a new era of human development. I call this the *Age of the New Individual.* Next, we'll see how that era may help us cheat catastrophe, from regional calamities to planet killers.

CHAPTER FIFTEEN

LESSONS FROM TOKYO— LEARNING TO PREDICT THE BIG ONE

We saw in an earlier chapter how a big earthquake in Tokyo could cripple the Japanese economy, and consequently the world financial structure. Today, thousands of Japanese geologists are profoundly concerned about this risk, but they're limited by socioeconomic constraints to what they can do. So let's take a speculative look into a future when a disaster has already struck: when the horrific results have shocked governments into action to make sure such a calamity never recurs. This is the world of a fictional geologist who, together with an international team, uses molecular technologies to unlock the secrets of the tectonic plates on which Tokyo restlessly sits.

AFTER THE QUAKE: THE DAWN OF DESKTOP MANUFACTURING

The quake leveled eight hundred thousand homes in Greater Tokyo, killing a quarter million souls and sending the world economy into depression when its largest lending capital was crippled. Needless to say, attitudes to disaster preparedness have changed. No longer do smug government bureaucrats pretend that everybody is ready to cope with a big tremor, as they did even after the 1995 Kobe quake exposed that fallacy. The far more devastating Tokyo catastrophe generated a revolution that swept away the old government. A fringe opposition party has been voted in, vowing to rebuild Tokyo in such a way that it could never again be reduced by a quake to the second rate power it is now. Moreover, sobered by the magnitude of the damage to Tokyo, the United States and the European Union are launching programs to retrofit New York, Frankfurt, and other financial centers, since it's clear they, too, are in quake zones that could annihilate them at any time.

For Japanese quake experts like Hiro Naguchi,* who survived by virtue of living in Kyoto, the megadisaster has been a blessing in disguise. He's gone from being just another engineer to deputy head of the largest reconstruction and disaster prevention project in history: a multitrillion-dollar transformation of the decimated city of Tokyo.

Hiro gets ready for work by donning his virtual-reality suit, which looks like a skintight body stocking. The prototype is a remarkable invention, still available to only a few, but soon to hit mass markets and give the fractured Japanese economy a badly needed jumpstart. Millions of sensors, nanochips, actuators, wireless communicators, and temperature controllers are embedded in the millimeter-thin breathable fabric that molds to his contours like a body sock. He's uncomfortably aware that one of its first applications will be for cybersex. It's made the real thing outdated by comparison, and eliminated disease risk. Right now, churches, governments, and ethics experts are busy debating the impact this might have on personal relations, but that's not his worry.

He walks to the next room of his apartment. The blank walls—molded to resemble the interior of an eggshell—come alive with 10 trillion nanodiodes and nanochips painted onto the surface: about ten thousand times thinner than a conventional coat of paint. The room array and his suit together suck as much energy as a lightbulb used to, and are powered by solar cells painted onto the windows of his apartment.

"Let's go to the office," he says. The room transforms into a place that exists only in cyberspace. On one side, images of his twenty colleagues show up at a virtual table.

The other half of the room is reserved for holographic images. It's blank.

"Good morning, everyone." He bows respectfully and everyone returns the greeting. "Yuri, what do we have today?"

Yuri Kopalov, cochairman of the group, is based in southern Russia, where, decades ago, tens of thousands died in an earthquake. Centralized Soviet designs from earthquake-free Moscow had been sent to a seismic region, dooming many to die in pancake collapses of apartment buildings. As an infant, Yuri had survived in the rubble.

"Well, I have exciting news," Yuri bubbles with a grin.

---

*Names used in this chapter are fictional and are not intended to refer to real people.

There's a short delay as translating software whizzes through a trillion operations to check the meaning of Yuri's reply in Russian, then translates it to the language of every other participant. The software has some way to go in real time, but it's improving by leaps. This group communicates fluidly in seven languages, instead of having to struggle with English as everyone's second language. It's transforming international working relationships. Individuals no longer have to shove themselves into the cultural mold of a foreign language. They can read, write, and express themselves with the nuances necessary for deep understanding.

Some had feared this would lead to cultural chaos, but the opposite has occurred as business relationships flourish, especially among smaller organizations that can't afford international language training. This instant ability has shrunk the world more than the Internet did so many years ago. Most important, the rate of social understanding is catching up with the rate of technological discovery that ran ahead of it for centuries.

"I think we've done it," Yuri announces. The other half of Hiro's workroom comes alive. A real-time three-dimensional image appears, taken from a thousand feet beneath the surface of Tokyo.[1] "We've got the tectonic tension calculations to where we're able to predict a surface slip, based on rock type, magma pressure, and tension at the face."

Everyone in the meeting knows this picture is sent by hundreds of millions of nano-sized sensors, injected miles beneath the surface where earthquakes are born. Tokyo has been chosen as the first site for this, because of the earthquake, along with the government's desire to show its electorate that something real is being done. Japan also has a commanding lead in the nanosensing field, due to years of earlier research. The sensors—able to withstand enormous pressure with their graphite nanotube construction—are pressure-injected into hundreds of drill tubes under the Sea of Japan. They communicate with each other via wireless frequencies from their molecule-sized transmitters that send signals just far enough through the rock to form a gigantic neural network that looks like the workings of a huge brain. The image of this brain rotates in the room for Hiro's scrutiny. He can zoom to the minutest detail of the three tectonic plates that meet beneath Tokyo's streets, then zoom out to get a thousand-mile radius.

It's a fantastic sight that no one in the world has seen before: a real-time three-dimensional picture of the earth's crust.

To make a greater impression, Yuri has programmed the heat sensors to relay a temperature increase to everyone's VR suits, so they can get into the feel of things. The hot magma is producing beads of sweat on their foreheads. It isn't real, of course, because everyone would be fried in an instant. Still, it lends a sense of immediacy.

"What this means"—Yuri's voice trembles with excitement—"is that we'll get at least a five-minute warning when the plates are about to move, and we may quickly get up to ten minutes as the calculations improve."

Everybody in the meeting knows what this means, so he does not belabor the point. With a five-minute warning, each ground transport, elevator, and electrical service in range of the quake could be shut down or automatically secured. Potential victims could protect themselves. It's the end of earthquake surprises.

For decades, quake prediction was stymied because geologists didn't have information. They knew generally what went on beneath the surface, but the cost of getting details was prohibitive. Then a joint venture between Lawrence Livermore Labs and the University of Kyoto came up with a nanosensor that had superlative qualities. It could withstand steel-melting temperatures and titanium-crushing pressure, and communicate half a meter through rock to the next nanite, thereby indicating its own position via relay to positioning satellites. With these attributes, it could do what it was supposed to: send temperature and pressure data to a central node.

Cost was key. Each nanosensor cost ten cents. The newly invented molecular assembler had made this possible. Millions of identical molecular machines could manufacture themselves using nothing more than carbon, hydrogen, oxygen, nitrogen, and a few trace elements as feedstock.

Sadly, the human price paid before governments would apply this technology to earthquakes had been horrendous: the destruction of Tokyo by a quake that everyone knew was coming but did little about. So many lives were lost, and the worldwide depression had been devastating.

The poignancy of this loss had weighed heavily on each of them, driving them forward. Everyone here today is quietly ecstatic that they finally have a solution.

"Congratulations," says Hiro. He knows the prime minister will be pleased.

CHAPTER SIXTEEN

THE LONG VALLEY CALDERA DEFENSE– AVOIDING A DARK AGE

LONG VALLEY, CALIFORNIA: A PRESENT-DAY SCENARIO

Half a trillion cubic yards of pulverized rock rip into the stratosphere, darkening the earth. One of nature's biggest time bombs, central California's Long Valley Caldera, has exploded. The mammoth ground wave triggers a quake in the Cascadia subduction zone that runs from Cape Mendocino along the coasts of Washington and Oregon up to Vancouver Island. The shock sends megawaves to Honolulu and Tokyo, submerging everything in sight of their shores. A hundred thousand die around the Pacific Rim, despite early warnings. Trillions of dollars evaporate in one day. Meanwhile, the U.S. Midwest has been buried in volcanic ash. A week afterward, the jet stream fills with dust. In a month, sulfuric acid rain falls on a quarter of the planet's surface. Crops fail as temperatures plummet. Diseases ravage the world when ecosystems are disrupted. The globalized economy crashes. A new dark age arrives–this time literally.

Too horrendous to contemplate? In northern California, the first warning signs of a potentially devastating event are already apparent[1] in the form of tree die-offs caused by heat from pooling magma. Long Valley may blow next year, or next millennium. We're still bad at making such predictions. Other potential ground zeros are at magma domes in Southeast Asia, New Mexico, Wyoming, and Italy. A much smaller event at Indonesia's Krakatau in the 1800s changed world climate patterns for years.[2] In 1958, as discussed in chapter 10, Alaska experienced a landslide-generated wave that exceeded several hundred feet.[3] Earthquakes of the size required to generate such waves hit Chile in the 1960s. On January 26, 1700, North America's West Coast was inundated by Cascadia-generated tsunamis that raged miles inland and went as far as Japan.[4] And eighty thousand years ago, the East Coast was submerged by similar hits. Today, there is new evidence to show that an unstable volcano in the Atlantic poses a new landslide/tsunami threat to that same coastline.[5]

Let's fast-forward back to the future again, to a time when we may have the tools to protect ourselves from such calamities, courtesy of the molecular revolution.

LONG VALLEY, CALIFORNIA, SOMETIME IN THE TWENTY-FIRST CENTURY

Thousands of nanosensors implanted deep in the great caldera detect a huge magma buildup thousands of feet below. A silent alarm goes off. On the surface, millions of grapefruit-sized subterranean robot excavators starting to replicate themselves on-site, in a digital fabricator, so that they can begin digging pressure-relieving tunnels near main vents.

Several cubic miles of material have to be removed, and as the machines bore deeper the work gets tougher. Weeks pass. Pressure in the volcano builds. Finally, the drillers approach their target, but as they do, a rock plug separating the relief tunnel from the magma bursts. The volcano explodes.

Crucially, the preemptive drilling has lessened the force of the blow, giving precious time to prepare for impact. Everyone in the area was long ago evacuated, thanks to dependable early-warning systems. Once-reluctant residents now leave such zones willingly because they trust these systems' accuracy. Homeowners also have the benefit of mass mobility, along with confidence in affordable reconstruction. High-risk communities have already moved—including houses, businesses, and services—thanks to disassembly and reassembly capabilities that arrived with deployment of nanotechnology.

Thus, mass deaths from such blows are now avoided.

Nonetheless, the explosion is still devastatingly strong, sending a giant tsunami across the Pacific at five hundred miles per hour.

Fortunately, Pacific Rim cities are each part of the tsunami warning network, and had been alerted weeks ago during the caldera-drilling phase. Seconds after the long-awaited Cascadia earthquake occurred, tsunami warnings were raised along all coastlines. Millions of vehicles were airborne in half an hour. Millions more potential victims made it to the carbon-reinforced transport tunnels that permeate cities and double as evacuation corridors. These tunnels were constructed by billions of computer-guided excavation nanites. High-speed, frictionless electromagnetic vehicles float on a thin magnetic cushion. These aren't like the old subway systems, where hundreds of thousands crowd into a single tube. Instead, these low-cost tubes run, spaghetti-like, underground. With this increased capacity and high speed, they have the capacity to carry large coastal populations from harm's way.

As the tsunami approaches Hilo and Honolulu, graphite nanotube-reinforced curtains lower themselves from a string of floating barriers that ring the islands. As the curtains unravel toward the seafloor, they position themselves to form a slope: an artificial shoreline (see Fig. 31). At an electronic command, their molecular structure transforms from flexible to rigid. When successive waves of tsunami energy hit this first line of defense, they deflect off of it. Some of the surface energy screams upward along the slope to roar over the barriers in hundred-foot mountains that dissipate harmlessly on the surface.

Still, a series of weakened pulses that pass through the barriers in the deep waters off the islands continue their charge, submerging Honolulu and Hilo up to the fifth-story level of downtown areas. If this had happened in 2010, great swaths of the cities would have been annihilated, with power systems destroyed and buildings knocked over or stripped clean of interiors. Sewage, oil, and other toxic pollution would have poured from ruptured tanks.

But not today. Buildings are impregnated with carbon nanotubes thirty or more times stronger than steel, rendering them virtually indestructible against wave action. Doors and windows are hermetically sealed, turning buildings into submarines for the duration. External power lines don't exist, having been replaced by transparent solar coatings on every building and street that relay energy to office buildings, homes, and vehicles. Fossil fuels are gone. Oil-based lubricants are gone. Engines have no moving parts because they draw power from a solar grid painted invisibly on every vehicle and building.

Those unlucky enough to be unable to escape the wave are saved by utility fog that jumps from the walls of homes and offices to enclose them in a protective, cushioning, waterproof envelope that contains enough oxygen to sustain them for a few minutes. If more air is required, nanobots extract it from the surrounding water.

Twenty-four hours pass and the flood subsides. Things are still ugly. Debris and mud blanket everything. Some underground passages are flooded because it's not possible to seal everything on such short notice. Yet before residents have returned to check the damage, an army of biological nanites—nano-scale mechanical-biological mites—emerges to start consuming the gunk, after mobile pumps have displaced the largest volumes of floodwater.

Because everything has its own DNA or other nano-scale label, robots sort the debris and return it to its designated location, or send it for recycling

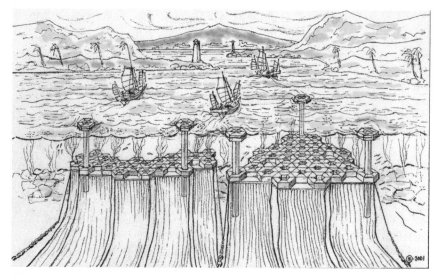

Fig. 31. A new concept for a tsunami barrier. Such barriers would be possible with new materials manufactured from carbon nanotubes. The curtain consists of a structural layer, molecular computers, and motors. This would be deployed from a semisubmersible platform, depicted above. Once the curtain was in place, it could receive an electronic impulse that would cause the material to transform from flexible to rigid. This would present a solid barrier to an incoming tsunami. Such "false shorelines" would be the first line of defense to protect seaside settlements from destruction. (Image courtesy of Eric Orbom)

if it was destroyed. In a few days, the gunk is gone. In a week, the city is cleaned and functioning again. Bloated nanites have been transported to forests, where they fertilize the soil with their load of nutrients.

Meanwhile, in San Francisco, the blast has triggered a big quake near the city. It's far less powerful than the cataclysmic 8.0 level that would have occurred without the nanite drillers in Long Valley. Still, it's stronger than the 1906 quake that wiped out the city. Even with detection technology, residents have only five minutes to get ready. Fortunately, much of the preparation has already been done. What little surface traffic there is (most has moved underground or overhead) comes to a halt, as detection systems in vehicles sense the oncoming tremor. Buildings constructed a hundred years ago still appear the same, but their exoskeletons contain an invisible nanotube framework designed to dissipate the shock. Liquefaction used to be a big problem for parts of the city built on landfill, but it's not anymore, thanks to these new materials.

Fire also used to be a threat. Now, though, homes have razor-thin, fire-resistant fiber reinforcing that keeps structures intact and stops fires before they start.

If injuries do occur, their severity is minimized because emergency services spring into action. In early years this would have been impossible because of power failures and blocked roads. Tremors killed many when emergency services couldn't reach them. Today, blockage is minimal and power systems are uninterrupted. Thus, thousands of deaths from shock and injury are averted because the injured can now be reached and saved through on-site emergency treatment.

Far away, in Japan, residents get enough warning to escape the coastline. Yet this is no ordinary tsunami. By the time it has traveled the five thousand miles to Japan, it's fifty feet tall when it hits shallow shoals. Nevertheless, the automatically activated tsunami curtains are able to stop flooding in most areas. Disease and death from contaminated water aren't threats, because water systems are rendered safe by reinforced piping and decentralized synthesizers that pull water vapor from the atmosphere.

Days later, a gigantic plume of volcanic dust has descended to earth, blanketing much of California and the Midwest states up to fifteen feet deep. Billions of speck-sized robo vacuums are busy sucking it up and transporting it to forests for fertilizer, or to the desert if it's too acidic. Loss of surface transportation isn't such a big problem because tunnels and aerocars still work.

Nevertheless, a far more sinister threat remains. The rest of the plume has entered the jet stream. It circles the earth, blotting out the sun and threatening to lower temperatures worldwide. This is by far the worst of the assaults on civilization. Yet, invisible to the human eye, a silent war is being waged in the stratosphere. Millions of tiny sky bots, launched immediately after the eruption, are consuming dust particles as feedstock, to replicate themselves in the trillions. They drop out of the jet stream as their bloated mass makes them too heavy to stay aloft. In less than half a year, the sky is clear. The ground is a bit dusty, but this volcanic washout is soon absorbed into the soil.

Thus, the catastrophe that could have changed the course of human history is a dud.

CHAPTER SEVENTEEN

HOW TO AVERT ARMAGEDDON— HAVE AN ASTEROID FOR LUNCH

BACK TO PRESENT-DAY REALITY

In the clear desert night of White Sands, New Mexico, a computerized telescope scans the skies every half hour, compares each image, and then picks out asteroids that may be heading our way.[1] This automated process is more efficient than the efforts of the few dozen astronomers who've dedicated themselves for years to finding rogue near-Earth objects. No one could be happier to learn that their manual methods are being rendered obsolete. They know that now, we may have a fighting chance at finding planet killers before they find us.

But there's still a problem. If we find one, there's absolutely nothing we can do about it. Despite what we may have seen on television, no weapon in our present arsenal is capable of destroying, or altering the path of, an incoming comet.

In their collaborative book *The Light of Other Days,* Stephen Baxter and sci-fi grand master Arthur C. Clarke describe a planet-killing asteroid that is discovered to intersect Earth's path in five hundred years, and the devastating psychological impacts this seeming death sentence has on civilization.[2] In reality, most scientists believe that the chances of such an occurrence in the near term are infinitesimal. Yet the 1994 Shoemaker-Levy 9 Comet collision with Jupiter, along with discoveries of many craters on Earth's surface, set astronomers to revising those odds. It is now broadly acknowledged that collisions with smaller projectiles a few hundred meters wide are more probable, and that they are capable of causing planetary havoc.[3]

Space agencies are gradually setting up detection systems to seek out these smaller projectiles, but we still haven't figured out a reasonable way to defend against them. The conventional wisdom is that our best defense is to throw up explosives that shatter such space mountains or throw them off course, but the flip side is that if we mess it up, we may face dozens of smaller and deadlier rocks—or inadvertently nudge the big one more directly into our path. As early as the 1980s, astronomer Carl Sagan pointed out that those who have the power to deflect such asteroids also have power to put them in our path, and hold the planet to ransom.[4] Moreover, we have problems getting far enough from Earth to intercept the projectile where it's still possible to push it off course.

Astronomers are constantly trying to figure out how to do this. They've concluded that our best defense is to detect these things as far away as possible. This might give us time to send out probes, map targets to find out what they're made of, and possibly attach solar sails—vast expanses of material that catch the sun's energy—to tug them slightly off course. Or we might knock them fractionally off their path with nuclear weapons. By the time they got near Earth, such a small variance in the trajectory would be sufficient to make them miss us.

The bad news is that, at the rate of current expenditures on such methods, we're still a long way from being able to do that. The good news is that nanotechnology may offer other methods that can be used much closer to Earth, and much more cheaply.

Virtually unnoticed in this public discourse is what I call evolution's back door—using the very small to influence the very big.

In the near future, it might work like this:

The Spaceguard Foundation disperses a million nanosatellites throughout the outer solar system to detect incoming rogue objects. Such numbers are possible because each satellite is smaller than a thumbnail and has been manufactured in the moon's low-gravity environment for energy-efficient dispersal. At the same time, a sister group of unmanned nanobot bases have been established in the asteroid belt. These bases are the jumping-off point for a fleet of "asteroid munchers."

The nanobots are only a few microns in diameter. Some are specialized to form solar sails that take them to their destination. Others—billions—hibernate in a delivery pod. Once on the surface of an incoming asteroid, they begin to replicate themselves, then assemble into extraction bots to strip-mine the projectile, layer by layer. Their nano-scale construction makes them superefficient, because every molecule in their structure is precisely placed to maximize faint solar energy from a distant sun as their fuel. Asteroids have varying composition. In some cases converting their contents to

Fig. 32. Delivery package to a planet-killing asteroid. A small satellite such as this could deliver self-assembling nanobots to the surface of an asteroid. From there, they'd use the asteroid's own resources to build millions of preprogrammed factories that convert the projectile into harmless, and possibly useful, raw materials for space-based colonies. This illustration shows the NEAR satellite about to soft-land on the Eros asteroid. NASA achieved this first-ever landing in 2001, suggesting that such a delivery strategy is feasible. (Image courtesy of Kees Veenenbos)

fuel may accelerate the process. This isn't essential, but in the case of asteroids that are detected close to Earth, it may help to speed disassembly.

Such asteroid disassembly solves several problems at once. The self-replicating capacity of mining bots eliminates the need for a heavy fuel load to get there. While weak or intermittent sunlight may make replication slow at the start, due to the tumbling of the asteroid or its orientation from the sun, this may be overcome by an exponential self-assembly rate that sees several billion operating by the time the asteroid nears Earth. Moreover, the one-shot risk associated with nuclear explosives would be eliminated, because destruction is controlled. If a few billion nanites fail to function, there are still others to take over. They might also convert a threat into a resource. By the time it approaches our orbit, the asteroid would be transformed into thousands of micro factories spewing out resources for space-based manufacturing. These smaller units could be nudged into a trajectory that puts them into orbit. Residue would be reduced to bite-sized meteorites that make a spectacular but harmless night show as they hit Earth's atmosphere.

Getting to such asteroids and landing on them is relatively easy, if we don't have to travel ten or more years out. The Near Earth Asteroid Rendezvous (NEAR) satellite already demonstrated this when it mapped, then soft-landed on, the Eros asteroid in 2001. It did this at relatively low cost, despite not being designed for such landings.[5] A satellite the size of NEAR could easily carry enough self-replicating bots to begin the disassembly process. Thus, the delivery package is already in our hands (see Fig. 32). We need only the molecular assemblers to put in that package.

This is just a brief outline of how molecular assembly might protect us from rogue near-Earth objects. Yet such brevity should not lead us to overlook the deep implications: there can be no more compelling reason to develop a molecular assembler, at the fastest possible pace.

In the final chapter, we'll discuss how to speed the invention of such an assembler, and in so doing protect ourselves from near-Earth objects.

In the meantime, we've seen so far that molecular defenses might protect us against everything from regional earthquakes to a comet that otherwise would annihilate civilization and rip the atmosphere from Earth, leaving a dead planet. These defensive tools may not be hundreds of years away; they may be right on our doorstep.

Thus, the question we face in the very near term is, How do we make such tools without annihilating ourselves in the process?

PART 4

GETTING FROM
HERE TO THERE

Our task—perhaps the only one that will save us—is to turn what we have dreamed into reality.

—Howard Bloom, *The Lucifer Principle*[1]

CHAPTER EIGHTEEN

THE RIGHT QUESTIONS

"1930 will be a splendid employment year."
–U.S. Department of Labor, 1929[2]

Everything about the future, including the short term, is speculative. Quotations such as the above pre-Depression forecast show how wrong we can be about the stability, wonders, or terrors of what's to come.[3] Therefore, it pays to hedge our bets, just as the brilliant scientist and computer pioneer John von Neumann did in 1949, with this statement about computing:

> It would appear we have reached the limits of what it is possible to achieve with computer technology although one should be careful with such statements–they tend to sound pretty silly in five years.[4]

Perhaps because predictions are so often wrong, or perhaps because we're born optimists, there aren't many precedents for wise preemptive actions based on predictions. Normally, we have to live through a disaster, unleash one on ourselves, or have our habits altered by technology to overcome our state of complacency.

Unfortunately, when a big catastrophe hits, we may not be left with the luxury to recover our inertia. Like those who built New York City's emergency command post in the World Trade Center complex– despite concerns about its vulnerability–we'll wonder how the unthinkable happened, and we'll gain no solace from those who sounded the warning. In the worst cases, chances are, we won't get a second chance.

Only informed forecasts, based on solid evidence from the past and present, might save us from repeating history.

WHAT EACH OF US MIGHT ASK OURSELVES

We're at the stage when we have more questions than answers. The challenge we face is to ask the *right* questions, then consider our options. One overriding question is, How do we use molecular technologies to avert natural catastrophes? To find answers, we may have to ask other questions. Each of us has to get an idea of how we might use such technologies. Then we have to consider the potential impacts of those applications. With these ideas in mind to frame our investigations, the primary questions might go something like this:

1. What impact might molecular assembly or artificial intelligence have on my life?
2. How could my community and I withstand natural disasters by using such technologies?
3. Who are the players in this adventure, and am I one of them?

Who should ask such questions? *Each of us.* Molecular technologies cut across economic and cultural boundaries. They are under development in thousands of agencies and companies. If you're one of the millions who work with them, or are their client, you'll be affected. You'll have a shot at affecting them.

Here's a refresher on the fundamentals of the many crucial matters touched upon earlier, after which I list the players, along with questions they might ask in relation to the sectors in which they work.

It's becoming clear that the more we learn about nature's extremes, the more we see that forestalling our perilous journey to a molecular age may relegate us to nature's dustbin. The odds are that we'll see a cultural tsunami as a result of molecular advances *before* a natural cataclysm hits. But this is not certain. We're vulnerable in the short term to disasters such as a massive Tokyo earthquake, a volcano sliding into the sea, or the eruption of the Vesuvius caldera. These risks remain, if nothing else, unpredictable.

What are the first steps to surviving? How do we control our wild technological creations so we can use them to prevent nature's occasional onslaughts from rendering us prematurely extinct?

Here, in alphabetical order, are the players in our scientific, economic, and cultural lives that may help us find answers to such questions. Each category has questions that each of these players might ask of themselves. Furthermore, those of us who may be affected by the actions of such players might ask the same questions of them as well.

Generally, these questions cover the impact of molecular technologies, and also how we might use such technologies as a defense against natural calamity. Sometimes they overlap.

Agriculture

Farming is already profoundly affected by both molecular science and equally, if not more so, by natural environmental extremes. Companies compete for the rights to genetic strains that do everything from producing medicine to growing forests more quickly. Meanwhile, climate changes are also challenging productivity. Yet if a molecular assembler is invented with the capacity to replicate food substances, much of today's agricultural industry may evaporate. In this context, there are two sets of questions for the agricultural industry.

First, what positive things might the agricultural industry do to show the benefits of molecular science for individual consumers? So far, some companies have made a public-relations mess by promoting technologies such as pesticide-resistant crop strains that appear to help their bottom line, but do nothing to help consumers. Some companies are trying to overcome that PR boondoggle by promoting vitamin supplement benefits and other genetic enhancements that may show a real benefit. Other, more cynical corporate decision makers have adopted a stealth approach, shunning the public debate and focusing instead on influencing legislators with special interest tactics. In either case, we still don't see an openly stated, transparent strategy for solving truly big problems—such as, for example, stopping pesticide pollution or eliminating vitamin deficiencies. We see piecemeal approaches, but no strategy that engages the public imagination. Moreover, the industry has no publicly understood strategy for using molecular science to help farmers cope with sudden global climate change. What might that strategy be?

In the more distant future, what happens to agriculture and agroindustry when each of us can conveniently grow our own food, or, ultimately, manufacture it? Are life sciences companies going to own the software that manufactures food, then make it available at an artificially high price? Such a strategy might backfire if intelligent machines are designing food fabrication software so quickly that patent legislation can't keep up.

How will government agencies such as the Food and Drug Administration deal with synthesized food? Who regulates the right of individuals to alter food formulas for their desktop fabricators? If someone wants to add an amino acid to his synthetic steak one evening, does he need FDA approval?

Agriculture has been with us for thousands of years, so it's hard to conceive that it may end for many of us. But that's precisely the possibility that we have to consider.

Chemicals

Nanotechnology and genetics *are* chemistry. Molecular assembly is built on the combination and recombination of atoms. Thus, the chemicals industry has a leading role to play in how molecular science unfolds. The chemicals industry is undergoing a revolution right now, because as bulk chemicals become generic products, these companies are turning to high-end specialty chemicals, along with pharmaceuticals and genetics, for their revenue base. Coatings, colors, lubricants, adhesives, pharmaceuticals, pesticides, and plastics—engines of the chemical industry's growth—will be based on nanotechnology. Many of the questions asked previously for the agricultural industry also apply to the chemicals industry, because of their intimate and often seamless partnership. How do they show the first positive benefits of molecular technologies for consumers, instead of developing shortsighted products that backfire and heighten consumer paranoia? More than any industry, the chemicals business has to look seriously at the ramifications of digital fabrication. What's the role of the chemicals industry when bulk manufacturing has no value anymore, and instead, chemicals are manufactured as integral ingredients of products by software-enabled digital fabricators?

In the field of disaster preparedness, might it be possible to develop defenses to demonstrate how nanotechnology is a lifesaver for the general population? How might the industry transform hazardous chemical closed-loop systems into ones that are resilient against huge natural forces such as giant tsunamis?

Climate Adaptation Researchers

One of the most promising vehicles to combine adaptive technology with natural disaster preparedness may be climate research agencies. Billions of dollars are spent trying to forecast what climate changes will take place, and how. Virtually nothing is spent studying how molecular technologies may improve our chances of adapting to such changes. Examples of such adaptation are discussed throughout this book. They include more resilient agriculture, buildings that self-adjust to altered climate, and energy transmission that's immune to interruption by ice storms, hurricanes, and flood. How might a link be established between climate adaptation and molecular technologies? One way is to bring climate researchers and molecular science researchers together. So far this hasn't happened. These communities are normally worlds apart. Exceptions are in the remote-sensing and climate-modeling fields. These may be invaluable windows for climate adaptation researchers to jump through and get a line on adaptive molecular technologies.

Computing and Corporate Disaster Recovery

Beneath the whole molecular infrastructure lies computing. Nothing is possible without it. Computers map the genetic sequences, generate the virtual reality that lets us see at the nano level, design the molecular machines, do the trillions of calculations, and perform thousands of other tasks that the human brain can't handle. How could we use computing to adapt to nature's worst attacks? What about redundancy that might allow us to lose half our computing capacity without missing a beat? How about a worldwide crustal sensor network that gives us a reasonable earthquake warning, so we might at least protect ourselves? This would not be a surface or shallow-crust network like those we have now, but rather would consist of sensors that penetrate deep into tectonic plates. Moreover, too few experts are considering what would happen to computerized systems in a wide scale natural disaster. For example, if a quake measuring 8.0 on the Richter scale were to hit San Francisco, much of the United States's Internet storage and networking capacity would be wiped out, because a good part of it is located in Silicon Valley. By the same token, if a tsunami were to hit Miami, much of Latin America's computer network would shut down, because its servers are based in Florida. Hundreds of millions of dollars are spent on corporate disaster recovery plans, but nothing on the scale of what would be required in these situations. It requires thinking on a new level; one conceivable only in a world of supercheap computing, multiple redundancy, and megabandwidth. Who might lead such a conceptual leap? Corporate disaster recovery teams are the best candidates, but they need to study lessons from big-scale catastrophes, then link these with nanocomputing concepts to gain the conceptual capacities for planning. Companies that survived the World Trade Center collapse might be excellent candidates for such work.

There is no doubt that the September 11 disaster resulted in a sea-change in attitudes toward disaster prevention,[5] but the underlying question is, Will such newfound awareness translate into natural megacatastrophe preparedness, or do we have to learn this the hard way also?

Construction

Extreme engineering is a prerequisite for adapting to severe environments. House, bridge, and tunnel construction may be revolutionized by nanotechnology. K. Eric Drexler describes a time when molecular manufacturing may require only carbon, nitrogen, oxygen and hydrogen in bulk quantities[6] that are available virtually everywhere. What happens when the prices of materials and labor drop to fractions, and houses are built by nanobots? What happens when architectural plans are delivered via the Internet then fed into construction bots, so that a fully furnished and landscaped home is erected in a day? What happens when megaprojects involve building space elevators from the earth to above the atmosphere? Engineering and molecular technologies go hand in hand. It's time for the construction engineers to explore this link.

Disaster Agencies

No disaster agency on the face of this planet is capable of dealing with a megacatastrophe. The main questions such agencies might ask themselves are: How might we plan to put ourselves out of business? How could we give every individual the power to detect, escape, and rebuild without having to depend on a big-brother disaster agency to protect them? What is the best way to study how molecular technologies could help us adapt? Today, the multibillion-dollar disaster industry is based on the idea of intervention when disaster strikes, or on coordinated government policies such as zoning and construction standards. The idea that individuals might have the resources to escape quickly from an impending disaster, then come back and rebuild their lives without government help is not in the cards: understandably so, because it's nowhere on the real horizon. Or is it? The conversation about a totally new infrastructure hasn't started yet, and might rightly be seen as too far down the road to be realistic. Yet how far is "too far down the road" when we talk about infrastructure planning that requires ten or twenty years? If molecular computers or assemblers are introduced in the next generation, they would relegate existing strategies for infrastructure protection to the waste heap. Instead of spending billions of dollars over the next years for conventional disaster planning, it might pay to put some of that money into technologies that obviate the need for approaches that have only limited success in any case. Yet it's doubtful that government agencies or nongovernmental relief agencies are able to do this alone. A better place to look might be corporations with long-term investment horizons that are looking at infrastructure loss in the case of a superdisaster.

Energy utilities are good candidates because of their long-term horizons, but also because molecular technologies may put their conventional sources of energy out of business. Furthermore, the damage recovery teams at big corporations—ones so big they don't bother with insurance—are qualified to do this, more so, perhaps, than disaster agencies or scientific investigators. They are also the ones developing molecular technologies to handle it. What might be the role of these corporations in adaptation technology? Probably substantial, and perhaps *profitable.*

Energy

If a tsunami hit the East Coast of the United States today, the entire energy system would be shut down for months: substations submerged, transformers soaked, and power lines torn apart. Molecular technologies might give us realistic tools to avoid this. Unfortunately, one barrier to adoption is the inertia of the fossil fuel industry. Oil executives get gray at the temples when asked about the effects of cheap, renewable energy on a worldwide fossil fuel network that includes production, pipelines, refining, and delivery. Molecular technologies are certain to disrupt these oligopolies. So far, the energy industry has run ahead of renewables by keeping the price of crude oil low and convincing governments to invest relatively little in renewables compared to fossil fuels. Molecular technologies may alter that equation. A renewable-energy era is about to dawn, when nano-scale materials make solar cells many times cheaper and much more efficient. Thus, the main question to ask is, What happens when the driving engines of our economy start to reduce their dependence on fossil fuels, the price of solar energy is limited to the cost of the hardware, and machines that use it cost nothing to produce besides the cost of software?

Another set of energy-related questions arises when we come to natural disasters. It's no good to replace one vulnerable technology with another. How can we develop solar, wind, and wave technologies that stand up in the face of natural calamities? How do we cheaply replace these new energy grids if they're destroyed? Right now, when knowledgeable environmental experts push for renewables, *they don't consider such questions.* For renewables to make us truly resilient against natural disaster, we have to start asking the uncomfortable questions now.

Ecologists and Environmentalists

These groups often overlap, but they are, by definition, distinct from each other. Ecologists are scientists who study the workings of the natural environment. Environmentalists are activists who use various means to press for positions based on their beliefs. Ecologists and environmentalists often work together, but sometimes are in conflict on key questions. Furthermore, the environmental movement is far from uniform, and virtually defies description these days. Therefore, in this section, when I refer to "environmentalists," I refer to the established environmental organizations that are broadly recognized, such as WWF–The Conservation Organization, Greenpeace, Sierra Club, Friends of the Earth, and the Nature Conservancy. I also refer to their active members and large foundations that fund their work.

As described throughout this book, much of the molecular future is fundamentally an ecological question, from the viewpoints of what molecular technologies do to ecology, and what ecology does to molecular technologies. Therefore, ecologists and environmental activists have special questions to ask themselves:

∞ As explained earlier, evidence suggests that earth is not a closed system. It interacts with the solar system via gravity, radiation, magnetism, comets, and asteroids. These have determining impacts on our ecology and our evolution as a species. This isn't a small issue; it's a big one that cuts to the heart of ecological theory and environmental-activist thought. How do we rewrite the principles of environmentalism to account for our newly understood relationship with the galaxy?

∞ How do we develop a new definition for "sustainability" that includes surviving catastrophic phenomena? Allusions to this are made in Agenda 21, adopted by world leaders at the Earth Summit in Rio de Janeiro 1992.[7] Although natural disasters and climate changes are discussed, adaptation to truly catastrophic phenomenon, such as a megatsunami, is not. How do we rewrite this part of Agenda 21 and incorporate it into environmental activism?

∞ What technologies might help us adapt to such extremes, so that besides trying to stop them from happening we might give society a chance to survive and prosper?

∞ How do we link up with disaster prevention experts and molecular technologists to give ourselves a deep understanding of big natural disasters and their solutions? FEMA, along with the European, Japanese, and U.S. space agencies, might be a good place to start, but they need a push from environmental and disaster prevention organizations. Some of the nongovernmental disaster relief organizations might be of help, especially those who recognize that disaster preparedness is the best form of disaster relief.

∞ What are the ecological implications of self-aware, intelligent machines? What might they do that we can't to clean up our old messes, while not creating new ones?

Finance

In 2000 it was estimated that hidden assets in offshore funds totaled about $5 trillion.[8] That was aside from the many trillions in regular offshore assets that fall under one tax regime or another. This "gray money" is greater than the net worth of many individual national economies. It falls outside the reach of governments. It moves nearly instantly at the command of computer trading programs. Some of these programs are now guided by artificial intelligence, and make more money at trading than their human counterparts do.[9] Such financial upheavals are only a small taste of what's to come in the molecular economy. With manufacturing of everything everywhere, and services provided to everyone from everywhere, there would be no need for many companies and individuals to keep their money in reach of national tax authorities. Furthermore, it may be necessary to reexamine the basis for taxation. Governments may gain the capability to undertake collective societal works by other means. For example, molecular assembly may give enormous power to governments to work for the collective good without having to tax citizens. This complex area requires much more thought than could be given here. Moreover, national currency regimes may be eliminated, as digital credits become the dominant international currency. Some questions are: What happens to the monetary system when everyone is able to satisfy his own basic material needs at very low cost? How would we use cash when digital manufacturing makes it impossible to differentiate a counterfeit bill or coin from the real thing? What happens to fiscal policy when digital information, moving at light speed, is the major commodity? How fast will monetary cycles move compared to, say, the ten- or twenty-year cycles of the late twentieth century, when products and patents go out of date in a matter of months instead of years?

Government

Federal, state, and municipal governments in the United States and around the world are unprepared for the molecular age. The list of policy, regulatory, and social issues associated with molecular science is all-encompassing, affecting most major branches of government from taxation to defense to health care. The first task is for political leaders and think tanks to undertake analyses of how the molecular age may affect the basis of government and democracy. What happens when we don't have to worry about trade or social services for our basic needs, because most of what we need is provided locally with digital manufacturing, and the biggest trade is in information? What happens when governments no longer have to spend time and money protecting energy supplies because we each have our own? How do we control the excesses of the ultrarich, the overabundance of the molecular assembler economy, and the challenge to intellectual-property laws created by intelligent, inventive machines? Who is going to govern vacation homes in orbit? Will Robo sapiens have the right to vote? What happens if half of all jobs are made redundant every decade? What happens to the War on Drugs when there's no import, export, or transport of contraband because drugs can be manufactured in a desktop machine using pirated software downloaded from the Internet? What happens to democratic controls when individuals can get as rich as small governments in a year or so? What kind of government and law enforcement will be practicable or necessary? Finally, why are governments spending so many hundreds of billions of taxpayer dollars subsidizing outdated, energy-guzzling, and polluting industries, when molecular technologies already show the potential to eliminate them?

Indigenous Peoples, the "First Nations," or Aboriginal Peoples

The Hopi, Australian Aborigines, Mongolians, Tibetans, and Inuit: these are among the cultures that others have tried to destroy. Now, they may hold some priceless assets in their histories: legends from a deep past. As we'll see later, such legends may tell us of cyclical natural calamities. How might everyone benefit from myths and petroglyph records depicting natural catastrophes that have been shown to strike repeatedly over the ages, and that may strike again? Elders and governing councils of these cultures might ask themselves, How do we reconcile the idea of "living in harmony with nature" with natural catastrophes that may decimate us? Did the ancients preach that we should live harmoniously with nature until the next volcanic caldera explosion destroys our livelihood, or did they learn a way for us to survive and progress? How do we reconcile the potentially negative impact of high technology, with its potential to cushion the blow of an asteroid hit? These questions are explored in the next chapter. Some pointers on where to look for answers may lie in age-old myths and wisdom.

It's up to the indigenous peoples of this world to begin reinterpreting their legends in this light, but it's also up to scientists to hear what they have to say. One organization that might consider bridging this gap is the United

Tribes of the Americas,[10] an international coalition of Native tribes that's emerging as a link between ancient peoples and the contemporary world. This organization, along with its affiliated group, the United Tribal Alliance,[11] has adopted the use of advanced technology as one of its strategic goals.[12] A joint project between the United Tribal Alliance and, for example, a futures institute or scientific organization, may help bring a fresh perspective.

Insurance

What's the relevance of insurance if many things are replaceable at very low capital cost, but liabilities from software are potentially unlimited? Is it feasible to have insurance against rogue fabricators, or robots that accidentally or deliberately kill large numbers of people or overrun ecosystems? In an era when software determines the content of most materials, is it possible to insure against defects in computer programs? In the life insurance field, some insurers voluntarily agreed for a time not to take into account the results of genetic tests when selling life insurance policies.[13] This was in response to criticism that they might discriminate against bad health-risk policyholders. Yet, as genetic screening becomes commonplace, such self-imposed bans are unlikely to last. On the other end of that risk scale, susceptibility to genetic defects may be reduced or eliminated with genetic therapies. How is the actuarial profession going to deal with such contradictions, between genetic tests that demonstrate a likelihood of reduced longevity, and new treatments that can fix genetic defects?

As far as big natural disasters are concerned, it's clear what happens to the insurance industry right now. First, the companies go bankrupt, then they turn to the government to bail them out, then they refuse to insure against such risks, and leave taxpayers to foot the bill as the insurer of last resort. This has worked miserably in California and Florida, where premiums from government-backed plans have skyrocketed while coverage limits have plummeted. Most average homeowners have grossly inadequate protection against hurricanes and earthquakes. The conclusion from this is that the insurance industry is virtually useless when it comes to coping with huge natural catastrophes. So what would it take to restore some sense or reasonable risk-assessment abilities to the situation, so that we might have reasonable insurance again? Telling people to move out of risky zones isn't the solution. The answers lie with molecular technologies that may give us more warning, more resilient communities, and cheaper ways to rebuild.

Investors

What should they invest in? By early 2002 the nanotechnology sector was entering the first stages of a venture capital investment bubble, similar to those that hit the early biotechnology and Internet sectors. Investors know by now what happened to those sectors in their nascent stage: their indexes collapsed. Furthermore, as explained earlier, "nanotechnology" seems to encompass everything from molecular assemblers to washing machines. Thus, money may start pouring into technologies that aren't truly nano. Some of these may produce profits, because there are many legitimate companies working on microtechnologies with great futures, but others will certainly produce a bust. The danger is that investors may grow dismayed with "nanostocks" that provide nano returns. This may lead to an investment downturn. In late 2001 Hewlett-Packard's director of quantum science, R. Stanley Williams, sounded the alarm on this risk at a conference of technology business leaders, when he remarked, "Ignorance and greed meeting in the marketplace is a recipe for disaster. . . . As a consequence, the field will lose credibility and momentum."[14] To avoid this, it would be helpful if a group of scientists began working with, for example, the Securities and Exchanges Commission to develop an investor guide to "What Is Nano?" They should also warn investment houses and brokers to avoid repeating the mistakes they committed while pushing questionable dotcom companies. Such preemptive actions might help preserve investor confidence in the legitimacy of the nanotechnology sector. Still, the best mantra is "buyer beware."

Here are a few potential breakout industries that investors may want to investigate, keeping in mind that the trick is to figure out when they come to market and how to take advantage of them, because if we look at the history of other breakout technologies—such as the car, telephone, radio, television, and computers—we see that the majority of companies that started them went broke. Sector index funds may be one way to avoid this, but these suffer from boom-and-bust behavior. Another challenge is how we define "sectors" of the molecular economy narrowly enough to assign tracking indexes to them. "Technology" or "biotechnology" or "nanotechnology" indexes may not be accurate enough. I leave that to the fund managers. In the meantime, here are some technologies and businesses to track.*

---

*This is not an offering for sale or an analyst rating and should not be construed as professional investment advice.

∞ Artificial and cloned organs
∞ Climate adaptation consultancies
∞ Coatings that prevent buildup of harmful bacteria
∞ Computer animators
∞ Crops with nutritional supplements or climate resilience
∞ Desktop digital fabrication
∞ Encryption and privacy security software
∞ Genetic programming and designing
∞ Genetic trait selection
∞ Guidance systems for small aircraft
∞ Imprinting technology (human features recognition)
∞ Intellectual-property consultants and lawyers
∞ Land somewhere in the middle of nowhere
∞ Life sciences companies—the big ones—but watch for "contingent lia-
bilities" provisions
∞ Mechatronics
∞ Nanobacteria-related tests and treatments
∞ Natural-disaster consultants
∞ Pharmaceuticals customized to treat diseases at a molecular level
∞ Pilotless-aircraft technology
∞ Psychology groups (everyone will need help to get through this era)
∞ Risk management consultants
∞ Robotic pets that are smart and sympathetic
∞ Remote surgical devices that eliminate invasive surgery
∞ Smart coatings that respond to the environment
∞ Smart dust and other sensors that survey the environment
∞ Software engineering academies and students
∞ Thermal photovoltaics
∞ Virtual-reality entertainment, including, but not limited to, sex suits
∞ Voice recognition (those who develop the algorithms)

And—some sectors to avoid, but with important caveats:

∞ Animal organ transplants, which may lose out to artificial organs
∞ Fossil fuel companies, although the short term looks great
∞ Genetic-liability underwriting
∞ Insurance industry generally, unless governments limit their lia-
bility, or the industry insists that clients adopt new disaster preven-
tion technologies as a precondition for coverage
∞ Utilities, unless they start manufacturing solar cells en masse

Labor

How should organized labor react when molecular assemblers and intelligent robots eliminate most manufacturing jobs? What is the nature of work going to be? How is organized labor going to cope with sudden job dislocation on an unprecedented scale?

Next, what are unions doing in order to plan for natural megadisasters? Do they have policies on them? In some cases, they do. The labor movement led the way in health care and worker safety. It has always been a big contributor to disaster relief. How might it contribute to pushing for initiatives on disaster preparedness?

Leisure

What happens when the only limit on leisure is imagination, when most people take vacations in their minds, and when alcohol is replaced with a cornucopia of molecular uppers, downers, and feel-gooders? What's legal fun? Given that recreational drugs constitute an enormous part of our lives, how are we going to cope with the tremendous power of a new generation of drugs that lack the somewhat self-regulating aspects such as hangovers or impairment? What happens when virtual sex is safer and more sensual than human sex? Does having sex with a holographic centaur that romps around your house count as bestiality? We need only look at the gruesome legalized violence of today's video games, compared to the illegality of displaying human breasts on a beach, to see that our legal system is already failing to cope with paradoxes of the virtual world. How are we going to redefine morality in that world?

Finally, in a world hit by big natural disasters the leisure industry might not do too well. For example, the beachside hotel industry is on the front lines of serious weather and tsunamis. How might the leisure industry plan to adapt? Invest in virtual reality? Stay-at-home vacations? Biodegradable resorts that rebuild themselves after being wiped out?

Management

Executives of big corporations today often hold more power than heads of government. As technology moves beyond the speed and understanding of governments, and management rather than shareholders control the big decisions, corporate managers effectively become a shadow government. They are making decisions, including scientific and technological ones that affect billions of lives. Yet, on the other hand, these executives are caught in the narrowing vise of the molecular economy: racing to beat obsolescence of their technologies, trying to keep employees from switching to competitors, looking over their shoulders if profit margins fall below expectations. Top spots at corporations have become revolving doors, as CEOs turn over at the fastest rate in history.[15] Moreover, although a few companies have managed to survive the past hundred years, the vast majority of corporations have gone under, been bought, been merged, or been divided up and sold off. Through all this, corporate concentration continues to increase. The life sciences, communications, and media companies that form the backbone of the new economy are consolidating chaotically. The classic example came when AOL, a company that started with seemingly minor services, such as holding customers' e-mail accounts for them, bought Time Warner, one of the world's largest news and entertainment companies. An ominous predecessor of that deal was the "merger" of Daimler-Benz with Chrysler, which later resulted in financial losses along with the departure of many top Chrysler executives.[16] Thus, on one hand, we see attempts at sectoral control, and on the other, instability in the form of revolving management. In such an atmosphere, the main question to ask is: Who's running the show? Who is sticking around long enough to see the results of their decisions? The question of control and responsibility may become more acute as machines take over more functions. While human managers come and go, computer networks retain the corporate memory and provide more continuity than staff. Thus, the day may not be far off when new managers get briefed by computers rather than by their human predecessors. What ethics should we instill in our robotic managers to make sure they give us good advice? The question of ethical judgment by machines will soon be with us in business. Are we ready for that? Is it a good idea? Are we going to let this happen by default, or take a proactive role in determining what tasks thinking machines should manage?

In the realm of natural disasters, managers face one big question that also constitutes an enormous opportunity. In a world where insurers are refusing to insure swaths of industry that lie in the path of hurricanes and earthquakes, *how can we use molecular technologies to construct a disaster-resilient industrial base*? Companies that learn how to do this will be the big winners when the next catastrophe strikes, or a relentless series of climatic extremes disrupts economic systems. Companies that don't may be out of business.

Medicine

Among individual sectors, health care now consumes the largest percentage of financial resources. As we get older, fears about how to pay for health care prey on us. Switch on a TV in America, and most channels are laden with ads from life sciences companies offering prescription solutions to conditions many of us haven't heard of. Billions are spent telling hundreds of millions of unqualified persons what drug to ask their physician to prescribe. What are the implications for the reputation of medical science? As we age, and become more dependent on medicine to maintain our quality of life, the role of the drug company and physician becomes more controversial. Moreover, these professionals may only be able to cope with the complexities of maintaining such life extension if they get help from intelligent computers for analytical work, and computerized machinery to keep us going. Therefore, as the population evolves toward artificially extended life spans, we face the prospect that civilization's existence may hinge on intelligent computing and computerized parts. We see the start of this with microprocessors implanted in our bodies as, for example, with prosthetics, heart pumps, retinal implants, pacemakers, and drug dispensers.

From this, the questions that health-care professionals need to ask themselves include: Are we ready to place our societal existence in the hands of computers that calculate our needs according to our personal genetic makeup? What's the role of robotic health care in a world where people switch body parts or bodies when they wear out, and switch genes to get new characteristics for babies? What happens when a new "child" emerges from an artificial womb fully developed as an adult, or we have the option to decide at what stage they come into the world? What happens to the growing field of holistic medicine when much of our body content is synthetic or genetically manipulated?

Ethical questions about research on fertilized human eggs are already high on the political agenda, so there's no point to belabor those here. They are precursors to aforementioned challenges. How we cope with them will say much about how we'll cope with the rest.

Military

Redefining military security in a digital world is a prerequisite. What are the military implications of having to fight individuals with the power of an army generated from a desktop factory, when biological weapons can be made and released at the flick of a switch? Looking at present trends toward civilians being the prime targets of warfare, what happens when genocide becomes the preferred method of warfare over battlefield conflicts? What happens when there are no more oil fields or trade routes to defend, but instead globalized information infrastructures that fall prey to digital terrorists? A few things become obvious. Tanks and aircraft carriers are useless.

Computers and genetics are the most potent weapons. How do we cope with these without creating an Orwellian world that undermines the democracy it's supposed to defend?

When we add natural catastrophe to this, the issue becomes infinitely more complex. How might we maintain civil order if a tsunami renders millions of people homeless in one hour? How might we cope with millions of refugees escaping a freak monthlong Arctic blizzard that incapacitates the northern United States and Canada? The military spends hundreds of millions of dollars on think tank scenarios, but in reality most of these are discarded and the infrastructure evolves at a relatively slow pace. That may have to change in the digital economy.

Nanotechnologists

This may be the most influential group of scientists the world has seen. In earlier chapters it was shown that there is no consensus right now among the top technologists about how their technologies should be controlled. Yet such technologies are just about here. Thousands of nanotechnologists are signing petitions[17] against what they fear are Luddites who want to stop their work. Yet by focusing only on criticizing such Luddites, many scientists seem to display reactive rather than proactive attitudes. Organizations—such as the Foresight Institute, the World Future Society, Coates and Jarratt,[18] the Copenhagen Institute for Futures Studies,[19] and the World Transhumanist Association, to name a few—are looking at how to get technologists thinking outside of their boxes. Yet the work of such organizations is greatly underdeveloped and underfunded, due largely to a lack of attention by the scientific community. Who might put resources into helping the social-responsibility side catch up, and how might we weave this into the basic fabric of how nanotechnologists are educated? How does the imbalance between male and female researchers in this field affect the outcome of research?

On the disaster preparedness side, nanotechnologists face a profound challenge. They have the power to alter the whole disaster prevention paradigm by making our defenses infinitely cheaper and more flexible than they are today. As explained throughout this book, the future of our postindustrial society, and our capacity to transcend to another level of civilization may rest on these defenses. Yet virtually no one is examining this. How do we tie the work of nanotechnologists to that of disaster prevention decision makers? Who at the Federal Emergency Management Agency, the Environmental Protection Agency, or the Red Cross is working on these relationships? Who in the disaster preparedness divisions of multinational corporations is considering this? How might such disaster preparedness cooperation help mend the growing rift between environmental activists and nanotechnology researchers? What are the first technological applications to focus on?

Natural-Resource Managers

Most natural-resource decisions have long time horizons. Mines and forestry operations are often planned over a ten- to thirty-year span. This means that hundreds of billions of dollars in investment and technology decisions that are being made today may start coming into play around the time molecular assembly gets going. The molecular assembler could eventually make conventional mining and forestry obsolete, because we'll be able to digitally fabricate products from materials that we extract locally from the soil, water, and air. On the other hand, molecular machines may be used to greatly speed mining and forestry until we gain the ability to synthesize such complex compounds. Who in the mining and forestry industries is looking at this? What are the environmental implications? What are the implications for nations such as Canada and Indonesia, which rely heavily on forestry and mining products? What are the implications for arid nations when they can manufacture endless supplies of water?

On the disaster preparedness side, we face a different set of questions for a different set of natural-resource managers. What might long-term ecological records, such as those preserved in trees, glaciers, and riverbeds, tell us about past disasters? How do we preserve them so they'll be able to reveal their secrets? These ancient ecosystems are being disturbed at record rates just at the time when they're beginning to divulge invaluable information about the earth's climate, its flood patterns, and effects from near-Earth-object collisions. We need to preserve these ancient monuments not just for their own sake, but also for the data they may give us about threats we face.

Philanthropists

How might molecular technologies transform the causes we give our donations to? How could molecular technologies be used to enhance the charitable work that we support? How might truly big natural catastrophes render our philanthropic contributions irrelevant, and what might we do to prevent that? How might molecular technologies be used to overcome socioeconomic barriers to disaster prevention? How do philanthropists get up to speed on this issue? For high-tech philanthropists, how might their technologies be used for climate adaptation and natural-disaster preparedness?

Real Estate

Does the adage "location, location, location" still apply if we can work from anywhere using broadband Internet? What happens to real estate brokers when intelligent machines can perform most of their work? Software is already accurately forecasting some housing prices, for example.[20] What happens to land prices when an individual can build a tropical farm under a bubble in North Dakota, and get there from New York in an hour? What happens to land prices when owners can construct palatial homes at low cost? And ... what happens to real estate if the coastline ends up under a wall of water from a tsunami?

Social Workers

What problems might we have when many of us don't have to work to provide our basic means of support? When some kids are manufactured, instead of born? When undetectable drugs constitute part of the human psyche? When we share the earth with other intelligent machines—perhaps smarter than ourselves?

Space Agencies, Space Organizations, and Aerospace Companies

The marriage between space and molecular science is absolute. The connection between surviving natural catastrophe and exploring space is equally so. Space technology and molecular technologies together expand our abilities for climate change detection, hurricane forecasting, damage assessment, and superstrong disaster-resistant materials. Moreover, life support, broadband communications, propellants, radiation protection, and everything else that makes space exploration possible depend on molecular technologies. More than other sectors, the space industry has a gaggle of technophiles who endlessly discuss questions about molecular science and natural-disaster detection. Unfortunately, the discussion often doesn't make it through to the policy makers who control the purse strings. Thus, the questions space workers might ask themselves include getting the right concepts into the minds of taxpayers and politicians:

∞ Space agencies have a line into what may be the strongest motivators for getting our government to invest in building a molecular assembler: rogue asteroids. How might we use molecular technologies to build a defense against such near-Earth objects? Could we build a molecular assembler that would let us "disassemble" them?

Right now, astronomers calculate that we'll need to find such big menaces many years out in space to have a chance of deflecting them. But why not consume them? Why not use self-assembling molecular machines with high solar-conversion efficiency to eat dangerous asteroids? Looking at calculations put forward by nanotechnology pioneers, we see that assembly and disassembly may have exponential mechanistic capabilities. Such machines may dismantle an asteroid in weeks or months. This may reduce the distance at which we have to intercept such projectiles, because it would take less time to neutralize them. Such a scenario may not be a precise depiction of how disassembly might work in practice, but as visionary Richard Feynman said, "The principles of physics, as far as I can see, do not speak against the possibility of maneuvering things atom by atom."[21]

∞ How do we get the general population excited about space again? Space agencies worldwide have often failed to connect the relevance of their work to the lives of citizens who pay their salaries with taxes. A tour of the Visitor Center at the Kennedy Space Center reveals the level of some NASA contractors' public-education capabilities, compared to what the agency has the capacity to do. For example, as late as January 2002 there was no mention of "nanotechnology" at the Visitor Center's Web site,[22] although other NASA Web sites deal with it extensively. Visitor Center exhibits seem to have a hard time keeping up with what's going on in space. For example, there are few if any *functional* robots for visitors to interact with. One question the agency might ask is: How might we upgrade our ability to fire the public imagination? Commercial media such as the Discovery Channel do a good job of transforming space program announcements into something that viewers might watch. An example of this occurred in 2001, after an amateurish announcement by an on-camera scientist about one of the most spectacular achievements in space history: the NEAR satellite had become the first man-made machine to land on an asteroid. Though it generated considerable press, the story fell off the front pages of most newspapers after a day. No one could show what such an event meant for our own lives. Fortunately, by embellishing the event with a story about asteroid threats, then combining it with NASA's own images, the Discovery Channel succeeded in resurrecting the event months after it occurred. Thus, good scriptwriters and computer animators make great partners. We need more of that.

∞ How do we build large, constructed environments? Right now plans are on drawing boards, deep in the cupboards of NASA, and more prominently on the Web site of the Mars Society,[23] for colonizing Mars and other parts of space. Central to these plans are constructed environments. Such environments could teach us much about preparing to survive if Earth's climate turns hostile. Molecular technologies will let us build such environments. These may let large populations survive temporarily devastating ecological extremes. Right now, our understanding of constructed ecosystems is still primitive. *Biosphere 2*, built in Arizona by a private foundation, was a bold test of our capability to construct ecosystems in enclosed spaces, but key support systems failed.[24] Instead of trying again on a larger scale, the owners turned it over to a university for "open-flow" rather than closed-systems energy and climate simulations. This may produce other useful results, but it was wrong to abandon the original goal. That goal requires renewed support. The Mars missions or other space habitat projects may take this over.

∞ How do we improve sensing and forecasting of big natural disasters? The National Oceanic & Atmospheric Administration (NOAA) and NASA have always been big weather partners. NASA and FEMA are cooperating on a program for remote sensing of potential disaster zones.[25] Yet these are far from the scope required to cope with megadisasters or for preventing molecular technologies from running amok. Such catastrophe scenarios need to be injected into the game plans of that cooperative arrangement.

∞ Will we populate the moon and planets with ourselves, Robo sapiens, or both? More than other groups, space agencies will be first to use artificial intelligence. They already use satellites and planetary rovers that make their own decisions. Therefore space organizations have a special responsibility to lead the AI discussion.

Transportation

What happens when everyone can go everywhere, whenever they want, and work from wherever they want? How do we manage air transport corridors? NASA is developing an air traffic control system for small vehicles now, but who determines exactly where those corridors go? Right now it's mostly the federal governments who do this. Do we want national governments telling us exactly where we can park our cars on our roofs? How do we, as individuals, have a say in it?

On the disaster preparedness side, we have to ask: What would happen to our transportation infrastructure if a once-in-a-thousand-year tsunami hit

the East Coast? The entire north-south highway and rail network from
Florida to New Jersey would be knocked out. How might we use molecular
technologies to prevent this? How might skyways or networks of high-speed
transport tunnels help mass evacuations?

∞   ∞   ∞

These are just a few of the many questions for some of our key economic
sectors. It's not just the experts that need to ask them. Millions of us work
in these industries and have a role to play in their decision making. We can
each take our own initiative to ask these questions, and introduce our
coworkers to them. It's beyond the scope of this book to provide definitive
answers to these questions, but the ensuing chapters contain suggestions
about how we might begin. They also set out some principles to guide us.

CHAPTER NINETEEN

OVERCOMING CULTURAL AMNESIA

"Our tragic mistake was not that we pursued the new.
It was that we neglected the old."[1]
—Princeton University researcher
Edward Tenner, author of *Why Things Bite Back,*[2]
analyzing how low-tech terrorists used
ancient techniques to slip through America's
high-tech security net and destroy the
World Trade Center, December 2001

PRINCIPLE 1: FIND WAYS TO REMEMBER OUR PAST IN THE FUTURE

Whether it's terrorist attacks or natural disasters, we are vulnerable to ancient ways. Therefore, one of our great challenges is to develop a more accurate picture of nature's past catastrophic behavior. Starting points include the vast underground calderas that might erupt in our faces; continental shelves that hold remnants of settlements drowned by rising sea levels; and vast expanses of desert that still hold artifacts from ancient civilizations. From there we may be able to figure out what to do if past great disasters recur. We might also get clues as to which molecular technologies need to be deployed first, to protect us.

To do that requires a prior step—recovering from *cultural amnesia.*

OVERCOMING SCIENTIFIC RESISTANCE TO CATASTROPHE

It's time for the scientific community to start improving its distinction between historic natural catastrophes and the pseudoscience of catastrophism. By so doing, we may transcend the current situation, where the mere mention of catastrophic threats is often met with a roll of the eyes.

September 11 may have changed some minds about the odds of seemingly improbable disasters, but complacency and skepticism still prevail regarding natural megahits.

Standing square in the way of the required clarity is a band of nuts: everyone from cultists to alarmists, chomping at the bit to prove that the world is about to end in a flameout. Moreover, catastrophism has underpinned religious beliefs for millennia. In so doing, it has invited welldeserved contempt from scientists who rightly seek concrete evidence. Yet this suspicion of religious-type catastrophism is so ingrained in the scientific community that it's been hard for natural-catastrophe researchers to make their voices heard in the mainstream. Thus, the job of separating myth, fact, and groundless fiction is daunting, but it must be done if we're to get at the truth and avoid arrogantly discarding history's messages.

CUSHIONING THE BLOW— AN AGELESS HOPI MESSAGE

For decades, pilgrims have journeyed to a petroglyph in the Arizona desert to study symbols scratched on a stone at least ten thousand years ago by Hopi Indians. Interpreters of this ancient language say the symbols accurately predicted the arrival of Europeans in the Americas, railways, automobiles, the world wars, and the launching of *Spacelab*. They also claim the prophesies are tied to our own time. They warn that if we fail to learn how to live in harmony with nature, civilization as we know it will be destroyed.

Most scientists rightly observe that prophecies are totally without scientific basis. They are skeptical of prophesies that have come true, remarking that the odds of at least some predictions coming true are quite reasonable; even a broken clock is right twice a day. Yet it's important to differentiate between prophesies that try to predict the future, and legends that may lead us to evidence of past events that can recur. Natural history repeats, though perhaps never in precisely the same way. Such legends may help us to understand how and why.

The Hopi may be among the first civilized inhabitants of North America. Their villages have been identified as among the longest continually occupied settlements on the continent.[3] Their myths about creation and destruction of earlier "worlds" were detailed by, among others, the late engineer and author Frank Waters, in his 1963 work, *The Book of the Hopi*.[4] Then, in 1997, Thomas E. Mails published *The Hopi Survival Kit*;[5] a book containing the summary by a hundred-year-old Hopi elder of his tribe's philosophy and covenants. The information contained in such books suggests that for millennia the Hopi kept sophisticated oral and petroglyph records of their relationship with the natural ecology, including lessons about how humanity might survive or perish in it.

Some Hopi critics contest the details and perspectives of such books, so until further investigations are undertaken, it's prudent to consider these works as merely guideposts for where to look, rather than literal interpretations about our natural history.

The Hopi are special in North American history because of their strong pacifist tradition. While other tribes fought over territory long before the Europeans arrived, the Hopi appear to have used nonviolent strategies, except when their own survival was threatened, for example by aggressive Spanish conquistadors and religious zealots. That nonviolent tradition continued into the twentieth century. As late as the 1940s Hopi elders were still imprisoned for passively refusing to sign documents that empowered the U.S. government to take Hopi children from their homes and educate them in government-run institutions.[6] By these measures, it's amazing that the Hopi still exist.

Their pacifist philosophy and persistence have drawn many to the Hopi legends. These pilgrims journey to the petroglyph and besiege the Hopi with so many questions that a special Web site has been set up to answer the basics.[7]

While researching his books about the fate of ancient civilizations, archaeological journalist and engineer Graham Hancock found himself drawn to the Hopi. It's no coincidence that his book on the topic of ancient cataclysms, *Fingerprints of the Gods*, ended with recounting his own journey to Hopi elders, who again repeated this warning:

> The first world was destroyed as a punishment for human misdemeanors, by an all-consuming fire that came from above and below. The second world ended when the terrestrial globe toppled from its axis and everything was covered with ice. The third world ended in a universal flood. The present world is the fourth. Its fate will depend on whether or not its inhabitants behave in accordance with the Creator's plans.[8]

The Hopi are far from alone in their traditions on these questions. Many indigenous tribes carry similar elements in their oral histories. Such recountings of apocalypse are also contained in the written works of Buddhism and Christianity.[9]

Whether we accept these myths or not, the Hopi legends offer valuable ideas for how we might cope with a future that's rushing at us. Hopi interpreters use two phrases that are yet to be adequately defined in contemporary terms, especially as they relate to extreme natural phenomena that we're now beginning to learn about.

The first phrase is "harmony with nature." Hopi legends come back to the theme that nature is going out of balance, and we must learn to live in harmony with it if we're to advance. Environmentalists have frequently referred to such indigenous legends to warn of impending disaster if we continue to be out of sync with nature.[10] Yet what if this harmony also alludes to something else? What if it relates to our capacities to adapt to vast natural upheavals?

In light of our new knowledge of natural catastrophes that visit the earth from time to time, we must ask, How do we live in harmony with a natural system that occasionally threatens our very existence?

More than many ancient cultures, the Hopi elders appear to have maintained the integrity of these legends despite endless assaults by Europeans, and by other Native tribes. Before that message is further diluted by the sands of time, it may pay to explore what lies behind these terms, perhaps, for example, by more closely examining the phrase "cushion the blow."[11] By seeking methods to cushion the blows of destructive forces, might we be able to live in harmony with nature, but on a different level than has conventionally been interpreted?

Hopi elders warn against using technology for technology's sake. They say such a path is the way to the end of this world. Yet as we've seen in previous chapters, evidence of past natural changes is mounting. Could it be that there are scientifically valid methods to establish a new type of harmony with nature; to cushion the blow in times of upheaval, while maintaining an ecological balance in periods of calm?

There has been some, but not much, analysis of Hopi and other indigenous legends in the context of surviving cyclical cataclysms. We may have something to learn from such legends if we reexamine them in this light.

AN IMMORTAL MEMORY

In some ways we're genetically programmed to forget. Human suffering and emotional impact grow dim after only a few years. This has the advantage of protecting us psychologically and helping us to recover, then move on. Yet in that defense lie the seeds of a disastrous cycle, because we also tend to forget the lessons we've been taught. Thus, today we build huge infrastructures in vulnerable disaster zones despite the lessons of history. Tokyo, Lima,

Mexico City, Los Angeles, and Naples are a few examples. We face the age-old curse of one generation forgetting the lessons learned by another. When it comes to remembering the impact of disasters such as the 1923 Tokyo earthquake or the flood that created the Black Sea, we're still in the dark.

If we have so much difficulty keeping track of what happened only five or fifty generations ago, how do we preserve our wisdom over five hundred? Are we doomed to repeat an endless cycle of forgetting, then having to relearn the experiences of past generations and civilizations? Of the many threats facing us, this may be among the most dangerous.

Thus, we need a mechanism that lets us vividly recall the past and apply its lessons to surviving the future.

In his book *Cultural Amnesia*,[12] Stephen Bertman, a professor of languages, literatures, and cultures at Canada's University of Windsor, describes how this applies to remembering democracy's hard-earned lessons. Bertman concludes that the loss of collective memory is one of the fundamental threats to a democratic society. He argues that if we ignore the historic sacrifices that made these freedoms possible, we're less likely to appreciate what's required to safeguard them for future generations.

The same can be said for natural disasters, although Bertman doesn't dwell on this point. Around the world, we see that only a few years after an earthquake, volcano, or flood devastates a region, people rebuild and make the same mistakes. Tokyo is the classic, though far from the only, example, as explained earlier.

Bertman attributes the chief causes of cultural amnesia to:

∞ The long passage of time, where memory grows dim as the distance between an event and our present lives increases. This is compounded by the fact that as civilization grows older, there is more history to remember. The scope expands as we do in time.

∞ The "urgent senses": survival instincts that let us discard short-term memory for information that we need to run our daily lives. This protective mechanism keeps us from being paralyzed by fear based on past experiences. It also makes us susceptible to the pleasures of the present that tend to cloud out the nastiness of the past.

∞ Technology that has led to the dominant idea that "new" is intrinsically superior to "old." As change accelerates, we're more likely to discard old information in favor of newer data, also using our ancient survival instincts.

∞ The lure of materialism in a society that has come to be dominated by planned obsolescence. This combines with increased mobility to create a psychological environment that dismisses the old as obsolete.

∞ Referring to the United States, he cites the newness of the nation, its foundation on the pursuit of happiness, and the values associated with democracy, each of which offers its allegiance to the present rather than the past.[13]

To overcome these, he says, we need to preserve our collective memory:

> Memory is not just a defense against totalitarianism. . . . It can also be the active means to our further liberation, a reservoir from which we can draw a renewed sense of direction and purpose.[14]

Much of our collective memory about deep history is being created and preserved by geologists who are unearthing evidence of past cataclysms that we must learn to avert. Will we preserve that memory?

There are other technical barriers to how we retain our memory.

In the 1990s British scholar Bruce Lloyd began a project known as the Wisdom of the World, in which he used the World Wide Web to gather ideas about the nature of wisdom.[15] A theme reflected in many postings was a need to link the past, present, and future to avoid repeating history's mistakes. From this, Lloyd defined wisdom as "useful knowledge with a long shelf life."[16] But how do we achieve this when the shelf life of most new information is growing shorter, and the means to remember it are becoming obsolete more rapidly? For example, try to transfer data from an "ancient" early 1980s-era ten-inch floppy to a rewritable DVD. In a few short years we've made the link hard to complete.[17] So how do we maintain continuity over many millennia?

Computer intelligence pioneer Ray Kurzweil says there may be a solution. He predicts that molecular science will give us the capability to download personalities to a chip, then insert their data into a cloned or synthetic body.[18] On a profound scale we're reaching toward the capacity to *back up* human beings. This is not something for the next century. Kurzweil believes it will happen in the next few decades. If he is right, human memories and personalities will evolve for centuries. Is this the ultimate insurance policy, or damnation to eternal life? Will our individual perceptions of the past become so fragmented from our own specialized experiences that we lack the cultural cohesiveness to move forward? There would be those who remember catastrophic changes from centuries ago. How much influence would they have? It depends on what percentage of the population they'd make up; whether they'd survive other hazards that wipe out those memories; and how they'd keep those memories high in the collective consciousness.

What is smart enough and tough enough to last millennia, store masses of data, and know how to talk to whatever variant of our species might be around by then? What materials should we use to send messages to the future? Almost everything from ancient civilizations except rock and astronomical configurations is gone. What happened, for example, to the drawings and tools for the pyramids?

In a modern-day homage to such monuments, Danny Hillis, one of the first developers of parallel computing, along with Whole Earth Catalog founder Stewart Brand and other members of the Long Now Foundation, are building a Clock of the Long Now[19] to last ten thousand years.

> What we're doing with the clock is to do even more for time what the photograph of the Earth did for space. Like understanding of the earthly environment as one whole thing—we're trying to understand a period of time reaching 10000 years into the past and 10000 years into the future as one containable thought. . . .[20]

One of the foundation's options is to spread clocks among locations around the world to guard against loss, as Brand explains:

> The problem with (putting such a clock in) cities—cities, forests and rivers, ocean edges—they're all so volatile. And cities especially have such a high metabolism of stuff going through them and they're the targets for wars and so on—you can't really count on many centuries in the cities—you get in a sufficiently barren desert, it's not going to turn into agriculture, probably is not going to turn into a bunch of things; it's more stable.[21]

The choice of locations for such a clock bears careful thought. Research by geomorphologists such as Mary Bourke has shown that over time, deserts do turn into agriculture and lakes, and vice versa.[22] Parts of the Sahara desert were green less than ten thousand years ago, and many dry spots in North America were inland seas. Furthermore, if we look at the potential capacity of our future civilization to build in remote locations, we see that present assumptions about where populations might live could be wrong. A clock built on a mountaintop might end up on Main Street of a future mountain resort. Conversely, if we put such clocks in plain view so everyone could pay attention to them, kids who travel the world with lasers in aerocars may vandalize them, zapping time clocks just for fun.

Brand also argues that although works such as the pyramids may have been intended to relay information across time, they didn't work well because they were vandalized or their messages were lost to antiquity. This argument seems odd, because archaeologists have learned vast amounts about the state of the earth as it was thousands of years ago from messages carved in those monuments then. In 2001, for instance, Professor Fekri Hassan from University College London, announced that he and a team of scientists had verified the accuracy of hieroglyphs in the tomb of a regional governor, Ankhtifi, which indicated that "all of Upper Egypt was dying of hunger. . . ."[23] Geological evidence from nearby lake beds, along with rainfall records gleaned from stalactites in caves, confirmed that the cause of the famine was a drought, thus verifying both the accuracy and the date of this message, sent four thousand years ago.[24] As we spend more efforts uncovering these messages, we find that some ancient civilizations were effective at sending them through time using languages that, while not used today, are still understandable. We need only scratch further beneath the surface to find more information.

These examples show some of the problems in devising ways to relay messages over extended spans of time. They also suggest that we should study more carefully how ancient societies relayed such messages, rather than assuming that our high-tech society has superior methods. We may be attempting to reinvent a time wheel that's right at our feet.

As human memories are increasingly overloaded by trillions of bits of new information that multiply exponentially, we face the prospect of forgetting an increasingly extensive past, thereby dooming ourselves to repeat our mistakes and be devastated by natural disasters. Yet molecules give us many possibilities for solutions. The "endless human" described earlier, whose consciousness might survive indefinitely in computerized neural networks, may be one way to establish a long-term collective memory. The human brain is still poorly understood, but we know that it contains far more memory and computing capacity than we use in our daily lives. Molecular technologies, especially genetics and nanotechnology, are unlocking these mysteries. As they do, we may gain the capacity to use far greater portions of our brain for conscious recall of memory. If the structure of our brain prevents this, we may have neural implants that let us call up memory on demand. Or we may transfer our memories and consciousness into an external brain based in a robotic body. This may be the ultimate "clock of the long now."

Such methods become possible only if nanocomputing lets us stuff teraflops of data into compact memory banks, then attach those to our brains without short-circuiting them. It's a daunting task. Yet once achieved, it may give us the capacity to keep the sum total of our experiences in our consciousness, thereby retaining the lessons from our past.

CHAPTER TWENTY

BYPASSING
THE ROAD TO HELL

*"I believe that we have done a poor job of explaining
the complexities and the importance of biotechnology
to the general public."*
–Norman Borlaug, agronomist
and Nobel laureate[1]

PRINCIPLE 2:
LEARN FROM THE DARK SIDE

In November 2001, when scientists announced that they had cloned a human embryo, reaction was chaotic. The White House said it was "100 percent opposed" to human cloning, while the president of America's Catholic Alliance, Raymond Flynn, said it was "a moral breakdown. . . . Human reproduction is now in the hands of men, when it rightfully belongs in the hands of God." Researchers countered by claiming that it would help cure human diseases. Skeptics criticized the announcement as a prematurely inflammatory alarm, because the embryos didn't live long enough to determine whether their cells could be used for disease research. Meanwhile, scientific publishers were divided. John Rennie, editor in chief of *Scientific American* magazine said it was "an amazing accomplishment. . . . You hesitate to describe it as a virgin birth, but it is sort of in that vein."[2] Editors of the *New Scientist* were more circumspect. "Far from heralding an age of wonder cures," they wrote, "cloned human embryos may be a dead end."[3]

By 2002 *Scientific American* and a host of other magazines had published articles describing the furor, while the experimenters themselves defended it as a step along the road to cloning, and that they'd published in the interests of scientific transparency.[4]

Thus, while scientists debated whether the event was an immaculate conception or a scientific sham, moral conservatives were vilifying it as a religious abomination.

Science helps us every day to discover, quantify, qualify, and cope with risk, then use that knowledge to learn where to draw the line. Yet, as the human cloning controversy vividly reveals, science seems to be heading further away from enjoying the public trust. Why?

THE ANIMAL CONNECTION

One of the most profound changes to society over the millennia has been a reduction in the percentage of us who kill to eat, or see animals being killed to feed us. This proportion has declined from being a majority down to a tiny fraction, while the numbers of animals being slaughtered has grown to tens of billions annually.[5] That inversely proportional relationship has been accompanied by the rise of a vigorous animal rights movement.

Since 1789 scientists have struggled with philosopher Jeremy Bentham's admonishment, "The question is not, can they reason? Nor, can they talk? But, can they suffer?"[6] Bentham believed that animals had the capacity to feel emotion. Yet it took two centuries until science was able to demonstrate that animals exhibit humanlike behaviors, such as mourning their dead.[7] This cataloging of animal emotions is relatively new. For centuries it was blocked by a scientifically wrong yet broadly accepted and religiously supported rationalization that animals don't have emotions. The conventional wisdom was that they exhibited cognitive responses to environmental stimuli along with learning behavior, but that these certainly weren't feelings. This conclusion paved the way for vast industries that often inhumanely raised animals for food and experiments, or annihilated them through habitat destruction. Now it seems we had it wrong. According to scientists, animals such as dolphins, apes, bears, lions, elephants, and other well-studied creatures have shown strong evidence of emotion.[8] When combined with studies of animal cognizance that demonstrated tool making among primates,[9] and sophisticated social behavior by farm animals[10]– hence greater intelligence than previously thought–these findings constituted an explosive new mix. That mix ignited in 2001 when millions of animals were slaughtered throughout Europe in attempts to control an out-

break of foot-and-mouth disease. The sheer scale of the cull of healthy animals had scientists pitted against each other in a tense discussion over
whether other preventative means could have been used.[11]

Such controversies reflect changes in public attitudes that have led to
attacks on scientists who utilize animals for developing drugs or other medical cures. Some activists, especially in the U.K., have undertaken high-profile actions against laboratory facilities in which animal experiments are
done. Others have proposed that primates such as chimpanzees be declared
"legal persons" and share some of the rights of humans, including freedom
from bodily harm.[12] In this atmosphere, scientists have been frustrated
about why the media and the public seem to underplay the many benefits
for humanity that came from animal testing, while attacking practices that
scientists undertook for years to produce such benefits.

But scientists themselves helped light this fuse—by continuing the quest
for truth. In this case, they did it by investigating animal emotions and
inventiveness. Resulting discoveries, combined with pressure from animal
rights groups, have caused a minor earthquake in the experimental field,
throwing a wrench into animal-testing practices.

Moreover, just as scientists are beginning to come to terms with this by
adopting new guidelines for the ethical treatment of animals, things are
about to get more complicated.

What will our position on testing be, for example, if we create self-aware
synthetic intelligence with greater intellectual capacities than ourselves in
some areas, for example, playing chess, and also with the capability to suffer?
The way we respond to that question may determine the direction our technologies take. To see how it might play out, let's look at how scientists have
reacted in the past to big ethical issues, and what have been the results.

Since the 1940s scientists who were horrified by the ultimate uses of
their nuclear research started international organizations such as the Pugwash Conferences, the Educational Foundation for Nuclear Science, and the
Union of Concerned Scientists. That, however, didn't stop our governments
from continuing to lie, for example, about bombing Cambodia or about illnesses involving nuclear workers.[13] Such travesties spawned yet more organizations in the 1980s, such as Physicians for Social Responsibility and
Computer Professionals for Social Responsibility.[14]

Many scientists have taken courageous stands on such issues, at great
risks to their careers. Yet the record of governments abusing science in the
name of national security is terrible, and the collusion of some scientists has
been extensive, sometimes overshadowing the courageous acts of their colleagues who stood up to oppose such acts. As we move into the molecular
era, this ethical dilemma takes on epic proportions. If we fail to make trans-

parent decisions about artificial intelligence, for example, a climate of fear may develop around these machines and the scientists who build them.

Furthermore, national-security questions that may have justified secrecy about, for example, nuclear weapons during World War II are far less applicable with AI, because open-source software is leading to the development of intelligent machines through vast international networks, rather than closed national programs. By trying to cover up or gag such networks, governments only stand to exacerbate public mistrust. Thus, the whole concept of "national security" may have to be rethought in the context of a networked scientific world. It can't happen too soon, because public cynicism about our institutions and leaders has been accelerating in the polls.[15]

Here are examples of contemporary books by a cross-section of authors regarding misuse or potential abuse of science and technology. The point of listing these works is not to show the nasty things scientists may be saying about other scientists; rather, it's to show that a whole genre of books written by respected authors–some of them scientists, but many who are nonscientists–seems to play a central role in shaping public opinion about science.

∞ *Science, Money and Politics: Political Triumph and Ethical Erosion,*[16] was written by investigative journalist and former *Science* magazine news editor Dan Greenberg. *Scientific American*'s reviewer calls it: ". . . better documented than most National Academy of Sciences reports . . . should be required reading for science policy makers, science journalists and any American who gives a damn whenever science–one of the nation's crown jewels–falls into irresponsible hands. . . . We need more Greenbergs."[17]

∞ *Betrayal of Trust: The Collapse of Global Public Health*[18] by Pulitzer Prize–winner Laurie Garrett, deals, among other things, with the failure of the medical and scientific establishments to preserve the public preventative health-care system in favor of supporting expensive cures for relatively small numbers of people.

∞ *IBM and the Holocaust: The Strategic Alliance between Nazi Germany and America's Most Powerful Corporation,* by award-winning author Edwin Black,[19] chronicles use of then-nascent computer technology–with apparent acquiescence of the United States's foremost computer corporation–to round up Jews and track their extermination.

∞ *Owning the Future: Inside the Battles to Control the New Assets; Genes, Software, Databases, and Technological Know-How That Make up the Lifeblood of the New Economy,* by MIT Knight Science Fellow Seth Shulman, discusses misallocation and abuse of patents that may be resulting in a stifling of innovation.[20]

∞ *White-Collar Sweatshop: The Deterioration of Work and Its Rewards in America,* by *New York Times* financial writer Jill Andresky Fraser, documents how the high-technology society has led to most Americans working longer hours for less money and reduced benefits.[21]

Together these works paint an unflattering picture of technology abuse. These are notable among the many books about technological ethics, because they come from a wide range of authors who are recognized in their respective fields.

Such works focus increasingly on who controls the process. Without exception, the authors conclude that voters don't and special interests do. They suggest widespread cynicism about who's calling the shots.

What does this have to do with the future of scientific discovery?

Once ideas grab hold in society, they take on their own life through self-replication. These ideas, sometimes called *memes,*[22] are apparent in financial markets, where every few decades a mania takes hold that pushes share prices up: investors still buy them even after the truth becomes apparent, until one day the stocks collapse under their own weight. For centuries financial bubbles have come and gone, displaying the same characteristics. We've just been through such a mania with Internet companies. This left scars on millions of pensioners who've had to forgo retirement because of their losses.

The result: the dominant meme shifted from overexuberance and confidence in the new economy, to a witch-hunt for the guilty. In the financial world, such swings are usually short-lived, but in other parts of our society they take more time to develop and recede.

Today we may be witnessing a similar shift in attitudes toward technology. Since the beginning of the industrial revolution and until a few decades ago, we based our economies on the idea that we could extract resources and then discard them without ecology biting back. Today that conceptual bubble is beginning to burst, even as it continues on its own inertia. The debate has become polarized, as smokestack industries cling to their subsidies, while environmentalists at the extreme edges of the movement oppose every potentially harmful technology. Furthermore, it's not just the extremists who are manifesting against technology. Mainstream politicians and religious leaders now find themselves squarely at odds with the cutting edge of science, as graphically demonstrated by the reaction to cloning of a human embryo. Such conflicts are not unusual in history, but the backlashes they have foreshadowed are not to be ignored. Thus, the balance between the perceived risks and perceived benefits of science may be shifting, although the outcome of the struggle between these perceptions remains uncertain. As the shrillness of the confrontation mounts, scientists and those who pay their salaries are among those being fingered as the culprits.

What could worsen this situation?

Sun Microsystems's chief scientist, Bill Joy, has notched up the controversy over the concept of *relinquishment,* and elicited strong reactions from technological and environmental leaders by stating:

> The new Pandora's boxes of genetics, nanotechnology, and robotics are almost open, yet we seem hardly to have noticed. Ideas can't be put back in a box; unlike uranium or plutonium, they don't need to be mined and refined, and they can be freely copied. Once they are out, they are out . . . we must act more presciently, as to do the right thing only at last may be to lose the chance to do it at all.
>
> . . . The question is, indeed, Which is to be master? Will we survive our technologies?
>
> We are being propelled into this new century with no plan, no control, no brakes. Have we already gone too far down the path to alter course? I don't believe so, but we aren't trying yet, and the last chance to assert control—the fail-safe point—is rapidly approaching.[23]

One of the solutions proposed by environmentalists to this Pandora's box is to keep it shut until we know for certain what's inside, and what the implications are of letting it out. This idea is embodied in the precautionary principle, explained earlier in chapter 14.[24] Its underlying philosophy has been brewing ever since Rachel Carson's classic book, *Silent Spring,* gave rise in the 1960s to concern over the premature release of untested pesticides into the environment. Today, parts of the precautionary principle are already entrenched in European and some U.S. state legislation, alarming American researchers and sparking an agitated debate around the world.[25]

THE CREDIBILITY GAP

Despite widespread readiness in industry and academia to blame environmentalists for opposing science and technology, commercial exploiters of genetics are often their own worst enemies. One of the biggest blunders occurred when a few life sciences companies decided to tell the European public what a great idea genetically modified (GM) organisms were—just as some experimental crops got out of hand, contaminating other farmers' fields and killing nontarget insect species such as cute monarch butterflies. That catapulted the environmental movement into an unprecedented position of prominence, leading to severe reductions in American corn exports, along with the introduction of bills in more than forty states to regulate genetically modified crops.[26] This genetic firestorm set back research in some areas and elicited a knee-jerk regulatory reaction that may hobble the industry for some time to come.

The question is, Will developers of molecular science pay attention to this ugly episode, or forge ahead and blame the environmental extremists for their troubles?

A few months after the original Bill Joy article appeared regarding the dangers of molecular technologies, I started exchanging information with nanotech researchers about the ecological implications of their work.[27] I found growing awareness of the debate over the promises and risks of nano-technology, but not much knowledge about how to evaluate the environmental benefits or risks of their own work. For example, most of those I spoke with said they did not know that commonly used chemicals in coatings, colors, and adhesives are known to cause health and environmental problems, and that nanotech might either solve or worsen these. They were surprised to hear that 3M Corporation withdrew its widely known Scotchguard brand of dirt protection coatings from markets, at a cost of several hundred million dollars in annual sales, due to "increasing attention to the appropriate use and management of persistent materials."[28] Many in the computing field were unaware about concern over use of gallium arsenide in chips, and the toxic pollution from their production and disposal.[29] Others were unaware of the positive potential of their work to fix problems. One such researcher at a conference I attended in the fall of 2000 gave a presentation about changes in coloration properties of materials at the nano scale. When I suggested using these properties to replace toxic pigments with physical coloration methods, he told me he'd been unaware that color was such an environmental problem in products,[30] but added that he'd look into it now that I'd told him. This suggests that researchers may be inspired to do such work if they get the right information through interdisciplinary communication.

Other nanotechnology researchers were surprised to hear that energy-efficient electronics might increase, rather than decrease, net energy consumption by accelerating our ability to manufacture billions of products that consume more energy in real terms. Nanocomputing energy efficiency was put forward in one conference talk as an environmental rationale to reduce consumption. Yet recent developments suggest an opposing trend. As energy-efficient technologies permeate the electronics industry, power consumption in Silicon Valley has behaved much in the same way that operating-system software has reacted to chip speed: consuming it, then demanding more.[31]

When I discussed this situation with one of the more environmentally enlightened scientists in the nanotech community, she explained that most of those I'd been talking to were basic researchers. They were the "R" part of the R&D field. Had I been talking to the developers, she said, I might have seen a different level of awareness, because they operate more in the real world and have to deal with, for example, environmental regulations. But this led me to ask, does that mean basic researchers shouldn't be aware of the potential applications or implications of their work? No, she replied, but basic researchers do tend to see their work as "amoral" in the sense that it may be used for either good or bad. They can't let themselves be stopped by such considerations; otherwise there wouldn't be progress. But then, I asked, isn't the rate of progress of basic research for certain disciplines influenced by whether the researcher is being paid to develop nanotubes for bomb making or to eliminate toxins from color coatings? To a degree, yes, she said, but the researchers may not see it that way. They're more interested in the technical challenge of the work. Then why, I asked, do many researchers talk about the great things their work might do for the environment–or complain about the nasty things that environmentalists say about them? Because, she replied, they want to feel like they're doing the right thing. We finished our discussion by agreeing that there was a certain incongruity between the "amoral" philosophies of basic research, and how researchers want to be perceived as human beings.

This sampling of observations and reactions is given only to show that researchers who hope their work leads to health and environmental advances, or are concerned about attacks on their work, are well advised to familiarize themselves with the motivations that drive environmental movements and the concerns of nonscientist citizens. It may also be time to extend ecological studies into realms such as nanoecology.

Nanotechnology may force us to reevaluate what constitutes an environmental risk. Risk evaluation often depends on the eye of the beholder. For example, if an environmental problem is defined as a serious threat to human health, then the biggest such problems worldwide are different from what we normally see on TV. They include cars that kill or maim millions; sewage that causes most of the deadliest diseases; and AIDS. Furthermore, if the present EPA practices of banning lethal environmental activities were applied to these killers, we'd not be driving, using tap water, or having sex. The list also contrasts sharply with the list of high-profile issues emphasized by news media, such as deforestation, wildlife preservation, and global warming. Thus, priorities depend on how environmental risk is perceived and by whom.

So do solutions. Each of the above-mentioned threats, with the exception of AIDS, has a short-term solution. Cars can be made much safer and drastically less polluting.[32] Sewage can be treated with existing, cheap technology.[33] The relative costs of fixing such problems, far from denting the economy, can spawn industries that generate new employment.[34] Further savings are achieved, and new industries arise, when subsidies to polluting industries are cut.[35]

Yet for the most part we're slow to do it. Why? And why, in cases such as biotechnology, does the meme suddenly seem to shift from passive acceptance to rapid rejection?

One of the key factors is perception of risk, and our perceptions of those who tell us about those risks. We can go on for years ignoring a risk, until one day, a difference in the *quality and quantity of information* we receive spurs us to acknowledge that particular risk–although on the surface, nothing in our lives seems to have changed. We're ready, for example, to pay a few hundred dollars less for an SUV that has a weak roof, having heard somewhere that this small saving leads to collapses in a rollover that may kill or injure our passengers. The chances that it might happen to us may seem small enough to warrant the risk. Besides, the vehicle *feels* safe, and statistics show that those who are more likely to get killed in an accident with another vehicle are the victims in the smaller car. So we make the purchase. Furthermore, when it comes to indirect health threats, such as pollution from cars, then risk aversion decreases because there's no smoking gun, and we get a lot of benefits from our cars.

But then something happens to the quality and quantity of data we receive. We find out that the company manufacturing the tires that our vehicles travel on hasn't been up-front about its safety records. We also find that thousands of passengers have died in rollovers.[36] Although the statistical chances of us being personally harmed by this are unchanged and still small, we lose trust in the brand because we don't like feeling that we're being taken for a ride. Also, talk show hosts and our neighbors' kids are making fun of our car.

How does this explain public angst about technology? Over lunch at a conference in Washington, D.C., a few nanotech researchers remarked to me that environmentalists know nothing about genetics and are dead wrong on genetically modified organisms. They blame the environmentalists for smearing a legitimate science and holding back benefits from humanity.

I brought up a few items they may not have fully considered: how GM organisms were released, then found afterward to be killing nontarget species; not just any species, but the monarch butterflies that are a favorite of many school children.[37] It was compounded when modified corn uncertified for human consumption jumped the fence, mingling with corn that then went into taco shells, high-profile products. This presented a risk that people who had allergies to the modified corn might eat it. Thus, the quality and quantity of information that consumers received about genetically modified organisms had changed radically. This occurred either because the scientists hadn't done their job adequately before releasing such organisms into the environment, or they had miscalculated the effects on the public psyche of what they may have thought to be a minor problem. The results were recalls that foreshadowed a 40 percent drop in U.S. corn exports.[38]

Hmmm, they remarked . . . interesting information . . . are you *sure* it's right? Now let's look at extracts from a petition signed by three thousand scientists:

We, the undersigned members of the scientific community, believe that recombinant DNA techniques constitute powerful and safe means for the modification of organisms and can contribute substantially in enhancing quality of life by improving agriculture, health care, and the environment.

The responsible genetic modification of plants is neither new nor dangerous. Many characteristics, such as pest and disease resistance, have been routinely introduced into crop plants by traditional methods of sexual reproduction or cell culture procedures. The addition of new or different genes into an organism by recombinant DNA techniques does not inherently pose new or heightened risks relative to the modification of organisms by more traditional methods, and the relative safety of marketed products is further ensured by current regulations intended to safeguard the food supply. The novel genetic tools offer greater flexibility and precision in the modification of crop plants.

. . . Current methods of regulation and development have worked well. Recombinant DNA techniques have already been used to develop "environmentally-friendly" crop plants with traits that preserve yields and allow farmers to reduce their use of synthetic pesticides and herbicides. The next generation of products promises to provide even greater benefits to consumers, such as enhanced nutrition, healthier oils, enhanced vitamin content, longer shelf life and improved medicines.

Through judicious deployment, biotechnology can also address environmental degradation, hunger, and poverty in the developing world by providing improved agricultural productivity and greater nutritional security. Scientists at the international agricultural centers, universities, public research institutions, and elsewhere are already experimenting with products intended specifically for use in the developing world.

We hereby express our support for the use of recombinant DNA as a potent tool for the achievement of a productive and sustainable agricultural system. We also urge policy makers to use sound scientific principles in the regulation of products produced with recombinant DNA, and to base evaluations of those products upon the characteristics of those products, rather than on the *processes used in their development* [italics are mine].[39]

This rational-sounding petition is instructive, because it shows how so many intelligent scientists put their signatures to a document that does not acknowledge *why* public paranoia is often *based* on fear of "processes used in their development."

The underlying problem is guilt by association. If one or several genetics experiments are found to be lacking after they've been deployed, a whole industry might get tarred with the same brush. Scientists increasingly run into this knotty problem, but they seem less certain how to cope with it than are, for example, damage-control experts in marketing.

Guilt by association can also rub off onto scientists from the conglomerates they work with. For example, cross-ownership in the food and tobacco industry is substantial. Some of the biggest food company owners also control tobacco companies that for years encouraged teenagers to start smoking, while denying that they were influencing them to do so.[40]

Independent doctors and scientists claimed for years that smoking was ruinous to your health, so consumer advocates are left wondering how an entire industry was allowed to get away with promoting it to kids for decades, while downplaying its harmful properties. Many scientists work in the tobacco industry. Yet, as scientist and whistle-blower Jeffrey Wigand found out, the price for exposing such secrets is job loss and career suicide.[41] His was a courageous move, but it also laid bare the years of silence by other scientists who were employed by the industry.

Some of these companies play a big role in our food industry. Therefore, one might ask, If tobacco company scientists kept quiet while their marketing departments pressed ahead, and it took the government so long to uncover what those companies really knew, then why should it be different when it comes to telling us about effects of DNA modification on our food? Scientists who work for the food subsidiaries might rightfully say they weren't responsible for tobacco. Yet to the many consumers who see a tobacco company logo imprinted on their food packages, and who may have friends or family members affected by tobacco-related diseases, the distinction might easily be lost. In such cases, it's easy to see why scientists face a credibility gap.

EMOTIONAL REFUGEES

Besides this, another profound psychology is driving a green backlash, one that nanotech researchers ignore at their peril.

Scientists, who are often the chief beneficiaries of technological power, are slow to recognize that this technology is producing *emotional refugees*. These are not people who've been forced from their homes; rather, they've become chronically yet indefinably ill, and their children are developing allergies in record numbers. In the U.K., for example, four in ten schoolchildren have at least one allergy. Similar trends are found in the United States, among adults.[42] What's to blame? Something in the air, food, or playground soil? Cumulative exposure to contamination, indoor air pollution, or stress? As thousands of chemicals with undefined degradation pathways enter the marketplace annually, then mix with a sedentary yet stressful lifestyle, medical science is unable to nail down a smoking gun. The technical term is a *nonpoint source*. People see bad things happening to themselves or their kids and wonder what caused it. In such cases, victims can't sue as they would for a defective tire. They can't vent their rage on manufacturers, because they don't know who caused it. To make matters worse, average consumers find themselves stuck with uninsurable health-care claims and unsympathetic HMOs. Victims experience a gnawing frustration. They have no one to blame and only costs to show. They sit with their families in worsening commuter traffic, breathing exhaust, knowing it might lead to their kid having an allergy attack but not having the financial means to avoid it. So they give the kids a pill to calm them and hope they grow out of it. This childhood pill popping is demonstrated in statistics that show substantial increases in tranquilizer prescriptions for kids.[43]

FOOD TOXIC ENVIRONMENTS

Ask physicians what the top "environmental" problem is and you'll often hear, "obesity."[44] The technological lifestyle of commuting, sitting at work, and eating bad food has caused an epidemic of fat. Many Americans and Germans are severely overweight and getting fatter. Countries such as Japan and China, where obesity and heart disease were unheard of until a few years ago, are getting fatter by the year. It's not surprising. Despite the advent of telecommuting, more commuters are sitting in more traffic jams on more highways than at any other time in world history, breathing each other's exhaust for hours every day. Office, factory, and professional workers spend their lives this way: We sit in cars or public transport; sit in traffic; go to an office cubicle, warehouse, or factory with windows that don't open; and sit in front of a screen for most of the day. Then we reverse the process to get home, where we eat a big meal then sit in front of the TV. If we're in the transportation industry we sit in a vehicle all day. The resulting weight gain has led to a large, but largely ineffective, diet industry. The downward spiral of overeating and overdieting produces prolonged stress syndromes. So when more than half of Americans look at themselves in the mirror they don't say, "I'm so happy science found a way to feed me." Instead, they wonder why science hasn't found a solution to their weight problem.

GREEN STRESS

Finally, we have the weather. For the first time in living memory, parts of Great Britain have found their cities underwater, not once, but many times. Parts of North Carolina, New Jersey, and Texas have found themselves in a similar situation. Factions of the environmental movement blame global warming, but other candidates for culpability include communities that expand into risky areas such as flood zones, or just normal changes in natural weather patterns. Try telling victims that science makes their lives better when their flood defenses are collapsing, and "Greens" are blaming technology for the weather that destroys their property values.

THE DANGERS OF DISMISSING
JOE AND JANE AVERAGE

Animal ethics, the credibility gap, emotional refugees, food toxic environments, and green stress. The point in describing these is neither to claim that society is disintegrating nor to say that science is bad, but rather to demonstrate that many of us are stuck with such new stresses and don't necessarily see scientists on the plus side of the equation.

When victims get no joy from governments or employers, they turn to those who offer an outlet: consumer and environmental organizations. The media is full of accounts from victims convinced that pollution caused their illnesses, but who can't find the smoking gun to prove it. Greater fury arises when smoking guns *are* found, but polluters and tobacco conglomerates fight to the end in court. A good example of such regrettable corporate resistance is the chromium poisoning case portrayed in the film *Erin Brockovich*.[45]

Scientists are well advised to consider the depth of these syndromes and not to dismiss them as reactionary or ignorant, despite the fact that they often lead to bad solutions. These are ordinary citizens trying to get to the next paycheck. They don't want to be told how science is improving their lives when their kids are zoned-out on drugs to combat undiagnosable diseases, they're sleep deprived from overwork in the new economy, and they can't fit into their jeans.

Molecular technologies undoubtedly have the potential to solve such problems, but whether they will is another question. For decades, we've known how to make vehicles safer, but we do so grudgingly despite minimal extra costs. Might this be, for instance, the destiny of nanotechnology: utilitarian but unnecessarily lethal?

Such a debate is not going to produce positive results until the scientific and technology communities are seen to more seriously acknowledge and address technology-induced misfortunes that fall upon ordinary persons and their children.

Furthermore, the scientific community has to address a persistent trend that has apparently benefited some in the short run, but may turn viciously against science when the aforementioned factors coalesce. In his earlier-described book, *Science Money and Politics*, Dan Greenberg chronicles decades of concern voiced by scientific leaders about the low level of public understanding of science, and the threat it poses to government support for research. Yet he also points out that amid this apparent sea of ignorance, the total amount of government and corporate funding for science has increased rather than diminished. Greenberg then documents how this dichotomy has led many scientists to effectively withdraw from the forum of public discussion, because they see no point in engaging a general public who apparently have little to do with funding scientific work. Instead, they cater to special interests, which are the driving force behind increases in scientific funding.[46]

Greenberg is not alone in describing the special interest phenomenon as it affects science. In late 2001 the venerable *Economist* magazine, a bastion of support for capitalism and scientific research, depicted a range of studies on widespread corporate-influence peddling in the peer review processes of scientific journals, especially where clinical trials of drugs are concerned.[47]

If such assessments are correct, this begs the question: How long are taxpayers going to put up with being excluded from so many science funding decisions?

The answer is, No one knows, but we do have parallels from other sectors of the economy to guide us. In equities markets, we've seen how small investors are happy to put up with someone else managing their money and telling them not to worry—until persistent loss of their investments causes enough pain to make them budge. At the point when investors perceive their interests being harmed, they react violently against the supposed guardians of their financial well-being: fund managers. A classic example was the bankruptcy in 2001 of the energy trading company Enron—once one of the largest companies in America—that led to loss of pension funds for thousands of retirees and workers. It precipitated congressional inquiries that embroiled the Bush administration, one of the largest auditing firms in America, and numerous investment funds that were caught off guard by the event.[48]

In the field of science, we haven't yet seen a scandal on the scale of Enron, and as such, taxpayers seem happy to go along for the ride. They seem to perceive more benefits than downsides, or at least they haven't grown sufficiently perturbed to do something that might upset the overall funding apple cart. But at what point does their perception begin to flip? When human cloning contravenes their religious beliefs? When their kids are persistently sick with one thing or another? When genetically modified food terrifies them? It is the convergence of such factors, rather than any one, that may cause the public to turn on science. Like the impoverished small investor, taxpayers are slow to anger but inconsolable once brought to a boil. This is what scientists have to watch for.

Here are some steps that molecular research institutions might take to help prevent such a shift in public sentiment:

- ∞ Don't lie to the public. Punish industry representatives who do. Every scientist who is paid to give expert testimony that he knows to be questionable is cutting the throats of the whole scientific research community.
- ∞ Acknowledge one of the psychological factors that serves as a breeding ground for attacks against science. That force comes from victims with real problems and no smoking gun to blame. Recognizing the motivations that drive this large group is the first step to finding solutions and combating unjustified attacks.
- ∞ Don't make promises you can't keep. There is a fine line between postulating that advantages—such as reduced energy consumption, widespread improved health, or lower pollution—may accrue, compared to promising that they will occur. Such a line must be carefully maintained by scientists, and forcefully relayed to the science media that serve as their interpreters. In this book, for example, an attempt has been made to avoid categorically claiming that a certain future will absolutely come to pass. Postulation is essential. Arrogant confidence is dangerous.
- ∞ Get out of the lab and into the real world. Spend much more time on public outreach; not just through slick television ads, but also via an extensive, concerted, personalized strategy that goes into our schools, municipal governments, and local media.
- ∞ Do things that produce broad public benefits. Identify projects that produce high-profile, positive results among large populations. Cure malaria. Help victims recover quickly from natural disasters. Make personal-computer software that doesn't crash.

NATURAL DISASTERS, WAR, AND PEACE

An entirely different set of dark-side challenges emerges when big natural disasters barge into the equation. Environmental extremes, such as those described in previous chapters, are going to confront governments with the specter of fighting the population they're supposed to protect. This is normal in a dictatorship, but catastrophic for a democracy. For example, if a tsunami wipes out a North American coast, military intervention may be the only way to keep order. No rational military commander wants to face such a situation.

Responses to natural disasters have been extensively studied by the military because they are similar in some ways to the aftermath of nuclear war. The military also participates in natural-disaster mitigation worldwide, from building emergency dikes in a flood to keeping civil order where conventional policing breaks down. It could be said that the U.S. Army Corps of Engineers is the world's largest disaster prevention agency, in view, for example, of the terraforming projects it has undertaken over the decades to prevent flooding.

From this experience, military and Army Corps planners know that a natural disaster of the scope described earlier would lead to martial law, with a painful road back to democracy. To prevent such an occurrence, it's in the interest of the military to support deployment of molecular technologies that help us guard against natural disaster.

Yet there are big military risks to deploying molecular science among the population. For that matter, there are big risks in deploying the technology among the military. Much has been said and written on both risks. Military leaders have testified in Congress and postulated publicly about nanotechnology's potential.[49] Technologists such as Joy, Kurzweil, and Hofstadter have debated the risks to civilians.[50] As early as 1985 visionaries such as Drexler explained them.[51] Meanwhile, research is proceeding at a breakneck pace. Defense agencies are pouring billions into its component technologies via, for example, an Institute for Soldier Nanotechnologies.[52] From this, one thing is sure: warfare will never be the same. War machines have a huge stake in the outcome of molecular science because it may destabilize them or make them obsolete. For example, once the price barrier to war is reduced, as it may be if molecular assemblers are able to manufacture many weapons anywhere, the potential may rise greatly for disaffected individuals, survivalist sects, and religious extremists to wreak havoc. For example, nuclear war has so far been averted in the religious struggles surrounding Israel, India, and Pakistan; but with nanotechnology, nanowar may be uncontrollable. In his essay *Molecular Nanotechnology and the World System*, Thomas McCarthy explains it this way:

> An unfortunate side-effect of molecular manufacturing is that it may contribute to the creation of a state of war regardless of the circumstances, whether they are ones of peace or of war. If MNT [molecular nanotechnology] makes weapons invisible, and does the same for factories, then the ability of one side to measure the capabilities of the other will be severely hampered, and per-

haps eliminated completely. This will be destabilizing in two ways. First, by making some weapons impossible to detect, it will prevent those weapons from fulfilling their role as deterrents in times when deterrents are needed, and thus will decrease the ability of states to dissuade potential aggressors from initiating military hostilities. . . . Secondly, the lack of armaments, which is necessary for not projecting hostile intentions and arousing suspicions during peacetime, will be meaningless. The lack of detectable armaments will not in itself be reassuring to other states, and even the true absence of armaments will be inadequate proof of commitment to lasting peace, when the tools of war can be generated cheaply on a few days' notice. . . .[53]

Molecular weapons may turn war by remote control into the prevailing battlefield strategy. That battlefield may also shift to civilian populations and corporate targets. With cheap, intelligent drones in the hundreds of millions, military strategists would have eyes and ears everywhere. So would terrorists. Thus, in the molecular age, small fringe groups may have greater power, while big military machines will have the awesome responsibility of controlling the capabilities that the lowliest field soldier would be able to unleash.

The first signs that centralized military command might have trouble coping with this new technology became apparent during field exercises in the year 2000, when, for the first time in history, soldiers in the field got a more advanced view of the situation than their remotely located commanders, because they were able to see their immediate environment *plus* the whole battlefield.[54]

Some technologists argue convincingly that the smallest number of nuts can wreak havoc with such technology. They may be right. But here's the dilemma: are we going to forgo these technologies and instead wait for nature's own disruptive ways to destabilize our civilization? Or are we going to go ahead with the risky tools that may save us?

The World Trade Center and Pentagon terrorist attacks are instructive. Terrorists, armed with relatively primitive weapons, took over passenger planes and flew them into buildings. Our high-tech air defenses and security were powerless to stop them. Should we stop flying or developing passenger planes because of this tragedy? No one would suggest such a thing. Yet it might be reasonable to suggest that we reinforce our buildings with affordable carbon nanotubes, so that rogue airliners can't knock them down. It might also be reasonable to install intelligent, encrypted override systems that prevent a hijacked plane from crashing into a building;[55] or install weapon detection devices that scan more accurately. While some of these measures are available now, molecular technology could make them more affordable and more dependable.

These measures could make our civil society more resilient against terrorist and natural attacks on our civil infrastructures, without invoking draconian security measures.

It's easy to argue that nature hasn't done as much damage as some nuts might do with new technologies, and therefore we should worry more about the nuts. Yet for those who watched Comet Shoemaker-Levy 9 slam into Jupiter, that argument has a hollow ring to it. Every time we come up against the relinquishment argument—i.e., relinquishing risky technologies—we have to remember that the universe is showing us that it's not what it seems. We're at risk from more than ourselves. Terrorists notwithstanding, we still have to develop the technology to cope with such natural risks.

That said, how do we minimize the chances of military destabilization?

In the powerful molecular economy, social disaffection is a one-way ticket to terrorism. This, in turn, leads to military intervention. Therefore, one way to avoid this is to stabilize the social and economic security of the individual: give each one of us a safe haven in this fast-changing world.

It has been argued convincingly that poor people don't start wars; rich ones do, as they argue over who controls what, or which religion is going to dominate. Yet the fodder for such conflict is inevitably the undereducated, young, low-income individual, and the intellectual, socially disaffected one. Without them, most wars in history wouldn't have been fought. Disaffected middle-class children, not poor ones, commit most high-school killings, so we have to deal with psychological poverty as well as the economic kind.

In the molecular age, it's improbable that we'd enjoy military security if economically, politically, or psychologically disenfranchised populations exist in significant numbers to be manipulated by those who control powerful decentralized weapons.

Nor is there much chance of avoiding this by applying the same restrictive regimes to molecular technologies that are applied to nuclear weapons. These new technologies may be too cheap and widespread to be controlled in such ways without a police state.

Still, guidelines for control are a necessity. More than a generation ago, visionaries such as Isaac Asimov saw the challenges of potentially Frankenstein-type technologies and devised principles to govern them, as shown in appendix A. In later years, visionary organizations such as the Foresight Institute have devised more detailed development guidelines, as in appendix B. These are partially based on principles adopted in the 1990s by the National Institutes of Health regarding genetic research.[56] Many claim the NIH guidelines have been extremely successful at preventing accidents, but this hasn't stopped a furor from erupting over the potential dangers of the technologies. The struggle to control them is just beginning.

Besides the issue of scientific controls, we are still left with one glaring conclusion: If we want the molecular age to succeed, we must concentrate on the security of the individual. Not just a few million individuals. Not just the majority. But every individual on the face of this planet.

Today, such attention to every one of many billions of individuals is improbable, but molecular science, by its character, may make it more practicable. This is not the "atoms for peace" illusion of the 1950s. Molecular technology, by its nature, may be infinitely customizable to meet the needs of the individual. It is this foundation—*this basic upgrading of the concept and application of human rights*—that we have to focus on, to generate broad support for going ahead. The rest of this book looks at how to do that.

A first step to achieving such security is to make sure that the enabling tools are broadly available for individuals to improve their own conditions. In the molecular age, those tools revolve around digital information, otherwise known as *intellectual property*.

CHAPTER TWENTY-ONE

USING OPEN SOURCE

PRINCIPLE 3: FAVOR SCIENTIFIC FREEDOM OVER INTELLECTUAL-PROPERTY RESTRICTIONS

In the case of molecular technologies, intellectual
property (IP) isn't just everything, it's the only thing.
For software, genetics, and nanotechnology, the first
copy of a product costs a fixed amount to produce, then
often costs virtually nothing to replicate. Thus, the property we have in our heads is the most valuable. To protect
it, an enormous legal web has been constructed that spans
the globe. Yet this web also ensnares innovative concepts,
preventing their spread among those who most often need
them. Therefore, in the "idea economy," the way we cope with
intellectual property may determine whether we succeed or fail
in the collective journey on which we're embarking.

One of the underpinning strengths shared by every digital technology is its most contentious asset: software. For instance, molecular
technologies operate at the submicroscopic level and, as such, require
interfaces between the user and the technology. That interface is virtual-
reality software that lets the user see what's going on. Moreover, other
software is used to control genetics and the development of artificial intelligence, just as our own genetic code is the software that controls our body.

Thus, as software takes on an ever-dominant role, the battle lines are
being drawn at intellectual property.

The *New York Times* gave this analysis of why software is so hard to
cope with when it comes to intellectual property:

... it is the peculiarly malleable quality of software that sets it apart from virtually every other industry. Unlike an automobile, where components are relatively fixed and separable, in the world of software the drawing of lines separating operating systems, applications, and features is a maddening and frustrating exercise, one that has been the basis of much of Microsoft's control over the industry.[1]

Why is this so important? Companies such as Microsoft, with monopolies on operating platforms for digital technologies, are in a position that affects the intellectual freedom of every scientist in the field of molecular technology. This is because scientists have to live by the rules of such restrictive platforms, without having much of a final say in how those platforms are developed.

It's easy to shoot at companies such as Microsoft because of their size, but thousands of other wannabe companies are ready to adopt strong-arm tactics, as evidenced by court records that show companies fighting each other or the government over control of markets.

Likewise, intellectual property is at the heart of lawsuits that rage over the practices of companies such as Napster and MP3, whose software lets everyone download, copy, and instantly reproduce music and other digital works.

In genetics, intellectual property has been at the heart of the vicious dispute between private companies and the National Institutes of Health over who owns the right to the human genome.[2] For example, in late 2001 the British company Oxford GlycoSciences announced that it planned to patent as much as one-eighth of the human genome—four thousand human proteins and the genes that code them—in one fell swoop. This generated outrage among gene-patenting opponents.[3] In pharmaceuticals, patents are among the key reasons why developing nations have not been able to gain access to their own underlying resources to carry out research on diseases that are not priorities for temperate nations.[4]

Intellectual property is one of the pillars on which the developed industrial nations have established and maintained their technological dominance, but it's also showing dangerous signs of getting in the way of progress. The Microsoft antitrust trial was only one of many warning signs. IP is being used, both purposefully and unwittingly, to stifle innovation and competition—just the opposite of what it was intended to do. For example, the trend toward secrecy in publicly financed health research was confirmed in a landmark study by Harvard University Medical School and the Massachusetts General Hospital, which found that 47 percent of scientists at 100 U.S. research institutions that receive public funding from the National Institutes of Health had been denied access to information about published research. The study revealed how—as the lines merge between academic and corporate research–secrecy makes it more difficult for scientists to verify the accuracy of published work, despite it being paid for by taxpayers.[5]

One of the worst elements of IP abuse is a flood of invalid patents. Some are "patent pending" but judged afterward to be unpatentable, while others are badly delineated due to pressures on underqualified patent examiners. As patent applications flood in and get more backed up, invalid patents emerge as a major impediment to progress.[6] They are used, unwittingly or deliberately, to hold up scientific developments. Thus, we are sinking into a patent quagmire that is blocking rather than serving one of its original purposes: to give inventors the power to develop their discoveries.

In the molecular age, to liberate each individual, we must give everyone access to ideas; but we must also make sure that individuals are given sufficient rewards to continue innovation. How do we balance these?

We may confront the dilemma sooner than we think. Pitted against several thousand patent specialists at the U.S. Patent Office in Washington, D.C., are machines that may render patent law enforcement more obsolete than it is now. Genetic programming, as described earlier, is liable to put the patent office out of business. Genetic programming occurs when a computer develops its own designs based on a set of conditions provided by programmers. Some of these designs are already beyond the capability of humans to predict, and they take only days or hours to develop. This is already resulting in a flood of patentable designs that themselves are being superceded in an ever-accelerating spiral of invention. Given that the U.S. Patent Office is already overwhelmed with patent applications despite having hired thousands of new patent examiners, it's clear who is going to win this contest: the computers.

One possibility might be for the patent office to hire computers to analyze these genetic designs. Then we'd be in the position of computers judging other computers' ideas. From this, we immediately begin to see where artificial intelligence might take over some of the foundations of our industrial infrastructure. The other option is to recognize the impossibility of enforcing patents and rewrite some rules, such as those that cover operating-system platforms, for example. Yet, in a capitalist society, this is the equivalent of standing up in front of Joseph McCarthy and admitting to being a communist. The world's largest companies are dead set against dilution of the patents regime, especially in the drug and software business, and with good reason. But the reality is that machines are developing patentable concepts exponentially faster than governments are able to recognize them, or patent holders are able to enforce them, or the courts are able to limit abuses of them. This contains the seeds for much of our economy to begin functioning outside the legal system, more so than the so-called grey economy does now. Anarchists may rub their hands at the prospect, but the implications for a civil society are serious.

For those who claim that patents are the only way to progress, we need only look at one example to see how cracks are appearing in the walls of that argument.

Open-source software has many definitions, but is generally acknowledged as software whose developers allow free redistribution—including the source code, along with modifications and derived works.[7] For years, Microsoft claimed that using open-source software was among the great threats to civilization, because it allegedly leads to a breakdown of the patent-enforceability regime.

To the surprise of many, though, it was then revealed by the *Wall Street Journal* that Microsoft had been using open source in its own programs, despite unequivocal claims to the contrary.[8] Thus, while Microsoft was leading the charge for years against open source, it was using such software on one of its more frequently accessed sites.[9]

Why so much fuss about open source? Here's why: software is the foundation of the molecular economy, and open source is the fly in the ointment of those who would seek to monopolize that economy. It is profoundly fundamental to our future.

In recognition of this, organizations such as the Foresight Institute have started open-source projects[10] to flesh out the issue. Moreover, the Patent Policy Working Group of the influential World Wide Web Consortium—which helps to determine policy for much of the Internet—has started grappling with the issue of licensing fees for technologies that are incorporated into Web standards.[11] Until 2001, much software for the Web was built on free platforms. The consortium's proposal to allow companies to collect fees for some of these elicited a barrage of protests from software developers who were concerned it would lead to information bottlenecks and constriction of the very innovation that built the Web. Yet these are only tips of the information-liberation struggle. The key challenge is how to bring software into compliance with the same underlying principles that have allowed science to move forward until now. Linus Torvalds, inventor of the open-source Linux operating system, succinctly summarized the argument this way:

> Open source code should be put in the tradition of sharing of information and intellectual freedom that has been practiced by Western science since the time of the Greeks.[12]

Ironically, regardless of how this discussion works out, intelligent machines may render it moot. Once they gain greater designing abilities, they may just go ahead and apply whatever they think of next. Who's going to stop them?

CHAPTER TWENTY-TWO

REDESIGNING DEMOCRACY FOR ARTIFICIAL INTELLIGENCE

*"The more the world is specialized
the more it will be run by generalists."*
—Marcel Masse, president,
Treasury Board of Canada[1]

PRINCIPLE 4: REDESIGN DEMOCRACY TO COPE WITH ARTIFICIAL INTELLIGENCE

I n a world where individuals may have the power to command trillions of robots that can terraform half a planet in a few years, or make a mockery of intellectual-property laws that underpin a capitalist society, the most important technologies may be those that empower us to make decisions collectively, instead of abdicating them.

Present primitive governing mechanisms probably aren't going to work. They're too slow to react, too dependent on vested interests, and too arbitrary to cope with the type of power we're talking about. For example, international collaboration on decisions that affect the global environment could be a prerequisite. Otherwise competitive wars would break out each time a nation, corporation, or individual tries to alter ecosystems to match their own requirements. We won't have time to argue for years over a global climate protocol. If Saudi Arabia decides to convert the desert into a forest, or Russia decides to convert the Arctic tundra to greenhouse-fed rain forests, this has implications for the rest of the world. The low cost of such terraforming would make our present decision-making structures obsolete. We wouldn't be able to control it with the types of dictatorships or democracies that we have today. We'd undoubtedly try, but that could lead to disasters that we'd be well advised to avoid in the first place.

We're at an inflection point, where democracy may either improve or slide into ecofascism. If we want to give citizens collective control, instead of migrating toward a world dominated exclusively by scientists and technocrats or malicious dictators, then we'll have to use molecular magic to upgrade the imperfect art of democracy.

Convincing arguments will be made, as they are by most dictatorships or big governments, that ordinary people lack the expertise to guide complex machinery. To counter this, citizens require an improved capacity to deal with complicated decisions. In this struggle, artificial intelligence becomes paramount. Technologies that enhance our capacities to input, interpret, and act on trillions of bits of information may be key to establishing a molecular democracy. Semi-intelligent personal Robo servers represent one possibility to help us assimilate and cope with such information overloads. Another possibility is to have neural implants for improved memory and collective communication. As with our other technologies, it depends on how we use them. On one hand, we may risk becoming part of the Borg, as depicted in *Star Trek*, where each biological entity is an unquestioning part of a compact, centralized authority. At the other extreme, our new neural capacities may transform each of us into superindividuals who independently make decisions: a culture of intellectual anarchy. Somewhere between those extremes is *independence* that relies on *interdependence*.

One thing seems certain: as individuals, we'll need more time, brainpower, or both, to act as citizens. Rather than running ourselves silly in a wired economy, we'll need technologies that free us to take on governing responsibilities. This may sound utopian, but the alternatives we face have been aptly described by a notorious author:

> First let us postulate that the computer scientists succeed in developing intelligent machines that can do all things better than human beings can do them. In that case presumably all work will be done by vast, highly organized systems of machines and no human effort will be necessary. Either of two cases might occur. The machines might be permitted to make all of their own decisions without human oversight, or else human control over the machines might be retained.
>
> If the machines are permitted to make all their own decisions, we can't make any conjectures as to the results, because it is impossible to guess how such machines might behave. We only point out that the fate of the human race would be at the mercy of the machines. It might be argued that the

human race would never be foolish enough to hand over all the power to the machines. But we are suggesting neither that the human race would voluntarily turn power over to the machines nor that the machines would willfully seize power. What we do suggest is that the human race might easily permit itself to drift into a position of such dependence on the machines that it would have no practical choice but to accept all of the machines' decisions. As society and the problems that face it become more and more complex and machines become more and more intelligent, people will let machines make more of their decisions for them, simply because machine-made decisions will bring better results than man-made ones. Eventually a stage may be reached at which the decisions necessary to keep the system running will be so complex that human beings will be incapable of making them intelligently. At that stage the machines will be in effective control. People won't be able to just turn the machines off, because they will be so dependent on them that turning them off would amount to suicide.

On the other hand it is possible that human control over the machines may be retained. In that case the average man may have control over certain private machines of his own, such as his car or his personal computer, but control over large systems of machines will be in the hands of a tiny elite–just as it is today, but with two differences. Due to improved techniques the elite will have greater control over the masses; and because human work will no longer be necessary the masses will be superfluous, a useless burden on the system. If the elite is ruthless they may simply decide to exterminate the mass of humanity. If they are humane they may use propaganda or other psychological or biological techniques to reduce the birth rate until the mass of humanity becomes extinct, leaving the world to the elite. Or, if the elite consists of soft-hearted liberals, they may decide to play the role of good shepherds to the rest of the human race. They will see to it that everyone's physical needs are satisfied, that all children are raised under psychologically hygienic conditions, that everyone has a wholesome hobby to keep him busy, and that anyone who may become dissatisfied undergoes "treatment" to cure his "problem." Of course, life will be so purposeless that people will have to be biologically or psychologically engineered either to remove their need for the power process or make them "sublimate" their drive for power into some harmless hobby. These engineered human beings may be happy in such a society, but they will most certainly not be free. They will have been reduced to the status of domestic animals.[2]

These words were penned by the Unabomber, Theodore Kaczynski, the lone terrorist who targeted dozens of high-level scientists and technologists, killing several with letter bombs. Interestingly, these precise words were also quoted by Sun Microsystems cofounder Bill Joy to describe the potential dangers of molecular technologies in the hands of maniacs who might want to terminate the technological world, or use it for other mad aims.[3] In quoting this passage, Joy explains, "Kaczynski's actions were murderous and, in my view, criminally insane. He is clearly a Luddite, but simply saying this does not dismiss his argument; as difficult as it is for me to acknowledge, I saw some merit in the reasoning in this single passage. I felt compelled to confront it."[4]

It's a chilling picture of what we may face if democratic institutions can't cope with decisions about how—or whether—to mold our environment. Is molecular science going to help us make collective decisions, or lead us to the era of technocratic priests who rule as the church did during the Middle Ages? This question is complicated by the impacts of nanotechnology and artificial intelligence on economic relations among, and within, states. These may be as disruptive as they are revolutionary. In his essay *Molecular Nanotechnology and the World System*, Thomas McCarthy describes conflicts that may buffet nation-states. He describes how "designer communities" may end up as the units on which our democracies are based:

> MNT [molecular nanotechnology] will change states from the inside out, altering their capabilities for industrial production, for military action, for power projection, for independent behavior. The capabilities of states that are thoroughly infused with advanced molecular manufacturing technology will be quite potent. But the changes that take place are not only meaningful in terms of absolute power. We must also take into account the changes in relative power both within and between states that may develop during the introduction of MNT into the world and during its spread. . . .
>
> MNT may take the bite out of independence by removing the need to capture economies of scale in order to prosper. By making the cost of capital (in the forms of factories and production equipment) negligible, "one-offs" and limited production runs of goods may have almost the same cost as their mass-produced counterparts. This would eliminate the need for participation in a larger market in order to capture the benefits of economies of scale; thus, trade, and with it the interdependence that accompanies trade, would shrink by an enormous degree. This opens up the possibility of small, autarkic states, ones that can form on a basis other than economic need, so that religious, ethnic, linguistic and any number of other reasons for association can take precedence. What may result is the formation of small, independent, autarkic communities for specific groups of people; these *designer communities* may be the state of tomorrow.[5]

Such designer communities, despite their potential economic and political independence in the molecular economy of tomorrow, may not be able to cope with globalized impact of terraforming, or of responses to natural extremes such as climate changes. A more sophisticated international apparatus than the ones we have today would be required to respond. We may see, for example, thousands of local designer communities participating directly in decisions on globalized environmental issues via Internet voting. The integrity of each vote would be protected by DNA-encoded smart cards that operate through secure networks. For those who question the infallibility of Internet voting, the argument could quickly become moot when it's pointed out that the financial future of each individual rests with millions of faceless computers that hold our life savings and send our personal data everywhere. Just as these computers let us check our bank account or credit rating, Internet voting could let us each check our individual voting account.

The determining factor in our ability to do this is the human-computer interface. The dissemination and interpretation of information to, and by, every individual will determine how well we deal with globalized decisions at a local level.

Today, information overload is already upon us. The challenge is to empower everyone to make information-enhanced decisions and to be able to confirm that they were carried out.

We may find, for example, that connecting ourselves to memory-enhancing chips is an effective way to cope with this tidal wave of material. Or we may find that a vast capacity is already locked up in the human brain and needs only to be liberated by designer drugs.

The memory and insight to help us assimilate such information may be available only if we're each able to instruct artificially intelligent computers to organize data to the point where our own human brains could make a reasoned decision. On the other hand, artificially intelligent minds may be assigned the job of making certain decisions by themselves, such as the placement of new communities based on their global ecological impact. The vast gray area that we'd have to struggle with is when to let artificial intelligence decide for us, and when to take over the controls.

Some scientists believe that scientific endeavors will go on regardless of the governing mechanisms we adopt. Nothing could be further from the truth. History is littered with examples of science being hamstrung by politics or religion. The ongoing struggle between creationists and evolutionists in U.S. schools is the classic contemporary example of this.[6]

Thus, the future of science rides heavily on societal perceptions of technology.

While some fear that we'll abdicate decision making to machines that are smarter than us, we need to consider whether this is better or worse than the present situation, where we abdicate decision making to politicians or non-elected bureaucrats and executives, who frequently deceive us or withhold the truth, as we saw so graphically in the Enron and other such debacles. Optimistically, as more technical decision making is left to intelligent machines, we may have the time and incentives to make collective policy decisions, instead of suffering the current situation. Right now, less than half of America's voting population elect representatives, who then spend much of their time being told what to do by special interests. As technology becomes more powerful, such a trend might lead to control of society and science by those special interests. One way out of this, although greatly challenging to implement, may be to empower the individual voter to cope with expanded decision making.

In an enlightened molecular economy, local decisions by voters may give us more control over our lives, because many means of economic independence may be in our communities. Physical manufacturing of consumer products might occur in the home or community, not half a world away. The software may come from somewhere else, but the decision about what to produce becomes local or individual.

This, in turn, may transform governing by letting us see the impact of local decisions on our daily lives.

On the other hand, international questions about ecology, space, and other issues affecting the global commons may demand a vastly increased intellectual effort. The human-computer relationship could inundate billions of individuals with collective decision-making requirements that today's human can't assimilate. Enhanced intelligence may be the only way to cope.

Nor does this type of decision making apply only to our planet.

Using the Moon or Mars as a testing ground for colonization is on the horizon. This gives our species the insurance policy of being on more than one planet, in case an improbable planet killer wipes us out on Earth (see Fig. 33). Yet the democratic control mechanisms for this may be immediately contentious. Kim Stanley Robinson, in his trilogy *Red Mars, Green Mars,* and *Blue Mars,* gives a scientifically credible outline of planetary terraforming and the political turmoil that accompanies it. He describes how naturalists take on terraformers in a guerilla campaign of terrorism that pits human Martians against their financial backers on Earth.[7] Robert Zubrin, founder of the Mars Society and a strong influence on methods for exploring the red planet, is the true-life embodiment of Robinson's terraformer. He's convinced that making Mars habitable is the next great human venture. Zubrin and Robinson collaborated on the Mars Declaration, which incorporates a rationale for going to the red planet and colonizing it.[8]

Robinson's eloquent and scientifically impressive portrayal of how democracy might play out with molecular technologies is a testament to the value of social science fiction in explaining choices among conflicting futures. For example, his work gives form to the debate on whether it's right to play God with one planet in the name of saving another. It helps to elucidate the discussion between high-profile technologists who say that playing planetary deity is just too dangerous, and other equally accomplished minds who discount that fear by claiming that we have no choice but to move forward.

Fig. 33. The planet killers. Far larger than the asteroid that wiped out the dinosaurs, these monsters are capable of tearing our atmosphere away from the surface of the earth. Some scientists postulate this may have happened to Mars. Nanotechnology may give us the defensive weapons to find and neutralize doomsday rocks before they reach us. (Don Davis [artist], courtesy of NASA)

One question arising from this discussion among intellectual titans is whether we mortal individuals are going to see some of our plain old everyday problems taken care of, or if we'll just watch from our onboard TVs in our traffic-bound SUVs, as the debates that determine our fate run ahead of us.

CHAPTER TWENTY-THREE

LIBERATING EACH ONE OF US

PRINCIPLE 5: USE TECHNOLOGY TO LIBERATE THE INDIVIDUAL FROM FEAR

In his essay "Is Life Really Getting Better?" author Richard Eckersley eloquently sets forth an issue that lies at the heart of why most of us get out of bed each morning and do the things we do:

> A central tenet of modern western culture is the belief in progress, the belief that life should be and is getting better–healthier, wealthier, happier, more satisfying and more interesting. Is this the case? If our answer is "yes" then we can assume society is on the right trajectory, requiring only periodic course correction by governments.
>
> If the answer is "no" then the most fundamental assumptions about our way of life need reassessing. The task we face goes far beyond adjustment of the policy levers by government: It means having a more open and spirited debate about how we are to live in the future and about what matters in our lives.[1]

Eckersley draws from a body of studies he says show convincingly that "wealth is a poor predictor of happiness. . . . In most countries the correlation between income and happiness is only negligible; only in the poorest countries is income a good measure of well-being."[2]

Hogwash, say optimists. Our wealth-generating technological society is helping everyone live longer and better. We're keeping up with population growth by expanding our economies and improving agricultural efficiency. Personal freedoms are expanding. Women and minorities are gaining more rights. Medical care is improving. We're cleaning up the environment. The Cold War is over, and we're on the edge of a sustained economic boom, despite the occasional recession. Crime is receding. Education is broadening. We're in the midst of a scientific renaissance that sees us uncovering the secrets of everything from genes to the rest of the universe. Naturally there are problems, but the good news is that we're finally admitting them, for example, by using information technology to expose injustices. Despots are being dethroned worldwide. Discrimination on the basis of sex and race has finally escaped from the platitudes of legislation to the hard glare of the news media and the courts. Child labor, dowry murder, selective abortion of girl babies, female castration: these are being exposed at last, everywhere. As democracy and mass media infiltrate cultures, we're beginning to talk about these formerly taboo topics. That is the first step toward solving them.

Such is the case put forth convincingly by technology enthusiasts who argue fervently that since the ideological war is over, we just need to continue with making the rest of the world a mirror image of the United States. Those who challenge this concept are dismissed as whining liberal naysayers. After all, if things aren't getting better, why are we doing this?

That's a good question, because, as most of us know, there's a flip side.

The vast majority of individuals in our world today—including many Americans—are still economically insecure, despite the enormous wealth of nations. How does one arrive at such a conclusion? The numbers may sound numbingly familiar, but they bear mentioning nonetheless. Three billion inhabitants of the earth—roughly half—live on *less than two dollars a day.*[3] Even for the other half, the digital workplace and the greatest decade of growth—the 1990s—have been far different from what e-brokers hyped them to be. In her book *White-Collar Sweatshop,* author Jill Fraser documents reality: the average American worked more hours for less real pay, fewer benefits, and lower job security than a generation ago.[4] Fraser isn't alone in her depiction of this. Other critically acclaimed books that document "cyberserfdom" have been popping up for the past decade.[5] Technoskills don't seem to solve this; while they help to improve some parts of our material situation, personal debt has been increasing steadily. Here are a few of the outstanding financial symptoms underlying it:

From 1983 until 1997, amidst the greatest sustained economic boom in the wealthiest nation on earth, the top 20 percent of Americans accounted for more than 100 percent of the total growth in wealth, while the bottom 80 percent lost 7 percent.[6] In the decade of the greatest affluence in history, the percentage of children living under the poverty line actually *grew* from

15 to 19 percent.[7] Some of us may be getting wealthier, but beginning in the 1990s, right into 2001, the overall rate of personal savings declined steadily.[8] Health-care costs increased severely.[9] The "rat on a treadmill" effect—running harder for fewer rewards—subsequently accelerated.

The reasons for this do not require sophisticated economic analysis to understand, although they are extensively documented. They were accurately forecast decades ago by business writers such as *Future Shock* author Alvin Toffler, who for years supervised *Fortune*, one of America's foremost business magazines. Toffler documented how mechanization and computerization require fewer people to produce more goods and services. As fewer workers are required to run the machines, ownership of production—whether physical or intellectual—is concentrated in fewer hands. Computerization and communication also let markets operate more efficiently, so that services can be derived from the cheapest source, be it in Thailand or Tennessee. While millions of new-economy jobs are created to replace old-economy jobs, these new jobs are themselves becoming outdated, in the ever-shortening cycle of a globalized economy that seeks the lowest price.

Today, we see that as global price competition keeps a lid on wages, competition for good housing and education increases. So while food, consumer goods, and transportation have stayed relatively cheap in the new economy, the costs for a good home, safe neighborhood, and education have skyrocketed. Those who do own homes own a smaller percentage of them, and as such are weighed down by big mortgages. The so-called wealth-effect that economists have praised has made big-ticket items unaffordable for many, while keeping the rest of us running ourselves silly to finance them.[10]

Thus, new entrants into the new economy face triple jeopardy: global price competition, rapid skills obsolescence, and inflated costs for big-ticket items. These damning trends show that while things may be better in some ways, they're getting worse in others. Some might argue that such insecurity is good, because it inspires us to work hard and build society. While this may have merit, beneath it lies a debilitating economic scourge: loss of purchasing power that eats at our collective ability to transcend to another level of civilization.

Many of us, in rich and poor nations alike, share similar insecurities: fear of becoming obsolete, of being unable to educate or protect our children, of ruinous family medical costs—especially to care for our parents and ourselves, now that we're living longer than ever before. These are just a few of the factors that overstress us. If, as Toffler warned, we do not give a greater monetary value to other types of work, such as teaching, child rearing, and other tasks that help us cope with this fast-paced world, then we may be stuck with an increasingly dysfunctional society.

Unfortunately, we haven't taken his sage advice very well. Subsequently, tens of millions of children are left undersupervised at home after school as working parents struggle to make ends meet–even on six-figure annual earnings. This contributes to a host of social problems among the middle class. Job rage, road rage, air rage, student rage, and customer rage are the symptoms. Prescribing mood-altering drugs for supposedly hyperactive children–who are often under the age of five–is commonplace. Teenage psychological counseling occurs regularly. Some kids from affluent homes are shooting their teachers and classmates. Use of hallucinogens among high-school students is on the rise.[11] This is because one of society's underpinnings–child rearing–is being undermined.

We live longer, but in what condition? As affluence spreads worldwide, a food-toxic environment follows it. Obesity is epidemic.[12] This fatness regime depends on raising billions of animals in mechanized factories, while incidentally killing millions of wild animals for seafood.

Toffler also predicted that our relationships would become temporary affairs, to match our temporary lifestyles. With that comes temporary loyalties.

> As we hurdle into tomorrow, millions of ordinary men and women will face emotion-packed options so unfamiliar, so untested, that past experience will offer little clue to wisdom. In their family ties as in all other aspects of their lives, they will be compelled to cope not merely with transience, but with the added problem of novelty as well.[13]

Various polls in America show that nine out of ten adults lie regularly; 38 percent of adults under thirty say that being corrupt is "essential" to getting ahead; and 60 percent say they are ready to spend years in jail and have a criminal record if it means getting several million dollars for their trouble.[14]

To cope with stress created by these shifting environments, and by diseases of aging that show up when our life span is extended, we've built medical and drug industries that represent one of the biggest economic sectors in the world.[15] Prescription tranquilizer, stimulant, and antidepressant use is a mainstay of everyday existence for many, and growing rapidly in scope.[16] Over-the-counter remedies are everywhere. The War on Drugs has been a self-confessed failure, as admitted by the nation's largest child drug-avoidance program in 2001.[17] This is no surprise, because our dependence on legal drugs sets us up for using the illegal ones. The advertising industry spends billions of dollars advertising prescription drugs to a population that has no expertise to judge which ones they might need.

Despite enormous expenditures on medicine, the public health-care systems of many industrialized nations are falling apart through a combination of underfunding and overstress.[18] This is partly because our medical systems are overwhelmingly treatment oriented and underwhelmingly prevention oriented. It is also caused by disintegration of the nuclear family alongside the inability or unwillingness of society to pay for social support that disappeared along with that nuclear family. This is complicated by the AIDS pandemic.

To blunt our senses against such realities, we entertain ourselves with movies and video games that contain endless acts of violence. Media owners continue to perpetrate the biggest myth of modern time by claiming that violence in the media doesn't influence people's behavior—but product advertising does. The entertainment industry that pervades global consciousness has, according to the *Los Angeles Times*, a high incidence of institutionalized lying.[19] Documentation of such standardized deception was made more powerful and disturbing when it was undertaken by a respected newspaper that operates near the heart of the American entertainment world.[20] Whether this is a cause, reflection, or exacerbation of our human tendencies is still debatable.

The way the United States deals with criminality is also worrisome. Close to one out of every 150 Americans is now behind bars.[21] In the 1990s, the era of the greatest supposed "improvement" of economic conditions in world history—which were also touted as a cause of reduced criminality—the rate of incarceration rose the fastest. Although some studies suggest this trend may be reversing, there remains no doubt that America's prisons hold a significantly greater percentage of its population than compared to other democratic countries.[22]

To further control each other's behavior, we're growing a vast web of regulations. Despite years of regulatory reform, the Code of Federal Regulations grew from 54,000 pages in 1970 to 134,000 pages in 1998, and by 2001 stood at the longest in history.[23] Thus, friction on business growth and innovation is increasing, as management spends more time dealing with regulations. Increasingly, we are attempting to regulate ourselves out of a situation characterized by passing loyalties and technologies that outpace the enforcement capabilities of governments.

Meanwhile, just to reiterate, more than 3 billion people still live on less than two dollars a day.[24] That's more in real numbers living in real poverty than the total world population in 1945. The polarization of wealth is increasing. The share of wealth owned by the top 20 percent of asset holders in the world went from 70 percent in the 1960s to 86 percent in 1999.[25] The standard argument to support this gross inequity is that the rich put their capital to work, so the population as a whole benefits from it. One could say, cynically, that it might make sense to give the rich 100 percent of the wealth so everyone could benefit from how they put it to work. This was precisely the argument used by feudal landlords. It seems to be the direction that our economies have been heading, despite efforts by Federal Reserve economists to convince us that the debt-driven wealth creation of the 1990s left everyone with a bigger piece of the pie. According to the most reputable financial press in America, as more companies merge and spin off, workers' benefits are slashed using pension plan loopholes that have left millions with reduced benefits, while their pension funds are traded as prized assets among companies.[26] Often, the game becomes how to keep pension plan funds away from fund beneficiaries so that company asset portfolios are increased.[27] And it's not the labor unions or left-leaning liberals who claim this: it's the business press.

Taken together, these symptoms suggest that many in our society are in a stealthily deteriorating health-care and economic situation driven by our own technology. As the world gets wealthier, we suffer more from economic disparities and materialist stress. Often our technology liberates us, only to trap us in other ways. Nor is this position being argued only by the disenfranchised. One of the planet's most successful capitalists, financier George Soros, has made a forceful case that fundamental changes to globalization are required, if we're to succeed socially.[28]

Similar messages are emanating from other quarters. Robert Pool, author of the best-selling book *Fat*, concludes that to fix its obesity epidemic, "America must remedy the unhealthy environment of its cities and towns."[29] The United States is not the only nation with this problem. As noted earlier, Japan, China, and Germany each display similar symptoms among their affluent populations. This trend suggests that as the world gets wealthier, we're getting fatter and more passive. To counter that obesity and passivity, teenagers—especially girls—are smoking in increasing numbers, despite heavily reported studies showing the consequences.[30]

Our suburbs are also getting fatter, gobbling up increasing portions of the natural environment. While there is strong disagreement on who is responsible for environmental problems and how bad they are, one fact remains undisputed: Suburbanization and farming are the principal causes of habitat destruction as forests and grasslands are built on, paved, or ploughed under.

However grim they may seem, these assessments are, by themselves, insufficient reason for rejecting the technological path. I have already pointed out that we must move forward with technology in order to find new defenses against the threats posed by natural phenomena. But if a societal breakdown occurs, our chances of building defenses against big natural disasters such as those described in this book are nil. Therefore, the question becomes, How might we apply our new technologies to improving our lives collectively and preventing such a decline? If we don't find answers to this question, then molecular technologies may serve as just another step in our race toward social disintegration.

How do we solve this? First, recognize the root of the problem. In an age of unprecedented wealth, fear still rules: fear of poverty, fear of being fired, fear of being obsolete, fear of being bankrupted by medical costs, and fear of being unable to care for aging parents. Whether in America or Bangladesh, many of us still live in an age of fear.

Thus, to transcend the present situation, we have to free the individual from insecurities that have accompanied or been worsened by the new economy instead of being cured by it.

Molecular technology could help us to make that transition, if we choose the right paths. Many of the unfulfilled promises about our technological potential may be realized in the future. Transport and virtual-reality systems may let us do things with less stress. The robustness of our youth could extend into old age. Memory may become expandable and we might share it with others, if we want.

Yet these alone will not guarantee our security. By themselves, they may make us less secure if the world accelerates and we feel less in control of what's going on. So what might help us establish security for the individual?

EMERGENCE OF THE NEW INDIVIDUAL

One place to start might be to restore the capacity to manufacture our own needs in our own neighborhoods or homes, but without the drawbacks of the nineteenth-century methods. Every household or local community may, for example, receive a basic software package that lets them build and customize each appliance and piece of furniture. Many years ago, engineering visionary Buckminster Fuller foresaw similar capabilities, based on his philosophy that "one individual's well being could not be fulfilled until every person on earth was provided with the necessities of life."[31] He spent his life designing things that might do that. Unfortunately, he didn't have nanotechnology to work with, but digital fabrication may help to achieve his goal.

Affordable, localized health care may be another liberating technology. Instead of depending on expertise based in centers to which we have to travel, we may be able to bring affordable health care to where *we* are. The Internet and augmented reality together could let physicians or intelligent machines provide competent diagnosis, treatment, and preventative care— remotely. Already, home diagnostics is a burgeoning industry. With nanomedicine, the home clinic might come in a package.[32] The lack of basic nutrition and exercise, which leads to so many health problems, could be rectified by robotic preparation of nutritious meals and virtual- or augmented-reality environments that are more conducive to exercise.

Local communities could also be more in control of their needs. By manufacturing goods locally, we might establish a new age of personal security. This might also liberate our imaginations, because the variety of things we could manufacture for ourselves would be limited only by what we'd want them to look like. Moreover, this independence would let each individual move from place to place without the insecurity and stress that accompany today's mobile society. For example, desktop manufacturing may not require a centralized superstructure around it. Therefore, individuals in underdeveloped economies may gain the same benefits as those in rich ones. This type of fundamental security may give each individual a solid base from which to get on with life.

However, this might only occur if we price software affordably, or designate some of it as open source. Such a possibility is far from guaranteed.

One of the *limitations* on individual freedom may be the type of consumptive overabundance discussed earlier. Environmental consequences from, for example, terraforming a swath of desert, would have to be regulated by international governing bodies. This would place limits on the freedom of individuals to produce whatever they liked. But it need not lead to police states or threaten the fundamental security of individuals. The concept of "world government" may come to mean something far more benign if each of us is imbued with greater intelligence that lets us cope with the broader implications of our decisions. If we look back to the Law of Requisite Variety discussed in chapter 14, we see an ecological rationale for such enhancement. As individuals gain greater intelligence, they might more easily comprehend the complexity of ecosystems. Thus, rather than trying to drag nature down to a primitive level, as we often do now, we may opt for more intricate solutions that are more compatible with the natural ecology. Our definition of the earth as a closed system may have to be revised to account for the more complex relationship of our planet with solar and galactic systems. This, too, may be easier if individuals are imbued with enhanced intelligence.

Furthermore, as we reach these heights, we might learn whether our enhanced intellects are able to control the base emotions that have historically led us to slaughter each other in the millions. As individuals gain more power, might they be able to control the murderous human tendencies that have been part of our evolutionary past? Could we genetically "breed out" or otherwise repress excessive aggression—and would we want to? Or might we require a collectively controlled policing system, with powers sufficient to cope with superintelligent individuals?

SUPERORGANISMS

These questions bring us to the role of *superorganisms*, the segments of society that are created from the sum of our individual actions: the regulations, codes of behavior, ideas, mega-structures, and environments that we collectively establish.

Despite the tremendous power that the new individual may have, we'll each still be part of superorganisms. These organisms could have the power to transform our planetary environment at a blinding pace. Huge tracts of land that are now empty may become occupied. The Sahara desert, the Siberian and Canadian tundra, and Antarctica: we could make these habitable, not by changing their climates, but by constructing artificial environments. Such environments may range from personal exoskeletons that protect us from our immediate environment, to bubble cities that establish a greater protective perimeter. The struggle could intensify between the new individual who wants to mold his, her, or its environment, and the superorganism that needs to balance competing interests. This, in turn, could provoke an ecological and political crisis, as wilderness areas come under population pressure. The environmental movement may press for conventions that ban colonization of some zones by human or robotic life forms. That struggle may spread to near space, the Moon, and Mars, as author Kim Stanley Robinson so aptly depicted in his sci-fi classics.[33]

Yet a surprising solution may help to mitigate green (or, in the case of Mars, red) conflict. Molecular technologies may let us engineer artificial environments that are virtually indistinguishable from their natural surroundings. Our settlements may blend into the ecology in ways that are technologically impossible today. Our cities may morph into forests of giant sequoias or desert rock faces. If that seems comically distant, just look back to the first primitive signs of this in 1997, when municipalities began requiring transmission towers to be disguised to blend in with local surroundings.[34] Are there hazards to such chameleon effects? Might someone accidentally cut down a treehouse? Might criminals evade authorities because they blend into the environment too easily, just as the Klingons cloak their battleships in *Star Trek*? The chameleon effect may have downsides, but, as described earlier, we could also have universal sensing to let us determine the characteristics of objects. Our machines may recognize and talk to each other, regardless of whether they look like trees or humans. Chemical sensors could alert them to each other's presence and composition. Since 1997 DARPA has spent millions of dollars on a dog's nose program to incorporate the sensing ability of a dog into machines.[35] One aim is land mine detection, but other potential applications reach across many domains. Thus, the components of this superorganism could communicate with their own ears, eyes, and noses. If we go to transplant a tree in our backyard, an embedded bar code may tell us its characteristics before we touch it.[36] It might be a natural tree, part of a communications network, or both. Moreover, trees may learn how to generate their own electricity or defend themselves against being cut down by humans.[37] Environmentalists might arm them with robotically enhanced exoskeletal defenses, such as carbon nanotube armor. This makes for intriguing scenarios about trees fighting back.

We may also see exaggerations of the contradictions that have characterized relations between the individual and society for millennia. Violent resistance to colonizing the wilderness may occur alongside technological integration with the natural environment. Superrich individuals may clash with terrorists who oppose their power, or with seemingly oppressive superorganisms that threaten their freedom to do as they please. We may see stealthy, encrypted individualists fighting against privacy invasion by big-brother bureaucrats. Nanospies may give governments and companies the power to see anyone, anywhere, anytime. Already, the U.S. military has dust-sized airborne spy robots known as smart dust,[38] virtually invisible to the eye. We can imagine their privacy invasion capabilities once they are shrunk to molecular size. Similarly, these same tools may make it tougher for superorganisms to keep secrets from the population. Furthermore, the same smart dust has already been used to reduce energy consumption and improve comfort in rooms, by monitoring temperature conditions in every part of a building.[39] In the future it may also help parents keep track of their kids, and receive warnings if they run into trouble.

Megacomputing, along with material and energy abundance, may let us take on projects such as space elevators and asteroid mining. These could terraform our world and others, if we let them. The ethical implications of such power will no doubt be heavily debated. This would have to be done at the international level. Moreover, building a network of international tunnels for passenger transport would require international agreements and infrastructures. Military security against terrorism would also require a worldwide network, as would the administration of satellite space communities. Each of these constitutes part of a superorganism that may help to support the security of—but also place restrictions on—individual freedom. In some ways, it may be no different than the tug-of-war we face today, except that the complexity could be greater. Hopefully, our personal confidence to cope with it would also be greater. Gaining such confidence may require greater intelligence than we have now, so it's easy to see why enhanced intelligence might pervade our governing regimes.

As they compete for dominance, the new individual and superorganisms of society may also contribute to helping us survive huge natural catastrophes. They could give us the capability to detect, live through, and adapt to sudden changes in the natural environment. When we reach the stage where every individual is able to survive and prosper after a natural calamity, we will have reached a new level of civilized resilience.

THE SPIRITUAL CHALLENGE

In a speech to a Unitarian Universalist congregation in 1999, artificial-intelligence developer Ray Kurzweil speculated on the meaning of life in an age where human consciousness might be implanted into artificial bodies, and robotic minds might think like us.[40] He pointed out that religious definitions of human consciousness might be upended, but on the other hand, they may be one of the few ways we could explain our existence in an age of spiritual machines.

Kurzweil went on to say that when he asked his five-year-old daughter about the meaning of life, she replied matter-of-factly that it was to be happy, have fun, and help people. The same definition might apply in the molecular-science age, he surmised, except that people would be indistinguishable from artificial life forms.

I'm told that this speech was well received by this particular congregation, but no doubt such viewpoints would stretch the credulity of many conventional religious institutions.

The spirituality question extends to the ecological realm as well. Environmental engineering is already attacked as an arrogant, unspiritual, "man over nature" approach. A war has begun between religions and those who advocate such applications. Right now, that war centers on the cloning of humans and genetic engineering of crops, but it is bound to broaden, as most everything becomes imbued with artificially modified qualities.

If we're having trouble with that, what happens when our machines start asking for rights, our organs and limbs are farmed from clones, and our memories are implanted outside our bodies? These may well occur in our lifetimes.

The looming confrontation between religious leaders and environmentalists, on one hand, and molecular scientists and corporations, on the other, is a potentially debilitating one. Moreover, a rift has also emerged in the confines of the scientific and technological community about our ability to prevent runaway technology.

One thing that might help to heal these rifts is to adopt the principle of mass survival as a fundamental human right.

MASS SURVIVAL AS A RIGHT

Unlike Cold War technology, precise molecular manufacturing may empower us to protect every individual from truly big disasters. The cost of protecting each individual could be much less than we've seen at any time in history. Through the methods described earlier, we may see extremely low-cost methods that empower each individual on earth to protect him or herself. Thus, mass survival in the worst of disasters may be practicable.

Despite this potential, there is a danger that we'll take the dark road.

If people aren't confident that they'll be among the saved in a disaster, then cooperation by large populations with their governments becomes less probable. As individuals grow more independent and informed, they become less likely to blindly accept assurances by the government that everything is going to be all right. There are good reasons for such skepticism. We've now learned that in the Cold War, military planners had a strategy of lying to the public, not only about survivability of a nuclear war, but also about the effects of manufacturing nuclear weapons on the health of the workers who made them. The most patriotic of workers—those who sacrificed their health to help win the Cold War—were kept in the dark for decades about their illnesses, and were denied their right to compensation.[41] This deception may have been effective at the time, but has since cost governments and scientists dearly in terms of credibility—especially because they invoked the veil of "national security" to deceive their own people. As explained earlier, such credibility gaps have been instrumental in allowing movements to arise that fan the flames of mistrust in science, whether warranted or not.

Similarly, if the general public gets the idea that we're entering the molecular age with some dark secrets for protecting an elite from its worst effects, there could be widespread resistance to the technologies.

The same goes for protecting us from big natural disasters. It may be comforting for some to have a Noah's ark strategy as a fallback, to protect the species from extinction in case of a planetary natural catastrophe. This may resemble a time when special bomb shelters were built to harbor our leaders in the event of nuclear war. Yet in an age in which we realize that global war or catastrophe may mean the end of civilized society, such arks may be regarded with suspicion. Strategies that emphasize survival of a few at the expense of many are unlikely to get broad support. If anything, they may prejudice us against expenditures on natural catastrophe preparedness. Many of us might ask, if I'm not going to make it, why pay for someone else to survive? It's a good question.

Furthermore, the attacks of September 11 demonstrated convincingly that it's difficult for the wealthy to isolate themselves from dangers that threaten the broader population. Many who died in the World Trade Center were high-income earners. Many more were exposed to toxic particles.

This event confirmed that if wealthier individuals want to continue to move about comfortably in society, they have to help to make that overall society—rather than just their immediate surroundings—safer for themselves to live and work.

Thus, our leaders have to do everything possible to demonstrate that we're in this together, not only as a way to get support for going ahead, but also to avoid a truly dark side that can emerge with the onset of molecular technologies.

The truly dark side is that we may not need many survivors to reprop-agate. As discussed earlier, Unabomber Theodore Kaczynski described a world in which an elite decides that the mass of the population is no longer required, and keeps them in a genetically manipulated state of subservience or eliminates them altogether. Molecular technologies could provide the means to do this. Thus, in an era of ethnic struggle that dates from Nazi Ger-many through to Pol Pot, Bosnia, Rwanda, and the Congo, we're left with that gnawing postholocaust feeling that somewhere, some powerful people are privately thinking that such mass extinctions might not be such a bad way to control our seemingly runaway population.

To advance as a species, we cannot allow ourselves to pass that way, and the only thing to prevent it is our own human determination. We must plan to survive together, by learning to use molecular science to save large populations. We might consider, for example, a variation on the constitutional right to life—the right to mass survival. *The Humanist Manifesto 2000*, endorsed by a broad cross section of scientists, artists, and Nobel prize winners, alludes to such a theme. The manifesto doesn't specify using molecular technology for mass sur-vival, but it does provide an ethical framework in which that might occur:

> The overriding need of the world community today is to develop a new Planetary Humanism—one that seeks to preserve human rights and enhance human freedom and dignity, but also emphasizes our commit-ment to humanity as a whole.[42]

When this is taken in the context of our vulnerability to big natural dis-asters, we see that right now, our commitment to humanity as a whole is seriously hampered by the inability of our earthquake-resistant, flood-con-trolling, and wind-resistant designs to cope with the scale of superthreats just being discovered. We are not capable of assuring mass survival under such conditions. Just think of Comet Shoemaker-Levy 9 plowing into Earth instead of Jupiter. Such a possibility is one of the main reasons we require an explicit doctrine of mass survival. Therefore, *it's time for humanists to con-sider such threats, and what we might do to deal with them.*

Furthermore, a new perspective on mass survival has to be considered. If, as it seems, we'll be inventing self-aware machines, we need to consider their survival. If we start manufacturing thinking or feeling machines, then switch them off as we please, our commitment to our own mass survival may be compromised. This would be especially true if the line between human and machine becomes blurred. The day is approaching when we share the earth with entities that, if not as intelligent as ourselves in every way at first, may at least possess many of our cognitive and emotional traits. Thus, defining "consciousness" may be one of our next great ethical challenges.

We'll also have to establish new principles for mass survival of animal species that we're only beginning to learn have emotions. Molecular technology may render obsolete the age-old practices of killing such animals. If we can synthesize food without slaughtering billions of animals, and synthesize drugs without causing suffering to billions of lab animals, then killing them en masse as we do now becomes a barbaric act of butchery.

Why are such ethical questions so important to our molecular future?

As molecular technologies begin to enhance our power, we may start to seriously consider the concept of populating this solar system, and perhaps traveling outside it. At that point, we must ask ourselves, Do we want to populate this part of the galaxy with species that, on one hand, have a brilliant technological capability, but on the other, kill unnecessarily? Moreover, as we create machines with high intelligence, do we want to imbue them with such traits? What's to stop them from concluding that we're just another part of the chain, and should be treated just as we treat other animals and thinking machines? We have to begin considering how alien species—not those from other planets, but those that we create—would view us.

REDESIGNING GREEN GODS
FOR MASS SURVIVAL

The molecular revolution and new discoveries about natural earth cycles may each force a rewrite of the book on "sustainability."

If an analysis of La Palma shows that miles of rock are about to fall into the Atlantic Ocean and send a giant wave across to North America and the Caribbean, how would we apply the precautionary principle? Would environmentalists try to stop preemptive bombing of the island because it threatens a marine sanctuary? There's not one practical game plan in the environmentalist arsenal for dealing with such threats. Governments and international disaster relief organizations are unequipped to deal with such environmental realities. Nor is it adequate to say to the millions of people who live in coastal tsunami zones, "Don't live there."

Thus, conceptual underpinnings such as the precautionary principle may require rethinking. We have to compare the relative risks of adaptive technologies with the risks of truly large natural catastrophes. This may require a revision of environmental risk assessment methods.

THINGS WE CAN DO NOW

Besides considering the questions asked and principles outlined earlier, here are suggestions to promote a doctrine of mass survival. These may help to instill confidence in our molecular future.

Find a Cure for an Endemic Disease and Make It Affordable

The discovery of nanobacteria infections, as described earlier, suggests that we may have breached the barriers of diseases that were once thought to be an inescapable fact of life. The jury isn't in yet, but the way that nanobacteria research progresses may have important consequences for health care, not only in advanced economies, but also in developing nations, where heart disease and stroke together are the greatest killers.[43] The relatively small amount of ink devoted to nanobacteria in this book should not detract from its potential implications. The outcome of tests for detection and cures is awaited by many.

Furthermore, few economic or health disadvantages weigh on tropical populations as does malaria. Unlike AIDS, it is a disease of warm climates. It represents a clear and present threat to everyone who visits such regions. Unlike heart ailments, it has received far too little attention, while it has decimated millions of lives. Being an undertreated affliction of developing nations, it is commonly referred to as an "orphan disease." Yet we now have the funding mechanisms in place, the genetic arsenal, and the growing public awareness to conquer this insidious enemy. By so doing, science could convincingly demonstrate that molecular technologies aren't only for rich, temperate climate countries: They can also produce "mass survival" benefits for tropical climate populations. Yet finding a cure isn't enough on its own. Drug companies, national governments, and international agencies need to make sure the cure is deployed rather than restricted by cost. This ethically sound step would go a long way toward proving good faith by those who control the technologies that have the power to cure disease.

Start a Transhuman Rights Movement

Finding a cure for malaria is a first step toward showing that all of us are in this together, but the next step is to redefine what "all of us" means. Should Robo sapiens have rights? What about Robo servers, who feel pain?

Governments are incapable of moving fast enough to cope with such questions. By the time they get around to holding hearings on a particular technology, it may have already achieved dominance. We cannot depend on national governments to show leadership in this field. Technology moves too fast, and much of the power lies outside the government domain at the multinational-corporate and local levels. Instead, we need to develop interest groups that influence governments and corporations at every level.

One candidate is the human rights movement, which is effective at combining practical goals with fundamental principles. Yet human rights principles also have to evolve to encompass the rights of artificially intelligent machines. The nongovernmental organizations are best positioned to begin thinking about this.

These principles also have to be broadened to include consideration of how megadisasters infringe on human rights. Some works have been published on impact of disasters on human rights.[44] These warrant further efforts. Collaboration among human rights organizations, disaster relief organizations such as the Red Cross, and forward-looking organizations such as the Foresight Institute, the World Future Society, and the Union of Concerned Scientists might start us on the path to defining such a transhuman rights movement.

At the same time, we might consider how to guarantee the right of survival for the whole of humanity, transhumans, and other life forms together, by addressing the greatest wild-card threat to life on Earth: rogue near-Earth objects.

Get Excited about Something Big: Build a Molecular Assembler

There is withering skepticism in much of the scientific community about whether the molecular assembler is feasible. Balanced against this is a growing group of technologists who are convinced it's realistic. Yet aside from a few entrepreneurs and academics, few have sat down and mapped out what it might take for us to develop an assembler in a short time. Those who have done this aren't being given the resources they need to prove their case. New funding initiatives for nanotechnology may help to remedy this deficiency, but it's just a first step.

To solve this, we might consider the approach that led to our ability to determine longitude, which in turn opened the world shipping lanes to global commerce.

In 1714, following the lead of several other governments, the British Parliament passed the Longitude Act, offering twenty thousand pounds—the equivalent of hundreds of millions of inflation-adjusted dollars today—to the inventor of a "Practicable and Useful" means to determine longitude. The incentive worked. And it was done by one man, John Harrison,[45] who built on a foundation of the many inventions that came before his.

One individual probably will not invent the assembler, but the incentive might inspire a team to achieve it. Today, several hundred individuals and institutions have the financial resources to offer such a prize: say, a billion dollars for the first functioning molecular assembler.

Furthermore, the test of such an assembler would be its ability to disassemble an asteroid that was similar to, but smaller than, the one that the NEAR satellite landed on. We have the delivery vehicle (see Fig. 32). We now need the assembler. This would convincingly erase one of the biggest point source threats to survival of every species on Earth: an asteroid hit. It would also trigger trillions of dollars worth of manufacturing applications, help to eliminate poverty at much lower cost, and contribute enormously to our intellectual capital.

As with the Cold War race to the moon, we have an identifiable enemy to inspire us: that deadly yet still undetected near-Earth object that has our planet's name written on it. This is the type of inspiration we may require to get to the molecular assembler.

Let's solve our student-scientist deficit by turning teenagers on to the ultimate challenge: Build the first molecular assembler and *truly* save the world.

The assembler challenge has the added attraction of involving potentially every basic scientific discipline, plus many nonscientific disciplines. It offers many pathways to the prize: chemistry, biology, physics, mathematics, ecology, computer sciences, virology, electronics, nanotechnology, and animation are just a few. Technology industry leaders might have the time of their lives chasing "the Prize," along with its many fringe benefits. Groups such as Physicians for Social Responsibility and the Union of Concerned Scientists might help.

Let's go!

Start New Branches of Economics

When digital fabrication becomes widespread, the whole economy could be transformed. Such manufacturing doesn't require molecular assembly to get going. It's already here. Companies such as those mentioned earlier already have the machines. Therefore, it's time to start a new division of economics that focuses on digital fabrication. One place for economists to begin is by talking to companies that are building the fabricators. They might graduate to more future-oriented companies, such as Zyvex, which work on molecular assembly. Such a discipline could then be used as an entry into another division of economics: superdisaster preparedness. How does an economy prepare for, or recover from, a big hit? By answering such questions, we may be able to devise economic theories for how to survive. Much work is being done on the economic implications of climate change, but it seems nothing is being done on the economic implications of, for example, a super-tsunami on the globalized banking system. Banks and risk management companies have already done work on the implications of earthquakes for centers such as Tokyo and New York. Much experience has also come out of the World Trade Center attacks. Thus, there is a foundation to build on, to plan for economic survival.

Redefine Earth As Being an Open Ecosystem

Science has shown us that Earth is not closed to the effects of matter from space. Near-Earth objects have changed our planet's history, by decimating some species and leading to the ascendancy of other ones. Moreover, the effects of gravity, radiation, and other forms of energy from the Sun, the Moon, planets, asteroids, and comets are becoming more apparent as our detection methods improve. Therefore, it's time to release ourselves from the conceptual straitjacket of closed systems. Earth is, and always has been, an open ecosystem. It is open to materials and energy from the universe. The fact that not much matter seems to leave Earth once it arrives should not distract us from other, open-ended characteristics. Such a revision to environmentalist thought might open our minds to a realization that the ecology that sustains us is an intimate part of space, rather than separate from it. Perhaps then we may be more open to exploring space, and more capable of developing technologies that protect us from its more devastating phenomena.

CONCLUSION

Robo sapiens, Robo servers, and *Homo provectus*. They may be on the way. Are we ready? Who is deciding the types of consciousness we create? When was the last time a genetic modification of a species was put to a vote? Some governments have banned human cloning, stem cell research, and other types of life tinkering, but who did the banning? Who took it upon themselves to make these incredibly basic decisions for the rest of humanity?

Moreover, what might stop us from getting to the point of such sophisticated technology? What may have led to climate flips millennia ago? What might we do to adapt to them now? Are conscious machines prerequisites for letting us cope with such changes, or are they threats to our species? How might we make these elephants in the room of our existence–disruptive technologies and natural disruptions–dance together so that we improve our chances of surviving them both?

Most of us are left out of decisions on what types of consciousness are being created. Our leaders are explaining them to us inadequately. The stage has been taken over by special interests and extremist factions. There is no coherent direction. We get only clichés, such as "It will cure disease" or "It will kill us all." Moreover, when was the last time we heard a right-to-life group say, "We support the right to life for artificial intelligence, and we're going to push politicians to define it." When was the last time we heard an environmental group say, "We think that slave robots with limited intelligence are a good idea if they help clean up the ecology." When was the last time we heard the animal rights movement try to explain "What is an animal," in the context of artificial intelligence?

Already, an elite group of *Homo sapiens* is in our midst whose power is enhanced by the analytical capabilities of supercomputers. This group, though it is by no means unified, manages financial, medical, and communications empires with tools that the rest of us don't have. Each of these tools influences our lives heavily.

Thus, we might be already branching off into superspecies, with our leaders asleep at the switch. No cohesive policy exists for answering the defining questions of our time: Do we want artificial intelligence, and if yes, for what? Are we ready to have alien beings that are smarter than us walking around? Are we ready to have another branch of computer- and prosthetic-enhanced *Homo sapiens* running our businesses and governments? Are we ready for the shock of decommissioning a pet robo dog that has greater intelligence than our flesh-and-blood canine? Would this new type of pet let us perform such a mechanistic termination, or might it first plead for mercy? If we refuse, might it escape to join other renegade robots, as did the boy robot David in the film *A.I.*?

In these discussions, no one is representing "We, the people." Our governments aren't asking these questions, or if they are, they don't seem to be sharing them with us. Companies are selling us the products and services that constitute this revolution, but such companies aren't structured to protect our interests. Most religious institutions are decades or centuries behind. Some powerful government agencies and environmental organizations say we should stop deploying technologies if they seem too risky. Most scientists are too busy inventing things to start considering how to govern their impacts.

Thus, in the absence of a coherent approach, it seems that we may have at least some of these life forms. Furthermore, we might have to risk their downsides if we're to find new ways of surviving the kinds of awful changes that nature hands us from time to time. We may need artificially enhanced intelligence to understand nature's mysteries and the complexities that we're unleashing upon ourselves. If we're too squeamish about that, let's remember that we have the beginnings of such enhanced intelligence, in the form of computers that help us run everything from cars to life support, and implants that replace defective parts of our bodies. Moreover, if we want to stop killing ourselves by the tens of millions in violent conflict, we're going to have to get much smarter, because our genes haven't been able to solve that by themselves.

This means addressing at least one probability: that the days of dominance by *Homo sapiens* may be numbered, either by nature's disruptive technology or our own. In a blink of the geological eye we may be surpassed by our own creations or annihilated by nature's. Right now, we're not ready.

We lack the policies and underlying discussions. We lack the spiritual and emotional preparation. We're misfits for what's coming and, if we don't watch out, we'll end up like Neanderthal–pushed off the evolutionary chart.

The offshoot of this lack of preparation is that we may wake up one day to find–oops!–we have some new form of life here. The computer that beat Garry Kasparov is now running for Congress. And our dog is begging us not to turn it off.

What might be the result of this lack of preparation? As in the movie *A.I.*, we might start killing our creations for fear that they will gain the upper hand. We may try to use them against each other as we've done for millennia with other weapons, but are they going to listen? When the Joint Chiefs tell an autonomous robot soldier to wipe out thousands of soldiers retreating along a desert road, will it perhaps refuse, or offer a better solution to humanely disable them–or, more disturbingly, agree that it's a fitting way to avenge what we may have lost?

Science fiction has for years described intelligent robots that take over management from humanity. Futurist writers envision everything from robot-human wars to benign dictatorships to peaceful coexistence. Yet most of those stories are still stuck in the science fiction pages or in relatively small groups of enthusiasts. They haven't migrated to the front pages of our collective mind.

It's time we had a serious public discussion about artificial life and what it means. It's time to take Asimov's Laws of Robotics out of the imagination and into the legislative agenda. It's time for political parties to develop an artificial-intelligence policy. And it's time for every one of us to wake up to this reality.

Then we must face the other side of the coin. We have to consider the absolutes that may prevent us from transcending to a more advanced stage by derailing us so severely that we get stuck in a repeating Dark Age loop. Natural disasters are such absolutes: the certainties that can toss us backward.

So we're running many races at once: We have to protect ourselves against natural onslaughts that we're learning to see, but don't yet know how to repel. We'll have to learn how to coexist with intelligent machines before they learn to control us. We may even have to learn how to become part of each other in order to survive.

The age of dominance by *Homo sapiens* on this planet may soon end. We may be superceding ourselves. This might not be a bad thing. It may be a good thing, because it might help solve some of our deepest problems. But we need a framework.

This book has suggested some questions to ask. Let's start with those. It also suggested some principles to start with:

Principle 1: Find ways to remember our past in the future.
Principle 2: Learn from the dark side.
Principle 3: Favor scientific freedom over intellectual-property restrictions.
Principle 4: Redesign democracy to cope with artificial intelligence.
Principle 5: Use technology to liberate the individual from fear.

Finally, this book has suggested some inspirational projects to get us started:

∞ Find a cure for an endemic disease and make it affordable.
∞ Start a transhuman rights movement.
∞ Get excited about something big: build a molecular assembler.
∞ Start new branches of economics.
∞ Redefine Earth as being an open ecosystem.

These are not principles by which to run a society. They do not preclude the development of other principles and projects. *They are only ways to help start us on the journey.*

The price we stand to pay for slowing our quest or ignoring its importance is dear. We need look back only a few generations, to when the Great Kanto earthquake changed world history, to see that civilization is at the mercy of natural phenomena. We must learn how to adapt to this seemingly endless natural cycle.

The link between how we cope with natural hazards and how we cope with artificial intelligence is strong. The way we deal with one is an indicator for how we'll cope with the other. If we use molecular technologies to guarantee mass survival in the face of nature's worst extremes, then we may be able to transcend our dark side and enter a new era. Let's begin to ask the questions posed in these pages. Rather than closing the cover, it might be worthwhile to go back to chapter 18–"The Right Questions"–and get started.

∞   ∞   ∞

For updates on the evolving technologies and debates described in this book, go to www.ourmolecularfuture.com on the World Wide Web.

APPENDIX A

ISAAC ASIMOV'S LAWS OF ROBOTICS

∞ **Law Zero:** A robot may not injure humanity, or, through inaction, allow humanity to come to harm.

∞ **Law One:** A robot may not injure a human being, or, through inaction, allow a human being to come to harm, unless this would violate a higher-order law.

∞ **Law Two:** A robot must obey orders given it by human beings, except where such orders would conflict with a higher-order law.

∞ **Law Three:** A robot must protect its own existence as long as such protection does not conflict with a higher-order law.[1]

EXCERPT FROM *FORESIGHT GUIDELINES ON MOLECULAR NANOTECHNOLOGY*

Original Version 1.0: February 21, 1999
Revised Draft Version 3.7: June 4, 2000

DEVELOPMENT PRINCIPLES

1. Artificial replicators must not be capable of replication in a natural, uncontrolled environment.
2. Evolution within the context of a self-replicating manufacturing system is discouraged.
3. Any replicated information should be error free.
4. MNT device designs should specifically limit proliferation and provide traceability of any replicating systems.
5. Developers should attempt to consider systematically the environmental consequences of the technology, and to limit these consequences to intended effects. This requires significant research on environmental models, risk management, as well as the theory, mechanisms, and experimental designs for built-in safeguard systems.
6. Industry self-regulation should be designed in whenever possible. Economic incentives could be provided through discounts on insurance policies for MNT development organizations that certify Guidelines compliance. Willingness to provide self-regulation should be one condition for access to advanced forms of the technology.
7. Distribution of molecular manufacturing *development* capability should be restricted, whenever possible, to responsible actors that have agreed to use the Guidelines. No such restriction need apply to end products of the development process that satisfy the Guidelines.[1]

Also, read the Foresight Position Statement on Avoiding High-Tech Terrorism, at: www.foresight.org/positions/proCoalition.html.

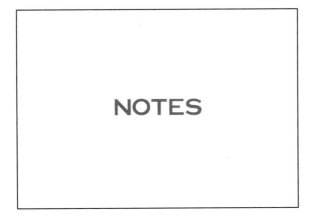

NOTES

A NOTE ABOUT THE NOTES

One of the great divides between scientists and laypersons is plain English. Most references cited in the pages that follow are from books, journals, and magazines that are one step removed from the highly technical realm of scientific journals. For example, references about E-Ink are from articles in technology magazines such as *MIT Technology Review*, company Web sites, and books written by the technology's creators. Underlying these publications is a vast supporting web of thousands of research papers that would take the length of this book to annotate. I deliberately stayed away from quoting these highly technical journal articles, not because they don't exist, but because I want lay readers to have access to easily understood background materials. If readers are interested to proceed beyond these, many of the annotated documents and Web sites contain references to supporting documentation in their own footnotes and bibliographies. In the section entitled "A Brief Sampling for the Scientifically Inclined Reader," I have included a brief list of technical works by some of the scientists referenced in this book. This list is very short, and does not include works that are already cited in the notes. Finally please note that Web page references sometimes go out of date because they are moved to other parts of a site, so it may require some detective work to locate older Web pages whose URLs are shown in the notes.

ABOUT THE TITLE

1. Alvin Toffler, *Future Shock* (New York: Bantam Books, 1974).
2. World Commission on Environment and Development, *Our Common Future* (Oxford and New York: Oxford University Press, 1987).
3. K. Eric Drexler, Chris Peterson, and Gayle Pergamit, *Unbounding the Future: The Nanotechnology Revolution* (New York: Quill William Morrow, 1991).
4. Theo Colborn, John Peterson Myers, and Dianne Dumanoski, *Our Stolen Future: How Man-Made Chemicals are Threatening our Fertility, Intelligence, and Survival* (London and New York: Little, Brown & Co., 1996).
5. Seth Shulman, *Owning the Future: Inside the Battles to Control the New Assets: Genes, Software, Databases, and Technological Know-How That Make Up the Lifeblood of the New Economy* (Boston: Houghton Mifflin, 1999).

INTRODUCTION: COLLISION OF FUTURES

1. Author's notes from a talk given at a Senior Associates' Induction session to nanotechnology chaired by Chris Peterson, president, The Foresight Institute, Palo Alto, Calif., September 8, 2000. Confirmed via e-mail correspondence with Chris Peterson, August 7, 2001.
2. "Honda's Robot Rings Opening Bell at NYSE," *Japan Times*, February 14, 2002 [online], www.japantoday.com/e/?content=news&cat=4&id=201219 [February 14, 2002].
3. More exact definitions for molecular and nanotechnologies are contained in chap. 2.
4. C. Burda et al., "Some Interesting Properties of Materials Confined in Time and Space of Different Shapes" (paper presented at the Eighth Foresight Conference on Molecular Nanotechnology, Bethesda, Md., November 3–5, 2000) [online], www.foresight.org/Conferences/MNT8/Abstracts/El-Sayed [August 10, 2001].
5. Dean Astumian, "Making Molecules into Motors," *Scientific American,* July 2001, 58, sidebar.
6. Edward Cornish, president of the World Future Society and editor of the *Futurist* magazine lists four types of continuity between the past and future: continuity of existence, change, pattern, and causation. This paragraph summarizes those types. The examples of each are my own. Edward Cornish, "How We Can Anticipate Future Events," *Futurist* 35, no. 4 (July–August 2001): 26.
7. Physicians are now training for surgery on virtual-reality machines that include virtual touch to replicate conditions during an operation. An example is at "UK Center for Minimally Invasive Surgery Begins Training," press release, University of Kentucky Chandra Medical Center, January 12, 2000 [online], www.mc.uky.edu/mcpr/news/2000/January/mis.htm [July 12, 2001].
8. Douglas Adams, *The Hitchhiker's Guide to the Galaxy* (New York: Pocket Books, 1981), p. 35. Arthur Dent, the main character, has a particularly bad day,

beginning with a notice that his house will be torn down, then deteriorating to a notice that Earth is going to be expropriated. This story is revisited many times throughout Adams's books.

9. See part 2 of this book for examples.

10. Many educational Web sites on the Internet describe Earth as a "closed system." Some explain that Earth is open to energy, but such energy is usually described as only emanating from the Sun, and being reflected by Earth. They usually make no mention of gravitational pull from the Moon, inflows of cosmic dust, or the significant input of energy from large meteor and asteroid collisions. Some sites mention meteors, but usually as having inconsequential effects on Earth as a "closed system." For example, Nike's Air to Earth site states that "the earth is a closed system: Practically nothing comes in, with the exception of energy in the form of heat and light from the sun, and nothing leaves except heat and reflected light." Air to Earth, "Lesson 1" [online], nikebiz.com/environ/lesson_1.shtml [September 8, 2001].

In its "Guide to Courses," PBS Adult Learning Services introduces a course segment with this statement: "In the past decade mankind has been made critically aware that Spaceship Earth is a closed system." PBS Adult Learning Services [online], www.pbs.org/als/guide/courselistings/courses/planet_earth/pltedescrip. htm [September 8, 2001].

"Kids for the Environment," sponsored by the Foster Foundation, Toyota, and Tattersalls, states that "A closed system means that all matter (physical things, rocks, trees, even you!), apart from a few rockets and meteorites, stay on earth. . . . Only energy, in the form of sunlight, enters the system called Earth and leaves it in the form of heat." "Build a Terrarium" [online], www.kids-for-the-environment.com.au/play/terrerium/pop_ terr1.html [September 8, 2001].

Moreover, none of these sites mention findings, such as investigations into the LINEAR comet, that buttress theories of how comets may have contributed to filling the earth's oceans: " 'The idea that comets seeded life on Earth with water and essential molecular building blocks is hotly debated, and for the first time, we have seen a comet with the right composition to do the job,' said Dr. Michael Mumma of NASA's Goddard Space Flight Center." "First Evidence that Comets Filled the Oceans: A Dying Comet's Kin may have Nourished Life on Earth," *ScienceDaily*, May 21, 2001 [online], www.sciencedaily.com/releases/2001/05/010521072649.htm [July 7, 2001].

See also "Sweet Meteorites. Scientists Have Discovered Sugars in a Meteorite, Adding to the List of Complex Organic Molecules That Have Been Found Inside Space Rocks," press release, NASA, Science@NASA [online], science.nasa.gov/headlines/y2001/ast20dec_1.htm [December 20, 2001]. A research team led by Dr. George Cooper at the NASA Ames Research Center in California discovered sugar-related compounds on meteorites that landed in Australia and Texas. Such compounds are building blocks for DNA and cell membranes. The discovery suggests that life on Earth may have been kick-started by space objects. See also Margaret Munro, "Sugar and Spice and the Whole Human Race, Meteorite Discovery Suggests Life Did Not Fully Begin on Earth," *National Post*, December 20, 2001, p. A3.

CHAPTER 1: SINGULARITY

1. Vernor Vinge, "The Coming Technological Singularity: How to Survive in the Post-Human Era" (paper presented at the VISION-21 Symposium sponsored by NASA's Lewis Research Center and the Ohio Aerospace Institute, March 30–31, 1993) [online], www-rohan.sdsu.edu/faculty/vinge/misc/singularity.html [July 7, 2001].

2. Gloria Chang, "World Chess Champion Loses Game 2 Against Computer 'Deep Blue' Amidst Media Frenzy," Discovery Channel Canada, May 6, 1997, [online], exn.ca/Stories/1997/05/05/01.asp [July 7, 2001]; interview with the *Globe and Mail* after Kasparov lost his second game in a six-game match against IBM's Deep Blue computer. Kasparov has made many other references to computers having alien intelligence.

3. In this scene the intelligent robot Gigolo Joe is explaining to the main character, child-robot David, why humans are trying to destroy them and why robots are successfully evading that destruction.

4. For an overview of some of the players in the Singularity discussion, see: Steve Alan Edwards, "Mind Children: Extropians," *21C–Scanning the Future* (April 1, 1997). Available online as "Surviving the Singularity" at members.aol.com/salaned/writings/survive.htm [August 27, 2001].

5. Ibid.

6. A list of scientists who've discussed some form of singularity occurring in our lifetimes is found at the Singularity Watch Web site developed by John Smart at members.home.net/marlon1/history_brief.html [July 7, 2001].

7. Ray Kurzweil, "The Age of Spiritiual Machines" (sermon to the First Unitarian Universalist Church of San Diego, January 23, 2000) [online], firstuusandiego.org/public/sermons/012300Kurzweil.ram [July 11, 2001].

8. Damien Broderick, *The Spike: How Our Lives are being Transformed by Rapidly Advancing Technologies* (New York: Forge/Tom Dougherty, 2001). Broderick argues that a "spike" in technological development could cause human obsolescence, transformation or transcendence when superintelligent, immortal machines supersede humanity. This theme was also explored at an International Futures Conference in Perth, Australia, in the year 2000, as described in an article by Richard Eckersley, "Doomsday Scenarios: How the World May Go On Without Us," *Futurist* 35, no. 6 (November/December 2001): 20–23. Here, the term "evolutionary pink slip" is used to describe the potential fate of *Homo sapiens* by the year 2050. See also Ben Bova, *Immortality. How Science Is Extending Your Life Span and Changing Our World* (New York: Avon Books, 1998). Renowned science and science fiction writer Ben Bova catalogues technologies that may cause us to redefine longevity. Together, these articles, conferences, and books suggest that the issue of transforming the human species is rapidly emerging from the scientific backwoods into popular consciousness.

9. Several challenges exist to the conventionally accepted time line. Some argue that *Homo sapiens* has been around for much longer and reached a high level of sophistication until various cataclysms sent us back to the stone ages. However, even if we accept this counterview, we still have relatively reliable indicators about the progress of technology for the past twelve thousand years.

10. Such time lines are more completely discussed in Ray Kurzweil, *The Age of Spiritual Machines* (New York: Penguin Books, 1999), pp. 261–80.

11. For an explanation of how and when this might occur, see Nick Bostrom, "How Long Before Superintelligence?" *International Journal of Futures Studies* 2 (1998), [online], www.systems.org/HTML/fsj-v02/superintelligence.htm [September 9, 2001]. For a description of the necessary preconditions for superintelligent AI, see The Singularity Institute for Artificial Intelligence, "General Intelligence and Seed A.I.: Creating Complete Minds Capable of Open-Ended Self-Improvement 2.3," May 18, 2001 [online], singinst.org/printable-GISAI.html [November 19, 2001]. This paper was written largely by prodigy Eliezer S. Yudkowsky, although the copyright has been assigned to the institute in the updated version 2.3.

12. Kit Sims Taylor, "Material Life, Markets and Capitalism," *Human Society and the Global Economy* [online], distance-ed.bcc.ctc.edu/econ100/ksttext/systems/economicsystems.htm [July 15, 2001].

CHAPTER 2: MOLECULAR BUILDING BLOCKS

1. Marcus Hewat, "Properties of Materials. Why is Atomic Structure Important?" [online], www.ill.fr/dif/3D-crystals/intro.html [July 11, 2001]. This Web site is recommended by the National Science Teachers Association. Hewat is a young French/Australian computer scientist specializing in three-dimensional imaging of atomic scale matter. He developed the aforementioned Web page as part of an educational Web site called Making Matter. The Atomic Structure of Materials, www.ill.fr/dif/3D-crystals/index.html [July 11, 2001].

2. The definitions shown here are general ones. There are many different technical definitions for each of these scientific disciplines, as we'll see later in this chapter. A precise, broadly accepted definition of robotics, for example, is yet to be established.

3. Richard P. Feynman, "An Outsider's Inside View of the Challenger Inquiry," *Physics Today* 41 (February 1988): 26–37.

4. Richard P. Feynman, "There's Plenty of Room at the Bottom: An Invitation to Enter a New Field of Physics" (lecture given to the American Physical Society, California Institute of Technology, December 29, 1959) [online], www.physics.umn.edu/groups/mmc/personnel/pete/There%20is%20plenty%20of%20room%20at%20the%20bottom.htm [July 7, 2001].

5. Ibid.

6. Ibid.

7. Ibid.

8. Some dates taken from Technanogy Web site chronology, www.technanogy.net/inspire/chronology.html [July 15, 2001].

9. K. Eric Drexler, "Molecular Engineering: An Approach to the Development of General Capabilities for Molecular Manipulation," *Proceedings of the National Academy of Sciences U.S.A.* 78, no. 9 (1981): 5275–78.

10. This sequence was described to me in an e-mail by Chris Peterson, August 7, 2001.

11. K. Eric Drexler, *Engines of Creation: The Coming Era of Nanotechnology* (New York: Doubleday, 1986).

12. K. Eric Drexler, *Nanosystems Molecular Machinery, Manufacturing and Computation* (New York: John Wiley & Sons, 1992).

13. K. Eric Drexler, Chris Peterson, and Gayle Pergamit, *Unbounding the Future: The Nanotechnology Revolution* (New York: Quill William Morrow, 1991).

14. Mitch Jacoby, "New Tools for Tiny Jobs," *Chemical & Engineering News* (October 16, 2000) [online], pubs.acs.org/cen/nanotechnology/7842/7842 instrumentation.html [August 15, 2001].

15. Gary Stix, "A Few 10^{-9} Milestones," *Scientific American*, September 2001, 36, sidebar.

16. Ibid.

17. "Researchers from Lucent Technologies' Bell Labs and University of Oxford Create First DNA Motors," press release, Lucent Technologies, August 9, 2000 [online], www.lucent.com/press/0800/000809.bla.html [August 15, 2001].

18. Maggie Fox, "Tiny Computers, Genome Project Top Year in Science," *Vancouver Sun*, December 21, 2001, A16.

19. "Scientists Build Tiny Computer Using DNA Molecules," *New York Times on the Web*, November 27, 2001, www.nytimes.com/2001/11/27/science/physical/27DNA.html [November 28, 2001].

20. Caroline Humer, "IBM Milestone Could Lead to Smaller Chips," *Los Angeles Times*, August 27, 2001, p. C2.

21. Richard Shim and John G. Spooner, "IBM Pixie Dust Breaks Hard Drive Barrier," CNET, May 21, 2001 [online], news.cnet.com/news/0-1003-200-5976693.html?tag=mn_hd [July 15, 2001].

22. "Bell Labs Scientists Build the World's Smallest Transistor, Paving the Way for 'Nanoelectronics,' " press release, Bell Labs [online], www.bell-labs.com/news/2001/november/8/1.html [November 8, 2001].

23. "Mitsui to Build World's Largest Nanotube Plant," Fiji Press English News Service [online], www.smalltimes.com/document_display.cfm?document_id=2814 [December 27, 2001].

24. "New World of Nanoelectronics May Arrive in the Near Future, AAAS Speakers Say," press release, American Association for the Advancement of Science, February 14, 2002 [online], www.eurekalert.org/pub_releases/2002-02/aaft-nwo 020602.php [February 15, 2002].

25. Lothar Kuhn, "Muehsame Fummelei–Forscher formen aus den kleinsten Teilchen der Welt Produkte mit wundersamen Eigenschaften," *Wirtschaftswoche* 52 (December 21, 2000): 143.

26. Will McCarthy, "Ultimate Alchemy," *Wired*, October 2001, 150–183.

27. Many billions of dollars have gone into investments for molecular chemistry, micro-electro-mechanical systems, biotechnology, and scanning tunneling microscopes, which together constitute the building blocks for molecular nanotechnology.

28. National Nanotechnology Initiative Web site, www.nano.gov/ [August 15, 2001].

29. California Nanosystems Institute Web site, www.cnsi.ucla.edu [December 31, 2001].

30. "NSF Will Establish Six New Nanotech Centers," *Foresight Update*, no. 46 (September 30, 2001): 1.

31. For a list of European Union nanotechnology research projects, see the CORDIS Web sites at www.cordis.lu/nanotechnology/src/sitemap.htm and www.cordis.lu/ist/fetnid.htm [August 15, 2001].

32. Tomoji Kawai, "Nanotechnology in Japan: Present Situation and Outlook" (paper given at the Foreign Press Center/Japan, March 9, 2001) [online], www.fpcj.jp/e/gyouji/br/2001/010309.html [August 15, 2001].

33. "Nanotechnology Takes Off Worldwide: Nanoscale science and engineering moves at the frontier of interdisciplinary research," National Science Foundation News Update, April 10, 1998 [online], www.eng.nsf.gov/engnews/1998_news/nanotechnology_takes_off_world.htm.

34. "China Should Temper Nano Enthusiasm with Sober Thinking, Researcher Says," *Asiaport Daily News* (Shanghai), July 19, 2001 [online], www.smalltimes.com/document_display.cfm?document_id=1706&section_id=53 [July 15, 2001].

35. Nano Nagle was born at Ballygriffin, Ireland, in 1718. Her foundation spread to become one of the most respected of the teaching profession and the cause of Nano Nagle for beatification is now being actively pursued. See www.mallow.ie/tourist/n-n.html [December 27, 2001].

36. The Foresight Institute Web site is at www.foresight.org [September 10, 2001].

37. "Molecular Nanotechnology: Thorough, inexpensive control of the structure of matter, based on molecule-by-molecule control of products and byproducts; the products and processes of molecular manufacturing, including molecular machinery." Drexler, Peterson, Pergamit, *Unbounding the Future*, p. 294.

38. Ibid., p. 293.

39. George M. Whitesides, "The Once and Future Nanomachine," *Scientific American*, September 2001, 78–83. Nobel prize laureate Richard Smalley is also a skeptic. In the same issue of *Scientific American*, he identifies the "fat fingers," i.e., too many arms that are too big to pick up atoms, and "sticky fingers," i.e., the tendency of atoms to stick to manipulators, as "fundamental . . . neither can be avoided." Richard E. Smalley, "Of Chemistry, Love and Nanobots," *Scientific American*, September 2001, 77.

40. As a rebuttal to Smalley and other skeptics, in the September 2001 issue of *Scientific American* K. Eric Drexler counters that "the need for such a large number of manipulators . . . has never been established or even seriously argued. In contrast, the designs that have received (and survived) the most peer review use one tool at a time and grip their tools without using any fingers at all." K. Eric Drexler, "Machine-Phase Nanotechnology," *Scientific American*, September 2001, 74.

After the *Scientific American* issue was published, a more substantive rebuttal to Smalley and other critics was published by the Institute for Molecular Manufacturing. This is one of the most concise arguments to date for the practicability of non-biological molecular assembly. K. Eric Drexler et al., "On Physics, Fundamentals, and Nanorobots: A Rebuttal to Smalley's Assertion that Self-Replicating Mechanical

Nanorobots Are Simply Not Possible," Institute for Molecular Manufacturing, 2001 [online], www.imm.org/SciAmDebate2/smalley.html [November 18, 2001].

41. The company is Zyvex Corporation. See company Web site at www.zyvex. com [August 15, 2001].

42. Fiona Harvey, "Advances with Tiny Computers," *Financial Times*, October 11, 2001 [online], globalarchive.ft.com/globalarchive/articles.html?print=true&id= 011011002640 [November 20, 2001].

43. Drexler, Peterson, and Pergamit, *Unbounding the Future*, p. 19.

44. James Von Ehr, interview by *Wall Street Reporter*, December 29, 2000. This interview was indexed and broadcast at the *Wall Street Reporter* Web site, www. wallstreetreporter.com/html2/interviewsa-z/archive_main.htm but had no transcript. It was transcribed by the author. Note: Calculations of the strength of carbon nanotubes vary, depending on what materials may have to be combined with them in structures. For example, some scientists calculate thirty times the strength of steel for carbon nanotube composites.

45. Drexler, *Engines of Creation*, p. 7.

46. Michael Roukes, "Plenty of Room, Indeed," *Scientific American*, September 2001, 55.

47. Chad Mirkin, "Nanotechnology: Fact or Fiction," *Chemical & Engineering News* (March 26, 2001): 185.

48. Aaron Hand, "Soft Lithography Prints TFTs on Curved Substrates," *Semiconductor International* (March 2001) [online], www.semiconductor.net/semiconductor/ issues/issues/2001/200103/six010307ln.asp [September 5, 2001].

49. These three processes were adapted from Introduction to Fabbers, a Web page of the Ennex Corporation, www.ennex.com/fabbers/fabbers.sht [July 15, 2001].

50. Ibid.

51. The Z Corporation Web site is at www.zcorp.com [September 5, 2001].

52. "Personal Fabrication on Demand," *Wired*, April 9, 2001 [online], www. wired.com/wired/archive/9.04/fab_pr.html [July 7, 2001]. Moreover, San Diego Supercomputer Center's Tele-Manufacturing Facility has a fabrication lab dedicated to academic model making along similar lines. See Stacey Smith Lang, "Skeletal Key," *Wired*, November 2001, 49.

53. "Personal Fabrication on Demand," *Wired*, April 9, 2001.

54. For a discussion about peer-to-peer sharing of fabrication software, much in the same way that peer-to-peer music sharing happens with file-sharing software, see Marshall Burns and James Howison, "Napster Fabbing: Internet Delivery of Physical Products" (Business Track Presentation at O'Reilly Peer-to-Peer Conference, San Francisco, February 16, 2001) [online], www.ennex.com/publish/200102-Napster/ index.sht [September 5, 2001].

55. Examples taken from Introduction to Fabbers–Fabber Applications, a Web site of the Ennex Corporation, www.ennex. com/fabbers/fabbers.sht [July 15, 2001]. See also Z Corporation case studies, www.zcorp.com/flash/index.html [September 5, 2001].

56. "Personal Fabrication on Demand," www.wired.com/wired/archive/9.04/ fab_pr.html.

57. Ibid.

58. Ibid.

59. See nanocell project description on James M. Tour's Web site, www.jmtour. com/info.htm [September 8, 2001].

60. Stephen Mihm, "Print Your Next PC," *MIT Technology Review* (November/December 2000) [online], www.technologyreview.com/articles/Mihm1100.asp [July 7, 2001].

61. "Xerox PARC and 3M Collaborate to Bring to Market Ground-Breaking Electronic Paper," press release, Xerox, June 29, 1999, [online], www2.xerox.com/ go/xrx/newsroom/T_release1.jsp?page=News+Item+2&release=Xerox+PARC+ and+3M+Collaborate+to+Bring+to+Market+Ground-Breaking+ Electronic+Paper [July 7, 2001]. Nicholas K. Sheridon is credited with developing one of the initial visions for electronic paper, but Jacobson's team also came up with a different version independently. See Steve Ditlea, "The Electronic Paper Chase," *Scientific American*, November 2001, 50–55.

62. Jim Battey, "Electronic Paper Gets Its Bearing," *Infoworld* (April 13, 2001) [online], www2.infoworld.com/articles/hn/xml/01/04/16/010416hnetrend.xml? Template=/storypages/printfriendly.html [July 7, 2001].

63. Steve Ditlea, "The Electronic Paper Chase," *Scientific American*, November 2001, 55.

64. "Xerox PARC and 3M Collaborate to Bring to Market Ground-Breaking Electronic Paper," press release, Xerox, July 13, 1999 [online], www.globalprint. com/eng/news/newsanzeigen.jsp?lang=en&news=2143; and "What is Electronic Ink?" E Ink Corporation [online], www.eink.com/technology/index.htm [June 7, 2001].

65. Steve Silberman, "The Hot New Medium: Paper," *Wired*, April 2001 [online], www.wired.com/wired/archive/9.04/anoto.html [December 31, 2001].

66. Drexler, *Engines of Creation*, p. 94.

67. Tom Fowler, "Professor Awaits Breakthrough; Nanotechnology Development May Drastically Alter Computing," *Houston Chronicle*, August 14, 2000 [online] amtexpo.com/nano/messages/1271.html [July 7, 2001].

CHAPTER 3: WHAT COMES FIRST AND WHAT DOES IT MEAN FOR EACH OF US?

1. Rachael Emma Silverman, "And Those Who Predicted What It Would Look Like Were Often Wrong," *Wall Street Journal*, January 1, 2000, p. R5. Silverman is quoting Asimov from *The Book of Predictions*.

2. Ibid.

3. Subsidies to the fossil fuel industry are broadly recognized inside and outside the environmental sector as a hindrance to the adoption of efficient technologies. For an executive summary see Roland Hwang, *Money Down the Pipeline: Uncovering the Hidden Subsidies to the Oil Industry—How tax breaks, government funding, and*

other subsidies benefit the oil industry at the expense of other energy technologies (1995) [online], www.ucsusa.org/vehicles/pipeline.html [July 7, 2001].

4. A large statistical base exists on barriers posed by subsidies to obsolete industries, but is aptly summarized by Maurice Strong, *Where on Earth Are We Going?* (New York: Texere, 2000), p. 375.

5. Simon Romero, "Once Bright Future of Optical Fiber Dims," *New York Times,* June 18, 2001 [online–*Note:* Web article access requires registration], www.ny times.com/2001/06/18/technology/18MELT.html?ei=5035&en=0e80925de 1525143&ex=993528000&partner=MARKETWATCH&pagewanted=print [July 9, 2001].

6. Ibid.

7. "Polaroid Files Voluntary Chapter 11 Petition, Receives $50 Million in New Financing," press release, Polaroid Corporation, October 12, 2001 [online], biz. yahoo.com/prnews/011012/nef010_1.html [October 30, 2001].

8. Ivan Berger, "ISP's Demise Brings a Bumpy Switchover," *New York Times,* January 10, 2002 [online], query.nytimes.com/search/abstract?res=F30713F6385D0 C738DDDA80894DA404482 [January 26, 2002].

9. Bill Keller, "Enron for Dummies," *New York Times,* January 26, 2002 [online], www.nytimes.com/2002/01/26/opinion/26KELL.html?pagewanted=print: "So where did Enron go wrong? . . . It figured if it could trade energy, it could trade anything, anywhere, in the new virtual marketplace. Newsprint. Television advertising time. Insurance risk. High-speed data transmission. All of these were converted into contracts–called derivatives–that were sold to investors."

10. Thomas L. Friedman, "Cyber Serfdom," *New York Times,* January 30, 2001, Op-Ed.

11. Patients are desperate for transplants. The first study of its kind released in October 2001 revealed that in Canada–one of the most advanced nations medically–about a quarter of patients that need new hearts die waiting for them. The study reveals that the problem is worsening. See "Status of Cardiac Transplantation Detailed in CCS Consensus Conference Report," *Info Cardio, The Official Newspaper of the Annual Canadian Cardiovascular Congress, October 20–24, 2001, Halifax, Nova Scotia,* p. 7 [online], www.ccs.ca/society/congress2001/infocardio/InfoCardio.pdf [December 3, 2001].

12. "What are Nanobacteria?" Nanobaclabs Web site, nanobaclabs.com/ [November 20, 2001]. The existence of nanobacteria has been broadly confirmed by medical laboratory testing. Numerous symposia and scientific studies about nanobacteria are referenced at this Web site.

13. Paul W. Ewald, *Evolution of Infectious Disease* (New York: Oxford University Press, 1994). Ewald has argued for years that many illnesses such as heart disease have infection at their roots. For a summary, see Steve Mirsky, "A Host with Infectious Ideas," *Scientific American Profile* [online], www.sciam.com/2001/0501issue/ 0501profile.html [January 25, 2002].

14. "What are Nanobacteria?" Nanobaclabs Web site, nanobaclabs.com/.

15. Ibid.

16. Dylan Tweney, "Smart Cards Are Finally Getting, Well, Smart," *eCompany*

Now, April 26, 2001 [online], www.ecompany.com/articles/web/0,1653,11515,00. html [July 7, 2001]. This was also discussed in a PBS interview with Sun Microsystems CEO Scott McNealy, broadcast as part of the program *Celebrity CEO's*, on KCET Los Angeles, June 1, 2001.

17. Greg Freiherr, "Providing the Power Behind Virtual Reality," *Medical Electronics Manufacturing* (Fall 1997): 60. Also online at www.devicelink.com/mem/ archive/97/10/007.html [August 15, 2001].

18. "Virtual Reality Organs Under Construction at UIC," press release, School of Biomedical and Information Sciences, College of Applied Health Sciences, University of Illinois at Chicago, March 1, 2000 [online], www.sbhis.uic.edu/E-news/ scangrant.htm; see also VRMedlab at www.sbhis.uic.edu/vrml/ [August 15, 2001].

19. "About the CAVE–The CAVE at NCSA," cave.ncsa.uiuc.edu/about.html [September 5, 2001].

20. Richard P. Feynman, "There's Plenty of Room at the Bottom: An Invitation to Enter a New Field of Physics" (lecture given to the American Physical Society, California Institute of Technology, December 29, 1959) [online], www.physics.umn. edu/groups/mmc/personnel/pete/There%20is%20plenty%20of%20room%20at% 20the%20bottom.htm [July 7, 2001].

21. "Materials Foresight on the Electronics Industry," *Report of a Working Party of the Institute of Materials (IoM) and the Institution of Electrical Engineers (IEE), London* (1998). Quoted from the Institute of Nanotechnology Web site, Section 11 [online], www.nano.org.uk/section11.htm [August 12, 2001].

22. Jessica Gorman, "New Method Lights a Path for Solar Cells," *Science News*, August 11, 2001 [online], www.sciencenews.org/20010811/fob4.asp [August 11, 2001]. See also Peter Fairley, "Solar on the Cheap," *MIT Technology Review* (January/February 2002): 48–53. This article explains that organic carbon-based materials could replace silicon-based solar cells and "can be dissolved to produce a photovoltaic ink, which an ink-jet printer could squirt in a thin film on a variety of surfaces."

23. Claudia Cattaneo, "Suncor Boosts Oilsands," *Financial Post*, November 14, 2001, p. FP1. Also, Carol Howes, "Oilpatch Poised for Spectacular Growth, Petro Canada in $5.8B Oilsands Expansion, Industry Giant Adds to $50B in Projects Already Unveiled," *Financial Post*, December 5, 2001, p. FP1.

24. "Materials Foresight on the Electronics Industry," section 10, www.nano. org.uk/section10.htm [August 12, 2001].

25. Ottilia Saxl, *Opportunities for Industry in the Application of Nanotechnology*, section 3, (2000) [online], www.nano.org.uk/contents.htm [July 10, 2001].

26. Ian Sample, "You Drive Me Crazy," *New Scientist* 171, no. 2300 (July 21, 2001): 24.

27. See the NASA Small Aircraft Transportation System (SATS) Web site at sats. larc.nasa.gov/main.html [August 15, 2001].

28. C. Burda et al., "Some Interesting Properties of Materials Confined in Time and Space of Different Shapes" (paper presented at the Eighth Foresight Conference on Molecular Nanotechnology, Bethesda, Md., November 3–5, 2000) [online], www. foresight.org/conferences/MNT8/Abstracts/El-Sayed [August 10, 2001].

29. Lia Unrau, "DOD to Fund New Rice-led MURI Initiative," *Rice University News* [online], riceinfo.rice.edu/projects/reno/rn/19990415/nanoshells.html [August 10, 2001].

30. Jennifer Darwin, "Spinning Big Plans for Tiny Technology," *Houston Business Journal*, December 1, 2000 [online], houston.bcentral.com/houston/stories/2000/12/04/story3.html?t=printable [September 3, 2001].

31. "ABB and Nanogate to Take Nanotechnology into the Realm of Quality-Assured Mass Production Using Robot-Based Automation," press release, ABB/Nanogate, April 4, 2001 [online], www.nanogate.de/datenbank/aktuelle_meldungen_e/meldung010.htm [December 31, 2001].

32. DeCorp Americas Inc. produces the FlatWire Ready product line, including DeWire. See the company's Web page at www.decorp.com/deProducts.shtml [August 12, 2001].

33. Earl Lane and Lou Dolinar, "The Keys to Finding Bin Laden. Experts Use High Tech, Humans," Newsday.com, September 26, 2001, www.newsday.com/technology/ny-usinte262384574sep26.story?coll=ny-technology-print [November 21, 2001].

34. Susan Scott, "Monk Seals Make Movies with the 'Critter Cam,' " *Honolulu Star-Bulletin*, February 24, 1997 [online], starbulletin.com/97/02/25/news/oceanwatch.html [August 20, 2001].

35. "Optical Glass for Instance–Crystal Clear Advantages," www.nanogate.de/_english/systempartnerschaft/schweizer_optics.htm [July 12, 2001].

36. Saxl, *Opportunities for Industry in the Application of Nanotechnology.*

37. "UC Berkeley Physicists Create Tiny Bearings and Springs Out of Carbon Nanotubes for Use in Microscopic Machines," press release, UC Berkeley, July 27, 2000 [online], www.berkeley.edu/news/media/releases/2000/07/27_nano.html [July 15, 2001].

38. "2000–Record Year For Robot Investment, Increase of 25%," press release, World Robotics Survey, United Nations Economic Commission for Europe, October 30, 2001 [online], www.unece.org/press/pr2001/01stat10e.htm.

39. Robocup Official Web site, www.robocup.org/ [July 7, 2001].

40. U.K.-based Armstrong Healthcare unveiled its "PathFinder" robot in September 2001: "The surgeon instructs the robot by marking a target and an approach path on the patient's scan. The robot carries a camera that automatically matches the scanner image to the position of the patient's head on the operating table." "Robots That Operate on Brains," *Wired News*, September 5, 2001 [online], www.wired.com/news/print/0,1294,46552,00.html [September 5, 2001]. Also, Brad Evenson, "Robotic Surgery a Boon," *National Post*, October 23, 2001, p. A9. This article describes how surgeons at London Health Sciences Centre in Canada have used robotics arms to conduct coronary artery bypasses on more than one hundred patients, resulting in less invasive procedures and leading to 40 percent less time in the hospital.

41. For more information on potential applications of nanomedicine, refer to Robert Freitas, *Nanomedicine* (1998–2001) [online], www.foresight.org/Nanomedicine/ [July 15, 2001].

42. Ibid.

43. The term "money trail" was used by, among others, Nobel laureate and maverick chemist Kerry Mullis, who has been critical of science-funding methods in his book *Dancing Naked in the Mind Field* (New York: Random House, 1998); see the back cover.

44. For a good overview of the drug resistance problem, see Laurie Garrett, *Betrayal of Trust: The Collapse of Global Public Health* (New York: Hyperion, 2000).

45. This quote was often dourly repeated to me by physicians at the Institut fuer Impfwesen und Virologie, Hamburg, Germany. In the 1990s our Germany-based research team lived in the tropics and often visited the institute to get immunization shots. Every six months when we visited, the doctor would ask, "Well, are you sick yet?" Finally, after the third year, team members got Dengue fever.

46. Ron Winslow, "One Patient, 34 Days In the Hospital, a Bill For $5.2 million," *Wall Street Journal*, August 2, 2001, p. 1.

47. Debora MacKenzie, "Special Report. Unaffordable Drugs. Protection Racket. Drug Companies are Tightening Their Hold on Third World Markets," *New Scientist* 171, no. 2300 (July 21, 2001): 18–20. In this article the author explains that a thousand children die of this treatable disease every day. Thus, in the thirty-four days it took to treat the American patient, approximately thirty-four thousand children died. Cost of one tablet in Uganda: seven cents. Cost in neighboring Kenya: more than two dollars. Total cost of treating thirty-four thousand children: less than treating one American patient.

48. J. L. Gallup and J. D. Sachs, "The Economic Burden of Malaria: Cross-Country Evidence," in *Health, Health Policy and Economic Outcomes*. Health and Development Satellite, Final Report. WHO Director-General Transition Team (1998). The quote itself is taken from the Global Alliance for Vaccines and Immunization FAQ page, www.vaccinealliance.org/ reference/health_growth.html [July 7, 2001].

49. Greer Van Zyl, "Africa Unites to Beat Malaria," *Daily Mail & Guardian*, (Johannesburg), March 29, 1999 [online], www.mg.co.za/mg/news/99mar2/29mar-malaria.html [July 7, 2001].

50. James Randerson, "Seeds of Destruction," *New Scientist* 172, no. 2319 (December 1, 2001): 19. A team led by Stephen Davis at the Commonwealth Science and Industry Research Organization, University of Canberra, Australia, have identified a way to genetically engineer male parasites whose offspring die. See also Stephen Davis, Nicholas Bax, and Peter Grewe, "Engineered Underdominance Allows Efficient and Economical Introgression of Traits into Pest Populations," *Journal of Theoretical Biology* 212, no. 1 (September 7, 2001): 83–98.

51. Andy Coghlan, "Parasite Wars, Choose the Wrong Vaccine and Malaria Will Get More Deadly," *New Scientist* 172, no. 2321 (December 15, 2001): 16. See also Sylvain Gandon et al., "Imperfect Vaccines and the Evolution of Pathogen Virulence," *Nature* 414 (December 13, 2001): 751.

52. Ibid.

53. "$100 Million for Malaria Research Pledged to Johns Hopkins University," *PR Newswire* 5, no. 6 (June 2001) [online], www.media-mark.com/prnewswire/biotech/ [July 15, 2001]; "Johns Hopkins Gets Pledge of $100 Million for Malaria Research," *Wall Street Journal*, May 7, 2001.

54. Anita Manning, "Plan Targets Global Diseases," *USA Today*, May 8, 2001 [online], www.usatoday.com/news/health/2001-05-08-global-usat.htm [August 10, 2001]. Also referenced at National Institute of Allergy and Infectious Diseases, National Institutes of Health, www.niaid.nih.gov/publications [July 13, 2001].

55. *High School and Youth Trends, National Institute on Drug Abuse*, 2000 Monitoring the Future Study [online], www.drugabuse.gov/Infofax/HSYouthtrends.html [July 12, 2001]; Donna Leinwand, "Club Drugs Sending More Young People to Hospitals," *USA Today*, July 25, 2001, p. 3A; also online at www.usatoday.com/usat online/20010725/3508804s.htm [July 25, 2001].

56. The volume of voice and data traffic monitored by the National Security Agency is documented in James Bamford, *Body of Secrets: Anatomy of the Ultra-Secret National Security Agency* (New York: Random House, 2000); see also www.random house.com/features/bamford/home.html [July 15, 2001]. One NSA program, ECHELON, intercepts more than 100 million transmissions per day. See John Lichfield, "Global Operation Combs the Airwaves for Business Secrets" *Independent* (London), April 11, 1998 [online], home.icdc.com/~paulwolf/independ.htm [September 8, 2001].

57. Bamford, *Body of Secrets*, epilogue.

58. Peter Weiss, "Light Shines in Quantum-Computing Arena," *Science News*, May 19, 2001 [online], www.sciencenews.org/20010519/ fob4.asp [August 10, 2001]; Indrani Rajkhowa, "Stopping Light in its Tracks," *Computers Today*, February 16–28, 2001 [online], www.india-today.com/ctoday/20010216/marvels.html [August 10, 2001].

59. Smart dust research Web site describing Defense Advanced Research Projects Agency (DARPA) project on microair vehicles, robotics.eecs.berkeley.edu/ ~pister/SmartDust/ [July 14, 2001]. See also the DARPA MEMS Web site at www. darpa.mil/mto/mems/summaries/Projects/index.html [July 14, 2001].

CHAPTER 4: WHAT COMES NEXT? MEGATRENDS THAT COULD ALTER OUR LIVES

1. At the time of writing these ads were broadcast by Motorola at www. motorola.com/home/ [August 13, 2001].

2. In July 2001 Kellogg's Toppas cereal boxes were carrying this educational material, in cooperation with the German-language science television program *Galileo*.

3. The world supersonic land speed record was achieved in October 1997. See "British Duo Sets First Supersonic Land Speed Record," CNN, October 15, 1997 [online], www.cnn.com/TECH/9710/15/brits.land.speed [August 28, 2001].

4. In 1900, 40 percent of the labor force was engaged in farming. "Farmers Revolt," Gilder Lehrman Institute of American History Exhibit, www.hfac.uh.edu/ gl/us25.htm [August 28, 2001]. By 1996, 3 percent of the U.S. population was engaged in farming. Although 21 million Americans are employed in industries associated with food and agriculture, just a fraction of those are growing the food. Patrick Plonski, "Current and Projected Future Employment Opportunities in Food, Fiber,

and Natural Resources," Minnesota Agricultural Education Leadership Council, www.maelc.state.mn.us/ua-employment.html [August 28, 2001].

5. Robotic prosthetics do this regularly.

6. Genetic programming is the technology. See www.genetic-programming. com for an explanation of genetic programming [July 15, 2001].

7. Ian Tattersall and Jeffrey Schwartz, *Extinct Humans* (Westview Press, June 2000). See also online at www.extincthumans.com/ [July 15, 2001]. An argument still rages over what species of human lived when, but this ten-year work effectively lays to rest the notion that one species of human gave way to another in a strictly linear fashion. The authors show, through examination of fossil records, that Neanderthal and *Homo sapiens* coexisted, and that up to fifteen different species of human may have existed over time. The authors challenge the conventional view of human history as a smooth continuum of change: ". . . the past few years have witnessed discovery . . . that clearly cannot be accommodated by the traditional notion of a single evolving lineage" (p. 40).

8. The term "Robo sapiens" was popularized in the book by the same name: Peter Menzel and Faith D'Aluisio, *Robo Sapiens—Evolution of a New Species* (Cambridge: Material World Press, MIT Press, 2000).

9. Indrani Rajkhowa, "Stopping Light in its Tracks," *Computers Today*, February 16–28, 2001 [online], www.india-today.com/ctoday/20010216/marvels.html [August 10, 2001].

10. Mike May, "I'm Just Flying Down to the Supermarket," *New Scientist* 162, no. 2188 (July 21, 2001): 24.

11. J. Storrs Hall, "Utility Fog: A Universal Physical Substance" (1998) [online], www.aeiveos.com/~bradbury/Authors/Computing/Hall-JS/UFAUPS.html [July 12, 2001].

12. The potential for electromagnetic vehicles to reach supersonic speed in low atmospheric pressure was discussed in R. M. Salter, "Trans-Planetary Subway Systems–A Burgeoning Capability" (1973); the abstract is available online at www.rand. org/cgi-bin/Abstracts/e-getabbydoc.pl?P-6092 [September 10, 2001].

13. "Fractal Robots," Digital Matter Control Web Site, Robodyne Cybernetics Ltd., www.stellar.demon.co.uk/new/contact.htm [July 7, 2001]. For a technical outline of fractal robots see Hans Moravec and Jesse Easudes, "Fractal Branching Ultra-dexterous Robots," Final Report, NASA Advanced Concepts Research Projects (January 31, 1999) [online], www.frc.ri.cmu.edu/~hpm/project.archive/robot. papers/1999/NASA.report.99/ [September 10, 2001].

14. "Artificial Lung on the Horizon, Reports University of Pittsburgh Researcher at International Society for Heart and Lung Transplantation Meeting," University of Pittsburgh Medical Center News Bureau, April 26, 2001 [online], www. upmc.edu/NewsBureau/tx/alung.htm [September 4, 2001].

15. Karl Ziemelis, "Going Up," *New Scientist* 170, no. 2289 (May 5, 2001): 24–27. This article describes the properties of carbon nanotube materials that make them suitable for space elevators.

16. Ibid., p. 25. The space elevator was first proposed in 1960 by Russian engineer

Yuri Artsutanov writing in *Pravda*, then popularized by Arthur C. Clarke in a 1978 novel *The Fountains of Paradise*. "Audacious & Outrageous: Space Elevators," Science@NASA [online], science.nasa.gov/headlines/y2000/ast07sep_1.htm [September 10, 2001].

17. The potential role of space elevators in colonizing Mars is aptly described by Kim Stanley Robinson throughout his trilogy, *Red Mars, Green Mars*, and *Blue Mars* (New York: Bantam Books, 1993–97).

18. Ziemelis, "Going Up," p. 26.

19. Neil Gershenfeld, *When Things Begin to Think* (New York: Henry Holt, 1999), p. 75.

20. "One of the central challenges of computer science is to get a computer to do what needs to be done, without telling it how to do it. . . . Genetic programming achieves this . . . by genetically breeding a population of computer programs using the principles of Darwinian natural selection and biologically inspired operations," Genetic Programming Inc., www.genetic-programming.com [July 15, 2001].

21. Isaac Asimov, *Robot Visions* (New York: Penguin Publishers, 1991), p. 417.

22. Genetic Programming Inc. is found at www.genetic-programming.com [July 15, 2001].

23. John R. Koza et al., "Automated Synthesis of Analog Electrical Circuits," in *Genetic Programming III: Darwinian Invention and Problem Solving* (San Francisco: Morgan Kaufmann, 1999) See also William Peakin, "Creative Computer can Invent to Order," *Sunday Times* (London), August 12, 2001 [online], www.sunday-times. co.uk/news/pages/sti/2001/08/12/stinwenws02005.html [September 5, 2001].

24. John Hockenberry, "The Next Brainiacs," *Wired*, August 2001 [online], www.wired.com/wired/archive/9.08/assist.html.

25. The first artificial heart was implanted in a human in 2001. See Rhonda Rowland, "Patient Gets First Totally Implanted Artificial Heart," CNN.com Health, July 3, 2001 [online], www.cnn.com/2001/HEALTH/conditions/07/03/artificial. heart/ [August 10, 2001]. Note: other artificial hearts were implanted in humans before this, but this was the first to be autonomously functional, allowing the recipient to move about independently.

26. Retinal Implants. See "Silicon Chips Implanted into the Eyes of Three Patients to Treat Blindness," press release, Optobionics Corporation, July 31, 2001 [online], www.optobionics.com/010730pressrelease. htm [August 10, 2001].

27. Hockenberry, "The Next Brainiacs."

28. Jennifer Kahn, "Let's Make Your Head Interactive. The Human Brain Project is combining wet anatomy with next-gen scanning, imaging, and networking to give neuroscience a revolutionary new tool–the globally accessible online mind," *Wired*, August 2001 [online], www.wired.com/wired/archive/9.08/brain.html [September 10, 2001].

29. H. G. Wells's *The War of the Worlds* was originally published in London (1898) and has since been reprinted numerous times by many publishers, including Bantam, New York (1988).

30. Arthur C. Clarke, *2001: A Space Odyssey* (1968).

31. Isaac Asimov, *The Bicentennial Man and Other Stories* (New York: Doubleday, 1976).

32. Merriam-Webster's Collegiate Dictionary Online: "Singularity . . . 4: a point or region of infinite mass density at which space and time are infinitely distorted by gravitational forces and which is held to be the final state of matter falling into a black hole," www.m-w.com [July 7, 2001].

33. These ideas are discussed more aptly by John Smart at his Web site Singularity Watch, members.home.net/marlon1/explore.html [July 7, 2001].

34. Vernor Vinge, "The Coming Technological Singularity: How to Survive in the Post-Human Era" (paper presented at the Vision-21 Symposium sponsored by NASA's Lewis Research Center and the Ohio Aerospace Institute, March 30–31, 1993) [online], www.rohan.solsu.edu/faculty/vinge/misc/singularity.html [July 7, 2001].

35. CERI Web site, www.ceri.com. For an example of the conflict with the FDA, see Dean Wolfe Manders, "The FDA Ban of L-Tryptophan: Politics, Profits and Prozac," *Social Policy* 26, no. 2 (winter 1995) [online] www.ceri.com/trypto.htm [July 7, 2001].

36. Genetic programming is further explained at www.genetic-programming. com [July 15, 2001].

37. A definition of "transhumanism" is found at transhumanist.org/index.html #transhumanism [July 7, 2001]:

Transhumanism represents a radical new approach to future-oriented thinking that is based on the premise that the human species does not represent the end of our evolution but, rather, its beginning. We formally define it as follows:

(1) The study of the ramifications, promises and potential dangers of the use of science, technology, creativity, and other means to overcome fundamental human limitations.

(2) The intellectual and cultural movement that affirms the possibility and desirability of fundamentally altering the human condition through applied reason, especially by using technology to eliminate aging and greatly enhance human intellectual, physical, and psychological capacities..."

This definition was further clarified in an e-mail from Nick Bostrom to the author, January 14, 2002.

38. These definitions are taken from the FAQ page of the Extropy Institute's Web site, www.extropy.org/faq/faq.htm [September 10, 2001]. As of the time of writing this page was being updated.

39. Ibid.

40. The Extropy Institute is found at www.extropy.org [July 7, 2001].

41. Ibid.

42. These definitions are taken from the FAQ page of the Extropy Institute's Web site, www.extropy.org/faq/faq.htm.

43. James Mallet, "Species, Concepts of," in *Encyclopedia of Biodiversity*, ed. Simon A. Levin (New York: Academic Press, 2001), vol. 5, pp. 427–40. Mallet is quoting E. Mayr, *Populations, Species, and Evolution* (Cambridge, Mass.: Harvard University Press, 1970).

44. M. Lynne Corn, "The Listing of a Species: Legal Definition and Biological Reality," report for Congress (Congressional Research Service, December 15, 1992).

45. "Researchers from Lucent Technologies' Bell Labs and University of Oxford Create First DNA Motors," press release, Lucent Technologies, August 9, 2000 [online], www.lucent.com/press/0800/000809.bla.html [August 15, 2001].

46. Human skin has already been grown outside of humans and is used in transplants. See Linda Moulton Howe, "Immortal Human Skin Cells, A Miraculous Answer for Burn Victims?" Earthfiles, December 17, 2000 [online], www.earthfiles. com/earth198.htm [September 8, 2001].

47. Supercomputing is being used at the Lawrence Livermore Laboratories to identify and mimic DNA. See Mike Colvin, "A New Kind of Biological Research, Advanced Simulations are Revealing the Exact Mechanisms of Key Biological Processes," *Science & Technology Review* (April 2001) [online], www.llnl.gov/str/April 01/Colvin.html [September 9, 2001]. Dr. Colvin leads the Computational Biology Group in Lawrence Livermore's Biology and Biotechnology Research Program.

48. For an introduction to how artificially intelligent machines may or may not replicate as a species, see Moshe Sipper and James A. Reggia, "Go Forth and Replicate. Birds do it, Bees do it, but could Machines do it? New computer simulations suggest that the answer is yes," *Scientific American*, August 2001, 35–43.

49. The issue of consciousness is explored by Ray Kurzweil in *The Age of Spiritual Machines* (New York: Penguin Books, 1999). See also Ray Kurzweil, "The Age of Spiritual Machines" (sermon to the First Unitarian Universalist Church of San Diego, January 23, 2000) [online], firstuusandiego.org/public/sermons/012300 Kurzweil.ram [July 11, 2001].

50. Tattersall and Schwartz, *Extinct Humans*.

51. James Martin, *After the Internet: Alien Intelligence* (Washington, D.C.: Capital Press, 2000).

52. Brad Lemley, "Computers will Save Us–The Future According to James Martin," *Discover*, June 2001 [online], www.discover.com/june_01/featsave.html [July 7, 2001].

53. Robert Kunzig, "The Unbearably Unstoppable Neutrino," *Discover*, August 2001, p. 40. Also online, www.discover.com/aug_01/featneutrino.html [September 8, 2001].

54. As explained in chap. 1, that group of award-winning scientists includes: Carnegie Mellon robotics director Hans Moravec, MIT artificial-intelligence explorer Marvin Minsky, Stanford medical informatics professor John R. Koza, nanotechnology pioneer K. Eric Drexler, speech synthesis developer Ray Kurzweil, and retired University of California computer scientist Vernor Vinge, among others. For an overview of some of the players, see Steve Alan Edwards, "Mind Children: Extopians," *21C–Scanning the Future* (April 1997). Available online as "Surviving the Singularity" at members.aol.com/salaned/writings/survive.htm [August 27, 2001].

55. Bill Joy, Ray Kurzweil, and Danny Hillis are just a few of the scientific entrepreneurs who've made money in the computing field by working on the enabling technologies for artificial intelligence.

56. "Satellite Trio to Test Artificial Intelligence Software," *Aviation Week*, May 30, 2001 [online], www.aviationnow.com/avnow/news/channel_space.jsp?view= story&id=news/ssat0530.xml [September 4, 2001].

57. Navy Center for Applied Research in Artificial Intelligence, www.aic.nrl.navy.mil/ [August 15, 2001].

CHAPTER 5: THE NANOECOLOGY REVOLUTION

1. David Perlman, "Toxic Clouds Billowed above Twin Tower Site: Expert Says Particles Were Worst Ever Measured," *San Francisco Chronicle*, February 12, 2002, p. A-1 [online], www.sfgate.com/cgi-bin/article.cgi?file=/chronicle/archive/2002/02/12/MNPARTICLES12.DTL [February 15, 2002]; "Trade Center Air Held Unprecedented Amounts of Very Fine Particles, Silicon, Sulfates, Metals, Say UC Davis Scientists," press release, UC Davis, February 11, 2002 [online], www.news.ucdavis.edu/default.lasso?-database=news_db.fmp_&-layout=news_detail&-response=search/news_detail.lasso&-logicalOp=or&-recordID=38684&-search [February 23, 2002].

2. "Zukunft der Nanotechnologie–Highlights der Hannover Messe 2001," Nano Web site, April 23, 2001, www.3sat.de/nano/cstuecke/18062/index.html [July 9, 2001].

3. "Nanotechnologie," Nano Web site, December 1, 1999, www.3sat.de/nano/bstuecke/02253/index.html [July 9, 2001].

4. "Nanostructured coating approved for use on Navy ships," press release, Office of Naval Research, April 5, 2000 [online], www.onr.navy.mil/onr/newsrel/nr000405.htm [July 9, 2001].

5. "Nanoworld–Bakterizide Beschichtungen," www.nanoworld.de/haus/produkt_wanne.html [July 9, 2001].

6. "Nanoworld–Waschbecken Duravit/Nanogate," www.nanoworld.de/haus/produkt01.html [July 9, 2001].

7. "Nano-Sonnencreme. Kleinste Partikel schuetzen vor schaedlichen Sonnenstrahlen," July 2, 2001 [online], www.3sat.de/nano/astuecke/20835/index.html [July 9, 2001]; and "Nanoworld–Sonnenschutzcreme," www.nanoworld.de/haus/ produkt_sonnen. html [July 9, 2001].

8. "Zukunft der Nanotechnologie–Highlights der Hannover Messe 2001," Nano Web site, April 23, 2001, www.3sat.de/nano/cstuecke/18062/index.html [July 9, 2001].

9. Ibid.

10. Lothar Kuhn, "Muehsame Fummelei–Forscher formen aus den kleinsten Teilchen der Welt Produkte mit wundersamen Eigenschaften," *Wirtschaftswoche* 52 (December 21, 2000): 143.

11. "New Biodegradable Nanocomposite Based on More Elastic Polylactide," *Hoover's*, November 2, 2001 [online], hoovnews.hoovers.com/fp.asp?layout=query_displaynews&doc_id=NR20011102670.2_d4f40000f8e46aef [December 5, 2001].

12. "China Takes Lead in Developing Nanometric Self-cleaning Glass," Xinhua News Agency, November 6, 2001 [online], library.northernlight.com/FA2001 1106620000141.html [December 5, 2001].

13. K. Eric Drexler, *Engines of Creation: The Coming Era of Nanotechnology* (New York: Doubleday, 1986), p. 94.

14. "NRDC Hails EPA Decision on Hudson River Cleanup As Triumph of Sci-

ence Over Cynicism," press release, NRDC, August 1, 2001 [online], www.nrdc. org/media/pressreleases/010801.asp [August 20, 2001].

15. Dennis L. Meadows et al., *The Limits to Growth* (New York: Universe Books, 1972); also online at www. clubofrome.org/archive/reports.html [August 12, 2001].

16. "E-mail Leads Workers to Use 40% More Paper, Study Says," *National Post,* November 19, 2001, p. 1. The story summarizes a study by the University of Surrey's digital world research center that found the amount of paper increased as each new communication technology was introduced.

17. The EPA, for example, started a program in 2001 to examine nanotechnology's ecological implications; see es.epa.gov/ncerqa/rfa/futures.html [August 15, 2001]. On August 24, 2000, a coalition of fourteen groups including the Earth Island Institute, Silicon Valley Toxics Coalition, and the Center for Media and Democracy placed a full-page ad in the *New York Times* entitled "TechnoUtopianism," quoting Bill Joy's darkest fears about nanotechnology's environmental impact and calling for public involvement. For a copy of the ad, go to www.turnpoint.org [September 8, 2001].

18. Michael Braungart et al., *Poor Design Practices—Gaseous Emissions from Complex Products* (Hamburg, Germany: Hamburger Umwelt Institut, 1997).

19. Drexler, *Engines of Creation,* p. 94.

20. Already today, electronic paper is being manufactured that has such energy-efficiency improvements over LCD displays. See Steve Silberman, "The Hot New Medium: Paper," *Wired,* April 2001 [online], www.wired.com/wired/archive/9.04 [July 15, 2001].

CHAPTER 6: REVIVING TROPICAL ISLANDS

1. The idea of virtual-reality characters that represent us was popularized, for example, by critically acclaimed science fiction writer Tad Williams in his best-selling book *Otherland* (New York: Daw Paperbacks, 1996). In this and other works, Williams describes a virtual world where every real-life individual has a virtual-reality counterpart in cyberspace. In 2001 this began to approach reality on the big screen with release of *Final Fantasy: The Spirits Within.* The virtually real heroine, Dr. Aki Ross, made the July 2001 cover of *Yahoo Internet Life* magazine. For examples of how seriously fans take these new virtual characters, see www.fffans.net [September 5, 2001]. Furthermore, in the emerging virtual-reality world of 2001, millions of online users already had personal "nyms": characters that they made up to enter online "chat rooms." Thus, the world of virtual-reality representatives is already dawning.

2. Ibid.

3. See the NASA Small Aircraft Transportation System (SATS) Web site at sats.larc.nasa.gov/main.html [July 10, 2001]. See also an analysis of the Moller SkyCar System and SATS at www.skyaid.org/Skycar/MollerEval_10_00.htm#_Toc 497038336 [July 10, 2001].

4. Benz patented the motor vehicle in 1886, inventors.about.com/gi/dynamic/ offsite.htm?site=www.toyota.co.jp/Museum/Tam/chart.html [September 5, 2001]. Seventy years later, horses were gone from urban streets in the United States. On

May 10, 1989, Paul Moller flew the M200X skycar prototype for the International Press. Since then the M200X has made over two hundred successful flights, www.moller.com/skycar/m200x/ [September 5, 2001]. Given the same seventy-year time line, this would put the aerocar into broad use by 2059.

5. NASA has been working on the system for some time. See the NASA Small Aircraft Transportation System (SATS) Web site at sats.larc.nasa.gov/main.html [September 5, 2001].

CHAPTER 7: WHO'S DRIVING THE MOLECULAR MACHINE?

1. Celera is found at www.celera.com. See also Michael D. Lemonick, "Gene Mapper," *Time*, December 17, 2000 [online], www.time.com/time/poy2000/mag/venter.html [August 11, 2000].

2. See the Nasdaq Biotechnology Index, www.nasdaqnews.com/news/pr2001/ne_section01_019.html [August 15, 2001].

3. Genetics are often used to alter the characteristics of enzymes for commercial applications. Genencor International is developing genetically modified enzymes for skin care products through a partnership with Procter & Gamble. Moreover, in an interview with *FTS Forum* magazine, Professor John Hotchkiss, director of graduate studies in the field of food science and technology at Cornell University, explained, "Procter and Gamble has put an enzyme system into a laundry detergent which breaks down cellulose so that it makes your clothes appear brighter because it takes away the fuzz that develops on cotton-based fabrics. If you can put that much cellulose in detergent and still sell it at a reasonable cost, through biotechnology it seems to me you can do almost anything." "The Guru of Active Packaging," *FTS Forum* (September/October 1998) [online], www.foodtechsource. com/emag/001/trend.htm [July 15, 2001].

4. For an introduction to how pervasive DARPA's role is in molecular research, go to its Web site and do a search of keywords such as genetics, robotics, artificial intelligence, and nanotechnology, www.darpa.mil [July 14, 2001].

5. For examples, see www.ri.cmu.edu/home.html [July 9, 2001].

6. For examples of search-and-rescue robotics, see the National Institute for Search and Rescue robotics Web site at www.csee.usf.edu/robotics/crasar/. For updates on use of robotics to search the World Trade Center wreckage, see www.aaai.org/AITopics/html/rescue.html [January 5, 2002].

7. For examples, see www.aibo.com [July 9, 2001].

8. For more information, see www.aibo.com/ers_310/ [December 30, 2001].

9. David Labrador, "Teaching Robot Dogs New Tricks," *Scientific American*, January 26, 2002 [online], www.sciam.com/explorations/2002/012102aibo/.

10. For examples, see www.honda.co.jp/ASIMO [July 9, 2001].

11. Kylie Taggart, "Canada Is the Lead Dog in Developing Robotic Surgery," *Medical Post* 37, no. 17 (May 1, 2001) [online], www.medicalpost.com/mdlink/english/members/medpost/data/3717/05B.HTM [July 14, 2001].

12. "FDA Approves New Robotic Surgery Device," press release, Food & Drug Administration, July 11, 2000 [online], www.fda.gov/bbs/topics/NEWS/NEW00 732.html [July 14, 2001].

13. See www.computermotion.com [July 14, 2001].

14. Tracy A. Schaaf, "Robotic Surgery: The Future Is Now," *MX*, March/April 2001 [online], www.devicelink.com/mx/archive/01/03/0103mx024.html [July 14, 2001].

15. The business of the NSA is, in essence, computers. It is totally dependent on computing to intercept, select, and decode messages. This computerized maze is documented in James Bamford, *Body of Secrets: Anatomy of the Ultra-Secret National Security Agency* (New York: Random House, 2000).

16. See www.genetic-programming.com [July 15, 2001].

17. For more information, see "A.I. in the News" at the American Association for Artificial Intelligence Web site, and also the Kurzweil AI Web site. The AI unit of JPL is at www-aig.jpl.nasa.gov/. The American Association for Artificial Intelligence news pages are at www.aaai.org. Kurzweil AI is at www.kurzweilai.net [December 5, 2001].

18. "Silicon Chips Implanted into the Eyes of Three Patients to Treat Blindness," press release, *Optobionics*, July 30, 2001 [online], www.optobionics.com/010730press release.htm [August 15, 2001].

19. Tom Fowler, "Professor Awaits Breakthrough; Nanotechnology development may drastically alter computing," *Houston Chronicle*, August 14, 2000 [online], amtexpo.com/nano/messages/1271.html [July 7, 2001].

20. Rick Overton, "Molecular Electronics Will Change Everything," *Wired*, July 8, 2000 [online], www.wired.com/wired/archive/8.07/moletronics.html [July 7, 2001].

21. Peter Weiss, "Light Shines in Quantum-Computing Arena," *Science News*, May 19, 2001 [online], www.sciencenews.org/20010519/fob4.asp [August 10, 2001].

22. Sun Microsystems, for example, developed a fast, easy language known as Java that most computers use to send programming instructions over the Internet for things such as Web pages. Sun and Microsoft have been locked in battle over Java and its use for many years. For more information, see "Microsoft Agrees to Settlement that Protects Future Integrity of the Java™ Platform" [online], java.sun.com/lawsuit/ [August 16, 2001].

23. Internet Architecture Board, www.iab.org/. For a description of various Internet standards organizations, see work.home.net/support/internetfaq-org.html# What%20is%20the%20IR [August 16, 2001].

24. Internet Software Consortium, www.isc.org/ [August 16, 2001].

25. Internet Engineering Task Force, www.ietf.org/overview.html [August 16, 2001].

26. Scansoft's Web site is at www.scansoft.com [December 5, 2001].

27. Lisa Guernsey, "Behind the Technology That Can Reproduce a Voice," *New York Times*, August 9, 2001 [online], www.nytimes.com/ 2001/08/09/technology/circuits/09VOIC.html?ex=9983931&pagewanted =print[August 11, 2001].

28. For a description of RealSpeak, see www.lhsl.com/realspeak/features.asp [August 16, 2001].

29. A description of the company Nanospectra is at the *Houston Business Journal* Web site: Jennifer Darwin, "Spinning Big Plans for Tiny Technology," *Houston Business Journal*, December 1, 2000 [online], houston.bcentral.com/houston/stories/ 2000/12/04/story3.html?t=printable [September 3, 2001].

30. For more on Evelyn Hu, see her biography at the California Nanosystems Institute, www.cnsi.ucla.edu. For a biography of Joe Jacobson and his colleagues at the MIT Media Lab, see www.media.mit.edu/people/faculty.html. For more information on James Ellenbogen and his colleagues at Mitre, see www.mitre.org/ research/nanotech/MITREnano_group.html. For a biography of Richard Smalley and a description of Carbon Nanotechnologies Inc., see cnanotech.com/4-4_people. cfm. For the NanoBusiness Alliance, see www.nanobusiness.org.

31. For a description of the robotic Canadarm and the company that built it, see www.mdrobotics.ca/pr12102000a.htm [August 16, 2001].

32. See the Mars Society Web site for a description of the artificial environment at Resolute Bay in northern Canada, www.marssociety.org [August 16, 2001].

33. See the Nanotech Investor newsletter, www.nanotechinvesting.com/ [August 16, 2001].

34. See the Technanogy Web site, www.technanogy.net [August 16, 2001].

35. See the MME Web site, www.mmei.com [August 16, 2001].

36. Brian Alexander, "Atomic Rulers of the World," *Wired*, June 9, 2001 [online], www.wired.com/wired/ archive/9.06 [July 13, 2001].

37. The NIST Web page is at www.nist.gov [August 16, 2001].

38. For a discussion of the relationship of the Napster case to nanotechnology, see Larry Berglas, "Nanotechnology and the Law," www.uslaw.com/library/article/ TNPCommercialCol090600.html? area_id=44 [August 16, 2001].

39. Margaret Kane, "David Boies Hired to Defend Napster," *ZDNet News*, June 16, 2000 [online], zdnet.com.com/2100-11-521581.html.

40. Foley and Lardner's Web site is at www.foleylardner.com [September 5, 2001], and they have a nanotechnology specialist on staff.

41. "France Funds Nanotech Incubator," *Foresight Update*, no. 46 (September 30, 2001): 5.

42. "European Commission and U.S. Launch Nanotech Collaboration," *Small Times*, December 3, 2001 [online], www.smalltimes.com/print_doc.cfm?doc_id= 2671 [December 3, 2001].

43. "Israel Line, Economic Briefs," Israel Ministry of Foreign Affairs, January 17, 2001 [online], www.israel-mfa.gov.il/mfa/go.asp?MFAH0j1d0 [July 7, 2001]. Also, "Tel Aviv Plans Nanotech Center," *Foresight Update*, no. 46 (September 30, 2001): 5.

44. "Taiwan Announces Five Year Nanoscience Program," *Foresight Update*, no. 46 (September 30, 2001): 5.

45. See the University of Melbourne's Nanoparticle Laboratory Web site at www.chemistry.unimelb.edu.au/nanoparticle/nano.htm [July 14, 2001].

46. "Canada will Establish Nanotech Research Center in Alberta," *Foresight Update*, no. 46 (September 30, 2001): 5.

47. Alexander Nicoll, "Research Set to Produce Miniature Weapons," *Financial*

Times, March 28, 2001 [online], globalarchive.ft.com/globalarchive/articles.html? print=true&id=010328001149 [July 14, 2001].

48. John McWethy, "Emerging Threat for U.S. Air Force," ABC News, June 14, 2001 [online], abcnews.go.com/sections/world/DailyNews/radar_stealth_010614. html [July 14, 2001]. For an example of how small nations are intensively researching military applications of molecular technologies, see the Romanian National Institute for Research and Development projects Web site, www.imt.ro/content-v/researchnew.htm [July 14, 2001].

49. 3sat–Nano Web site, www.3sat.de/nano.html [August 15, 2001].

50. *Frankfurter Allgemeine Zeitung* English-language edition Web site, www.faz.com/IN/INtemplates/eFAZ/default.asp [August 15, 2001].

51. *Japan Today* Web site, www.japantoday.com/e/?content=news&cat=4 [August 15, 2001].

52. *Small Times* Web site, www.smalltimes.com/ [August 15, 2001]; *Nano-Technology* magazine Web site, www.nanozine.com/ [August 15, 2001]; Nanotech Planet Web site, www.nanotechplanet.com; Nanotech News.com Web site, www.nanotechnews.com; NanoInvestorNews Web site, www.nanoinvestornews.com; Nanomagazine. com Web site, www.nanomagazine.com [January 27, 2002].

53. Kim Stanley Robinson, *Red Mars, Green Mars,* and *Blue Mars* (New York: Bantam Books, 1993–97).

54. Tad Williams, *Otherland* (New York: Daw Books, 1998).

55. *Chemical & Engineering News* 79, no. 13 (March 26, 2001): 98, 144, 200.

56. "Will Spiritual Robots Replace Humanity by 2100?" (panel discussion held at Stanford University, April 1, 2000). The transcript is available at www.technetcast.com/tnc_play_stream.html?stream_ id=271; the overview Web page is at www.technetcast.com/tnc_program.html? program_id=82; and the symposium page is at www.stanford.edu/dept/symbol/ Hofstadter-event.html [July 12, 2001].

57. Douglas Hofstadter, *Gödel, Escher, Bach: An Eternal Golden Braid* (New York: Basic Books, 1979).

58. Hans Moravec, *Robot: Mere Machine to Transcendent Mind* (Oxford University Press, 1998); and *Mind Children: The Future of Robot and Human Intelligence* (Cambridge: Harvard University Press, 1990).

59. Ray Kurzweil, *The Age of Intelligent Machines* (Cambridge: MIT Press, 1990); *The Age of Spiritual Machines* (New York: Penguin Books, 1999).

CHAPTER 8: ARE WE GETTING MORE OR LESS VULNERABLE?

1. Peter D. Ward and Donald Brownlee, *Rare Earth: Why Complex Life Is Uncommon in the Universe* (New York: Springer Verlag, 2000), p. 287.

2. Gerard Fryer, University of Hawaii, e-mail message to author, July 7, 2001. Fryer is part of the Tsunami Civil Defense Team and a recognized tsunami expert. The U.S. National Oceanographic and Atmospheric Administration (NOAA) is installing

seafloor pressure gauges to remedy this problem, but they are few and far between and do not solve the problem for much of the world. The main barrier is the high cost of such instruments and the low level of funding for finding such rare hazards. Regardless of that, nanotechnology may help solve this with cheap, ubiquitous sensors.

3. Antonio Ermirio de Moraes, the head of Brazil's largest industrial conglomerate, Grupo Votorantim, quoted in "Brazil Imposes Power Rationing," CNN.com, May 18, 2001; and "Poll: Most Brazilians Blame Gov't for Energy Crisis," CNN. com, June 3, 2001 [online], www.cnn.com/2001/WORLD/americas/06/03/brazil. politics.energy.reut/ index.html [July 12, 2001].

4. Clive Ponting, *A Green History of the World–The Environment and the Collapse of Great Civilizations* (Middlesex, England: Penguin Books, 1991), pp. 260–61.

5. Antonio Ermirio de Moraes, quoted in "Brazil Imposes Power Rationing" and "Poll: Most Brazilians Blame Gov't for Energy Crisis," www.cnn.com/2001/WORLD/americas/06/03/brazil. politics.energy.reut/index.html.

6. Bill Kaczor, "Scarcity of Water Increasing Political Tensions," *South Florida Sun-Sentinel,* April 26, 2001 [online], www.sunsentinel.com/news/local/florida/sfl-426drought. story [August 5, 2001].

7. "Baking the Big Apple," Earth Observatory Web site [online], earthobservatory.nasa.gov/Study/Adapting/adapting_2.html [July 14, 2001]; "Power Restored in New York As Heat Wave Subsides," CNN, July 7, 1999 [online], www.cnn.com/weather/9907/07/heat.wave.02 [August 30, 2001].

8. "Mayor Giuliani and Corporation Counsel Michael Hess Announce Lawsuit against Con Edison on behalf of City of New York, its Residents and the Board of Education," press release #288-99, July 15, 1999 [online], www.nyc.gov/html/om/html/99a/pr288-99.html [August 19, 2001].

9. Jesse H. Ausubel, "Mitigation and Adaptation for Climate Change: Answers and Questions," *Bridge* 23, no. 3 (1993): 15–30 [online], phe.rockefeller.edu/mitigation/ [July 7, 2001].

10. Nick Bostrom, "Existential Risks Analyzing Human Extinction Scenarios and Related Hazards" [online], www.nickbostrom.com/existential/risks.html [July 7, 2001]. As of the date of writing, this paper was unpublished except by the author on the World Wide Web. It is quoted and paraphrased here with the permission of Dr. Bostrom. According to Dr. Bostrom, the paper is being reviewed by several publishers. He has published a related paper, "The Doomsday Argument, Adam & Eve, UN, and Quantum Joe," *Synthese* 127, no. 3 (2001): 359–87. He is also working on a book, *Anthropic Bias: Observation, Selection, Effects in Science and Philosophy* (New York: Routledge, 2002). His dissertation was selected by editor Robert Nozick as one of the seven best philosophy dissertations of the year, to be published in a Routledge series. Bostrom did his Ph.D. at the London School of Economics, and prior to that studied mathematical logic, philosophy and artificial intelligence at the University of Gothenburg, Sweden.

11. Ibid.

12. Ray Kurzweil, *The Age of Spiritual Machines* (New York: Penguin Books, 1999), pp. 13–14.

13. Ibid.

14. Patrick Moore, *Pacific Spirit: The Forest Reborn* (Vancouver, Canada: Terra Bella, 1995), pp. 12–13.

15. Nick Bostrom, "Existential Risks Analyzing Human Extinction Scenarios and Related Hazard." This segment is my own abbreviation of a more lengthy explanation in his paper. In a telephone conversation on August 10, 2001, Dr. Bostrom reiterated that items in each category are in descending order of probability.

16. Ibid.

CHAPTER 9: SHAKING UP TOKYO AND THE GLOBALIZED ECONOMY

1. For a more detailed description of this, see Peter Hadfield, *Sixty Seconds That Will Change the World–The Coming Tokyo Earthquake* (Boston: Tuttle, 1991; revised, 1995).

2. Peter Hadfield, "Disaster Quake Wins Grim Place in Record Books," *New Scientist* 145, no. 1965 (February 18, 1995): 5.

3. Ibid.

4. "A South Kanto Area Earthquake and the Japanese Economy," *Tokai Monthly Economic Newsletter*, no. 122 (March 1989).

5. *What If the 1923 Earthquake Strikes Again?* (Menlo Park, Calif.: Risk Management Solutions Inc., 1995).

6. Hadfield, *Sixty Seconds That Will Change the World*, pp. 30–37.

7. Peter Hadfield, telephone interview with author regarding seismic evidence of crustal tension under Japan, December 14, 1999. This was reconfirmed in an e-mail dated August 7, 2001.

8. William J. Broad, "Earthquakes: A Matter of Luck, Mostly Bad," *New York Times*, September 28, 1999, p. D1.

9. Kenichi Ohmae, "Fixing Japan May Break U.S. Economy," *Daily Yomiuri*, May 3, 2001 [online], www.billtotten.com/english/ow1/00476.html. This article indicates that at least $550 billion in U.S.-held assets and debt could be quickly repatriated to Japan. Eric Helland, "Are U.S. Treasuries At Risk of Japanese Repatriation?" Risk Center, July 18, 2001 [online], www.netrisk.com/news/dailynews/2001_07_18_3274.htm [September 5, 2001].

10. Phred Dvorak et al., "Frayed by Recession, Japan's Corporate Ties Are Coming Unraveled," *Wall Street Journal*, March 2, 2001, p. A1.

11. Laurie Garrett, *Betrayal of Trust: The Collapse of Global Public Health* (New York: Hyperion, 2000), chap. 4. For an example of how earthquakes cause the spread of disease in poor nations, see an excerpt from Garrett's book, describing the spiraling health effects of the 1993 Latur earthquake that destroyed a million homes, at Amazon.com, www.amazon.com/exec/obidos/tg/stores/detail/-/books/0786865229/excerpt/ref=pm_dp_ln_b_3/107-6579546-2578162 [September 8, 2001].

12. For an example of the ongoing effects of Hurricane Mitch, see American Friends Service Committee, "Report on AFSC's Hurricane Mitch Relief and Recon-

struction Program As of December 31, 1999" [online], www.afsc.org/emap/projects/
mich0100.htm [September 8, 2001].

13. Usha Lee McFarling, "State at Epicenter of National Quake Damage
Survey," *Los Angeles Times*, September 17, 2000, p. 1. This story summarizes the first
national survey of earthquake risk in the United States by FEMA. It puts New York
State high on the list for potential losses.

14. Ibid.

15. U.S.-Japan Cooperative Research in Urban Earthquake Disaster Mitigation
Web site, ce.ecn.purdue.edu/~vail/jtcc/ [July 7, 2001].

CHAPTER 10: SO, YOU'RE BORED BY DOOMSDAY?

1. Dan Bruton, "Frequently Asked Questions About the Impact of Comet
Shoemaker-Levy 9 with Jupiter," Institute for Scientific Computation [online], www.
isc.tamu.edu/~astro/sl9/comet faq2.html#Q3.1 [August 29, 2001].

2. John R. Spence and Jacqueline Mitton, eds., *The Great Comet Crash: The
Impact of Comet Shoemaker-Levy 9 on Jupiter* (Cambridge: Cambridge University Press,
1995); David H. Levy, *Comets* (New York: Touchstone Press, 1998), p. 207.

3. Roy A. Gallant, "Tunguska; the Cosmic Mystery of the Century" [online],
www.usm.maine.edu/~planet/tung.html [August 29, 2001].

4. *Voyage to Atlantis: The Lost Empire*, ABC Television, June 10, 2001 (produced
by Disney).

5. See introduction, n. 10.

6. There is much discussion over how much material hits the biosphere of
Earth from space.

"Solar system bodies are not isolated from each other. Earth, for example,
encounters a suite of objects—interplanetary dust particles (IDPs) and small bodies
commonly referred to as the interplanetary debris complex. These bodies range
from fine dust that is delivered continuously to objects several tens of kilometers in
size or larger, including comets and Earth-approaching asteroids. . . ." National
Research Council, *Evaluating the Biological Potential in Samples Returned from Planetary
Satellites and Small Solar System Bodies, Framework for Decision Making* (Washington,
D.C.: National Academy Press, 1998), chap. 2 [online], www.nas.edu/ssb/sssbch2.
htm [July 7, 2001].

For evidence that bombardment was responsible for much of the water on
Earth, see "First Evidence That Comets Filled the Oceans: A Dying Comet's Kin
May Have Nourished Life on Earth," *ScienceDaily*, May 21, 2001 [online], www.
sciencedaily.com/releases/2001/05/010521072649.htm [July 7, 2001].

Regarding the amount of matter that hits earth now:

"Extraterrestrial Delivery of Organic Molecules to the Earth. . . . Dust collection,
both above the terrestrial atmosphere by Love and Brownlee (1993), and in the
Greenland and Antarctic ice sheets by Maurette (Maurette et al. 1995; Maurette,
1998) show that the Earth captures interplanetary dust as micrometeorites at a rate of

about 50-100 tons per day. About 99% of this mass is carried by micrometeorites in the 50-500 µm size range. This value is about 2,000 times higher than the most reliable estimate of the meteorite flux (about 0.03 tons per day) estimated by Bland et al. (1996). This amazing dominance of micrometeorites already suggests their possible role in delivering complex organics to the early Earth 4.2 to 3.9 Ga ago when the micrometeorite flux was enhanced by a factor of 1,000 (Anders,1989). . . ." A. Brack, "Life In The Solar System," Centre de Biophysique Moléculaire [online], www.edu.polytechnique.fr/Scolarite/Conferences/ASR22-2.pdf [September 8, 2001].

7. On March 13, 1989, the whole hydroelectric grid in Quebec was shut down in four minutes by a solar storm. See "Advanced Geomagnetic Storm Forecasting Capabilities Developed for the Electric Power Industry," press briefing American Association for the Advancement of Science, February 19, 2000 [online], www.meta techcorp.com/aps/AAAS_Press_Brief.htm [September 8, 2001].

8. Impacts of the weakening of Earth's magnetic field are still debated in agencies such as NASA, which is tracking the field fluctuation with its IMAGE project:

"We still don't know if the decline is just a natural ripple or a portent of something far more sinister," says Sten Odenwald, a researcher on the IMAGE Project. James Green, another IMAGE researcher, says, "There's going to be a long period of time–possibly many generations–when we're going to have to find a way to deal with all this extra energy. I don't know that anyone's done a proper scientific investigation of what will happen. It's certainly one of those things we should be looking into."

"Great Balls of Fire," *New Scientist* 171, no. 2305 (August 25, 2001): 28.

The Sun shifts its magnetic field every 11 years, and it has already happened for this solar cycle. The Earth's magnetic field flip is much more erratic and has happened approximately 25 times in the last 5 million years. It's been about 740,000 years since the last flip, however, so we're long overdue. There is evidence that we may be heading towards a reversal, but we can't predict when it would happen. Depending upon how quickly the field reversal happens, it could cause problems for things like electric power lines and oil pipelines, and if the field goes to near zero, it might cause a higher background radiation at the ground, but there is no evidence that previous reversals have had any major biological effect.

NASA Cosmic and Helioscopic Learning Center [online], helios.gsfc.nasa.gov/qa_earth.html [September 4, 2001].

Moreover, in the first astrophysical paper to deal with a biological subject, John Scalo and Craig Wheeler of the University of Texas at Austin have estimated that gamma ray bursts (GRBs) from massive star collapses have bombarded Earth at least a thousand times in its history, causing large-scale mutations of life forms by eroding the ozone layer and exposing organisms to UV radiation from the Sun. See

Marcus Chown, "Life Force, Blasts from Exploding Stars Can Change the Course of Evolution," *New Scientist* 172, no. 2321 (December 15, 2001): 10. See also John Scalo and J. Craig Wheeler, "Astrophysical and Astrobiological Implications of Gamma-Ray Burst Properties," accepted by *Astrophysical Journal* Oct. 2001; pre-print version [online], xxx.lanl.gov/abs/astro-ph/9912564 [October 30, 2001].

9. Peter D. Ward and Donald Brownlee, *Rare Earth; Why Complex Life Is Uncommon in the Universe* (New York: Springer-Verlag, 2000), p. 238.

10. Mel Waskin, *Mrs. O'Leary's Comet!* (Chicago: Academy Chicago Publishers, 1985), p. 21. This particular assertion is still speculative. Yet there is no doubt that asteroids and comets have their orbits altered by gravitational pull from planets. Moreover, William Bottke's research team at the Southwest Research Institute in Boulder, Colorado, has calculated that the Sun may be sending asteroids into Earth's path by degrading their orbits via heating and cooling. Their findings, published in *Science* 294, no. 1693, were reported by Justin Leighton, "Crash Course," *New Scientist* 172, no. 2319 (December 1, 2001): 25.

11. "First Evidence That Comets Filled the Oceans," www.sciencedaily.com/releases/2001/05/010521072649.htm.

12. The Anthropology Glossary, University of Santa Barbara Department of Anthropology Web site, www.anth.ucsb.edu/glossary/glossary.html [July 14, 2001].

13. Ward and Brownlee, *Rare Earth*, pp. 235–40. The authors give a good overview of Jupiter's role in cleaning the solar system of large objects than may have threatened Earth.

Conversely, some planets seem to draw asteroids into the path of Earth, as summarized by Donald Smith:

"Smaller asteroids–less than 12.5 miles (20 kilometers) wide–sometimes migrate to unstable areas of the main belt, known as resonances. Once within the resonances, asteroids are vulnerable to the gravitational pull of nearby planets–Mars, Jupiter or Saturn–which can elongate an asteroid's orbit. The change is sometimes enough to swing the asteroid onto a path that crosses Earth's orbit, setting up the possibility of a future collision." Donald Smith, "Getting Serious about Asteroid Strikes," *National Geographic News*, December 20, 2000 [online], news.nationalgeographic.com/news/2000/12/1220_asteroid.html [September 8, 2001].

14. David Morrison, "Is the Sky Falling?" *Skeptical Inquirer,* May/June 1997 [online], www.csicop.org/si/9705/asteroid.html [July 14, 2001].

15. Ibid.

16. Richard Huggett, *Catastrophism: Asteroids, Comets, and Other Dynamic Events in Earth History* (London and New York: Verso, 1997), p. 3 quoting Sherlock.

17. "The Alvarez Asteroid Impact Theory," Dinosaur and Paleontology Dictionary [online], www.zoomdinosaurs.com/subjects/dinosaurs/glossary/Alvarez.shtml [August 15, 2001].

18. "Impact Cratering On Earth," Natural Resources Canada [online], gdcinfo. agg.emr.ca/crater/paper/index_e.html [August 13, 2001].

19. *Tsunami!* Discovery Channel, April 10, 2001.

20. A. R. Nelson et al. "Radiocarbon Evidence for Extensive Plate Boundary

Rupture About 300 Years Ago at the Cascadia Subduction Zone," *Nature* 378 (November 23, 1995).

21. Robert Johnson, "Looking for Floods and Hearing Tales in Unlikely Places," *Wall Street Journal,* November 28 1999, p. 1.

22. "Ballard and the Black Sea. The Search for Noah's Flood," *National Geographic,* October 2000 [online], www.nationalgeographic.com/blacksea [August 13, 2001]. See also Linda Moulton Howe, "Evidence of 7000 Year Old Flood and Human Habitation Discovered Beneath Black Sea," Earthfiles Science/Medicine, September 17, 2000 [online], www. earthfiles.com/earth174.htm [July 14, 2000]; see also "Ancient Relics Evoke Noah's Flood," *San Francisco Chronicle,* September 13, 2000, A7.

23. "Mega Tsunami: Wave of Destruction," transcript, *BBC Horizon,* BBC2, October 12, 2000. The Lituya Bay tsunami is a broadly recognized and well-documented event, as described, for example, at the University of Southern California Tsunami Research Group Web site, "The 1958 Lituya Bay Tsunami" [online], www. usc.edu/dept/tsunamis/alaska/1958/webpages/index.html [August 14, 2001].

24. Ibid.

25. Ibid.

26. Tristan Marshall, "The Drowning Wave," *New Scientist* 168, no. 2259 (October 7, 2000): 26.

27. Christina Ward, "Tsunamis May Be Fueled by Sources Other Than Tectonic Shifts, Scientists Discover," Disaster Relief News Stories, July 19, 2000 [online], www.disasterrelief.org/Disasters/000719tsunamis/ [August 31, 2001].

28. Marshall, "The Drowning Wave."

29. Ibid. The team also includes Benfield Greig researcher Simon Day.

30. Bill McGuire, *Apocalypse: A Natural History of Global Disasters* (London: Cassell & Co., 1999), pp. 137–54.

31. Ibid.

32. Ibid.

33. BBC News report, October 11, 2000 [online], news.bbc.co.uk/olmedia/965000/video/_966968_heap_edit_0500_vi.ram [July 12, 2001].

34. Cynthia Long, "New Seismic Hazard Map Will Help Mitigate against Earthquake Damage," Disaster Relief News Stories, March 19, 1999 [online], www. disasterrelief.org/Disasters/990315Seismicmap/ [August 31, 2001].

35. "540 AD," transcript, *The Science Show,* ABC Radio National Australia, August 12, 2001 [online], www.abc.net.au/rn/science/ss/stories/s181594.htm. In this interview Baillie discusses the findings that are detailed in his book, *A Slice Through Time: Dendrochronology and Precision Dating* (London: Routledge, 1995).

36. R. B. Stothers, "Mystery Cloud of 536 A.D.," *Science Frontiers,* no. 33 (May–June 1984) [online], www.science-frontiers.com/sf033/sf033p19.htm [August 12, 2001].

37. According to Prof. Claus Hammer, a Danish specialist in ice core sampling and a member of the multinational team that has been testing ice cores in Greenland for more than a decade, ice cores from Greenland and Antarctica each demonstrated a large spike in sulphates around the 535 period, with no attendant iridium, thereby

supporting the theory of a massive volcanic eruption. Hammer's conclusions were aired in a PBS documentary, "Catastrophe," PBS, May 2000, but are still contested by other scientists, especially by Baillie, as to the accuracy of their dates.

38. Joel D. Gunn, ed., *The Years Without Summer: Tracing A.D. 536 and Its Aftermath* (Oxford: Archaeopress, 2000). Gunn is at the Department of Anthropology, University of North Carolina at Chapel Hill.

39. Joel D. Gunn, The Years without Summer–Tracing 536 A.D. and Its Aftermath, Web site introducing the book by the same title, www.ad536.org/ad536/default.htm [August 15, 2001].

40. David Keys, *Catastrophe: An Investigation into the Origins of the Modern World* (New York: Ballantine Books, 1999), p. 245.

41. Keys's hypothesis about a volcanic superexplosion is supported by, among other scientists, Los Alamos Laboratories volcanologist Ken Wohletz:

> . . . according to a volcanologist at the Department of Energy's Los Alamos National Laboratory . . . an eruption in the Indonesian archipelago could have produced a 150-meter-thick cloud layer over the entire Earth, triggering a chain of climatic, agricultural, political and social changes that ushered in the Dark Ages. Evidence supporting the catastrophe includes tree-ring and ice-core measurements, indications of a huge underwater caldera, and ash and pumice in the same area.

"The Dark Ages Really May Have Been Dimmer," *Daily News Bulletin*, Los Alamos National Laboratory [online], www.lanl.gov/orgs/pa/News/122000.html [September 5, 2001].

42. David Keys, *Catastrophe*, p. 5.

43. Richard B. Alley, *The Two-Mile Time Machine: Ice Core, Abrupt Climate Change, and Our Future* (Princeton, N.J.: Princeton University Press, 2000).

44. Richard B. Alley, "Abrupt Increase in Greenland Snow Accumulation at the End of the Younger Dryas Event," *Nature* 362 (1993): 527–29.

45. Alley, *Two-Mile Time Machine*, p. 3.

46. Ibid., p. 4.

47. Gordon Hamilton, "Pine Blight Beyond Human Control," *Vancouver Sun*, November 29, 2001, p. 1.

48. M. D. Max et al., "Sea-Floor Methane Blow-Out and Global Firestorm at the K-T Boundary," *USA Abstract* 18, no. 4 (1999): 285–91.

49. Richard Monastersky, "Can Methane Hydrates Fuel the 21st Century?" *Science News*, November 14, 1998 [online], www.sciencenews.org/sn_arc98/11_14_98/Bob1.htm [July 14, 2001].

50. *Ocean Mysteries*, Discovery Channel, June 7, 2001.

51. Ibid.

52. Ibid. See also Alexandra Witze, "Fiery Ice Offers Hope of a New Kind of Fuel but Scientists Fear Frozen Methane Will Fizzle," *Dallas Morning News*, December 4, 2000) [online], www.dallasnews.com/science/229632_hydrates04dis..html [July 14, 2001].

53. "Gas Methane Hydrates–A New Frontier," U.S. Geological Service Web site, marine.usgs.gov/fact-sheets/gas-hydrates/title.html [August 15, 2001].

54. Ibid.

55. In 2001 the Yellowstone Caldera was added to the U.S. list of official volcano observatories. This is relevant because it validates existence of a risk that, according to the *Los Angeles Times*, ". . . is forcing many geologists to rethink the very definition of hot spots and how they work." The increased vigilance results from use of sensor technology of the type discussed in this book. See Usha Lee McFarling, " 'Beast' Is Alive in Wyoming," *Los Angeles Times*, August 21, 2001 [online] www. latimes.com/news/nationworld/nation/la-082101erupt.story [August 22, 2001].

56. "Super Volcanoes," transcript BBC Horizon, February 3, 2000 [online], www.bbc.co.uk/science/horizon/supervolcanoes.shtml [July 9, 2001].

CHAPTER 11: AN ELEPHANT IN THE ROOM OF ENVIRONMENTALISM

1. "Destructive Storms Drive Insurance Losses Up–Will Taxpayers Have to Bail Out Insurance Industry?" Worldwatch News Brief, March 26, 1999 [online], www.worldwatch.org/alerts/990325.html [July 9, 2001]. Moreover, a stark example of the exposure of insurers to single-event disasters and resulting increases in insurance rates occurred with the World Trade Center attack. See "40% Additional Premium Income Will Offset WTC Losses, Says Lloyd's Chairman," press release, Lloyd's of London, November 27, 2001 [online], www.alex anderforbes.com/afuknews/PressArchive/LL8301.doc [January 23, 2002].

2. "Destructive Storms Drive Insurance Losses Up," Worldwatch News Brief, www.worldwatch.org/alerts/990325.html.

3. "World Disasters Report Predicts a Decade of Super-Disasters," press release, International Federation of Red Cross and Red Crescent Societies (June 24, 1999) [online], www.fema.gov/nwz99/irc624.htm [July 9, 2001].

4. "Reality Check Needed on International Aid Efforts, Says the Red Cross/ Red Crescent's World Disasters Report," press release, International Federation of Red Cross and Red Crescent Societies, June 28, 2001 [online], www.ifrc.org/docs/ news/pr01/3901.asp [July 9, 2001].

5. California Bureau of State Audits, "Department of Insurance: Recent Settlement and Enforcement Practices Raise Serious Concerns about Its Regulation of Insurance Companies" (October 2000), p. 18 [online], bsa.ca.gov/bsa/pdfs/2000_ 123.pdf [August 16, 2001].

6. Virginia Ellis, "The Fall of Insurance Commissioner Chuck Quackenbush," *Los Angeles Times* series, March 26 and 28, April 2, 4, 7, 10, 20, 21, 25, and 27; May 12 and 17; June 7, 16, 29, and 30; July 6, 7, 12, and 26; August 4; October 2 and 8; and November 30, 2000. This series won the George Polk Award for Political Reporting. See online abstract at notes.ire.org/ireresources.nsf/92d9c722054975d686256708006 7db05/4ac14111a3a4620c862569fb0071d964?OpenDocument [August 16, 2001].

7. "At the moment the emergency plan set up by the Italian government assumes that six hundred thousand people could be evacuated within seven days on the 40 trains that run daily. This plan also assumes that volcanologists can predict an eruption within 20 days. There is some concern as to the feasibility of this evacuation plan. A former professor of volcanology, Flavio Dobran is quoted as saying that 'the government's plan is preoccupied with trying to predict an eruption, when it should be concerned with disaster management and risk communication.' " Ellie du Celliee Muller, "The Volcanic Hazards of Mount Vesuvius," [online], www.brookes. ac.uk/geology/8361/1997/ellie.html [September 1, 2001].

8. World Commission on Environment and Development, *Our Common Future* (Oxford and New York: Oxford University Press, 1987), p. 43.

9. *Report of the United Nations Conference on Environment and Development*, Rio de Janeiro, June 3–14, 1992, chap. 7, sec. F [online], www.agora21.org/rio92/A21_html/A21en/a21_07.html [August 14, 2001].

10. These priority areas are quoted by Simone Rones in "Comments on Technology Foresight and Sustainable Development Article," *Futures Research Quarterly* (winter 1999): 69–70, table 1.

11. Strong, *Where on Earth Are We Going?* (New York: Texere, 2000), p. 387.

12. John Maddox, *What Remains to Be Discovered* (New York: Touchstone Press, 1998), pp. 332–33.

13. Karin Jegalian and Bruce T. Lahn, "Why the Y Is So Weird," *Scientific American*, February 2001, pp. 56–61.

14. Maddox, *What Remains to Be Discovered*, p. 333.

15. Ibid, p. 366.

CHAPTER 12: LOST MESSAGES FROM ANCIENT TIMES

1. *Skeptical Inquirer* is published by the Committee of the Scientific Investigation of Claims of the Paranormal, Amherst, New York.

2. Graham Hancock, *Fingerprints of the Gods* (New York: Crown Trade Paperbacks, 1995).

3. Robert Bauval and Adrian Gilbert, *The Orion Mystery: Unlocking the Secrets of the Pyramids* (New York: Crown Trade Paperbacks, 1995).

4. Rand and Rose Flem-Ath, *When the Sky Fell: In Search of Atlantis* (New York: St. Martin's Press, 1997).

5. Christopher Dunn, *The Giza Power Plant: Technologies of Ancient Egypt* (Santa Fe, N.M.: Bear & Co., 1998).

6. Charles H. Hapgood, *Maps of the Ancient Sea-Kings* (Kempton, Ill.: Adventures Unlimited Press, 1996).

7. Jane B. Sellers, *Death of the Gods in Ancient Egypt: An Essay on Egyptian Religion and the Frame of Time* (New York: Penguin Books, 1992); a revised version is

available at www.mightywords.com/browse/details_bc05.jsp?sku=MWTDJ7&private Label=false [August 16, 2001].

8. Peter James and Nick Thorpe, *Ancient Mysteries* (New York: Ballantine Books, 1999), pp. 58–76.

9. See documentation about the *BBC Horizon* scandal at the Official Graham Hancock Web site, www.grahamhancock.com/horizon/default.htm [August 17, 2001].

10. Excerpt from the Official Graham Hancock Web site, www.graham hancock.com/intro.php [July 14, 2001].

11. Andrew Cawthorne, "Canadians Find 'Lost City' Off Coast of Cuba," *National Post*, December 7, 2001, p. 1. See also Linda Moulton Howe, "Update on Underwater Megalithic Structures Near Western Cuba," *Earthfiles*, November 19, 2001 [online], www.earthfiles.com/earth303.htm.

12. David Keys, "Lost City Found Beneath Cuban Waters May Be a Trick of Nature," *Independent*, December 8, 2001 [online], www.independent.co.uk/story.jsp?story=109015 [December 9, 2001].

13. Clive Ponting, *A Green History of the World: The Environment and the Collapse of Great Civilizations* (Middlesex, England: Penguin, 2001).

14. Michael Cremo and Richard Thompson, *The Hidden History of the Human Race* (Los Angeles: Bhaktivedanta Publishing, 1999).

15. Ibid., see "Adverse Criticism from Establishment Scientists," prior to the title page.

16. Erich Von Daeniken, *Chariots of the Gods: Unsolved Mysteries of the Past* (New York: Putnam & Bantam, 1968).

17. For a summary of the criticisms of Von Daeniken's work, see Robert Todd Carroll's Skeptic's Dictionary at skepdic.com/vondanik.html [July 12, 2001].

18. The most noteworthy example of this occurred in 2001, when the Pentagon bought exclusive rights to *Ikonos* satellite images of Afghanistan to prevent them from falling into enemy hands. See Robert Remington, "U.S. Locks Up Rights to Spy Satellite Images," *National Post*, October 20, 2001, p. A7. This is also available online at www.spiescafe.com/aleksey/aleksey011023.htm. Other information about efforts at restricting Space Imaging activities came from Mark Brender, Washington Director for Space Imaging, telephone conversations with author (February–May 1999).

19. For more information, see Space Imaging's Web site at www.spaceimaging. com.

20. Andrew Cawthorne, "Explorers View 'Lost City' Ruins Under Caribbean," Arizona Central, December 6, 2001 [online], www.azcentral.com/azc-bin/print. php3: "American companies are prohibited from operating in Cuba by the long-running U.S. embargo on the Communist-run island." (Note: Arizona Central is a news Web site sponsored by the *Arizona Republic* newspaper and KPNX-TV).

CHAPTER 13: WHY GO THERE?

1. John Fiske, *The Meaning of Infancy* (1884). Fiske was known for his strong support of Darwinism.

2. For examples of such "business as usual" subsidies, see Roland Hwang, "Money Down the Pipeline: Uncovering the Hidden Subsidies to the Oil Industry— How tax breaks, government funding, and other subsidies benefit the oil industry at the expense of other energy technologies" [online], www.ucsusa.org/vehicles/ pipeline.html [July 7, 2001].

CHAPTER 14: TOOLS FOR DEFUSING TIME BOMBS

1. A large body of literature alleges UN New World Order conspiracies. A good summary of the theory is found in "The United Nations Role in the New World Order," at the Jeremiah Project Web site, www.jeremiahproject.com/prophecy/ nworder05.html [July 12, 2001].

2. "Warming Up: The National Academy of Sciences Report Concludes the Earth Is Getting Warmer and Humans Are Helping Cause It," *PBS News Hour*, June 7, 2001 [online], www.pbs.org/newshour/bb/environment/energy/warming_6-7. html [July 8, 2001].

3. Jan F. Feenstra et al., eds., *Handbook on Methods for Climate Change Impact Assessment and Adaptation Strategies Version 2.0* (Vrije Universiteit, Amsterdam: United Nations Environment Program and Institute for Environmental Studies, 1998) [online], www.vu.nl/IVM/research/climatechange/fb_Handbook.htm [August 1, 2001].

4. This is the author's paraphrase of a summary given by Ian Burton, Joel B. Smith, and Stephanie Lenhart in "Adaptation to Climate Change, Theory and Assessment," in ibid., pp. 5-2–5-4.

5. W. Ross Ashby, *Introduction to Cybernetics* (London: Chapman & Hall, 1956).

6. Stafford Beer, *Brain of the Firm* (New York: John Wiley & Sons, 1995).

7. "The Law of Requisite Variety," Principia Cybernetica Web, pespmc1.vub. ac.be/ASC/LAW_VARIE.html [August 15, 2001].

8. "Adapting Urban Areas to Atmospheric Change," Environment Canada Adaptation and Impacts Research Group, www.smc-msc.ec.gc.ca/airg/vertical_ gardens.htm [August 15, 2001].

9. This paragraph is based on a note sent by Brad Bass to the author (August 9, 2001), as a suggested modification to this segment of the text.

10. "The classification of adaptation responses . . . bear the losses, share the losses, prevent effects . . . etc." comes originally from I. Burton, R. W. Kates, and G. F. White, *The Environment as Hazard* (New York: Oxford University Press, 1978).

11. "The worst flood disaster in world history occurred in August, 1931 along the Huang He River in China and killed an estimated 3.7 million people." Søren E. Brun, "Atmospheric, Hydrologic and Geophysical Hazards in Coping with Natural Hazards in Canada: Scientific, Government and Insurance Industry Perspectives, Part 2: Natural Hazards In Canada, Chapter 2.0," Environmental Adaptation Research Group, Environment Canada and Institute for Environmental Studies, University of Toronto, June 1997 [online], www.utoronto.ca/env/nh/pt2ch2-3-2.htm [August 15, 2001].

12. Most generating equipment for the Three Gorges Dam is manufactured by dozens of Western companies. For a list, see www.probeinternational.org/probeint/threegorges/who.html [July 14, 2001].

13. John J. Kosowatz, "Mighty Monolith–China Builds the World's Largest Dam," Extreme Engineering, supplement to Scientific American, winter 1999, 14–24.

14. William K. Stevens, "Everglades Restoration Plan Does Too Little, Experts Say," New York Times National Desk, February 22, 1999.

15. Committee of Experts Appointed by Hydro Quebec's Board of Directors, Report on January 1998 Ice Storm (July 1998), pp. 9–10, executive summary [online], www.hydro.qc.ca/publications/r980727e/pdf/report. pdf [August 15, 2001].

16. "Project Impact," FEMA Web site, www.fema.gov/impact/ [July 12, 2001].

17. "Statement of the Independent Insurance Agents of America, Subcommittee on Housing and Community Opportunity Committee on Financial Services United States House of Representatives," July 19, 2001 [online], www.house.gov/financialservices/071901fw.pdf [September 5, 2001]; Marc Herman, "Rain Check: Flood after Flood, Taxpayers Still Subsidize People Living in Harm's Way," Mother Jones, March/April 1998 [online], www.motherjones.com/mother_jones/MA98/herman.html [August 31, 2001].

18. "Senator Murray Announces Restoration of Funds for Project Impact," press release, News from U.S. Senator Patty Murray, August 2, 2001 [online], www. senate.gov/~murray/releases/01/08/2001802D25.html [September 5, 2001]. See also "Bush Budget, Released on Day of Quake, Would End Disaster Program," KOMO News 4, February 28, 2001 [online], www.komotv.com/specialcoverage/bush_budget.htm [September 5, 2001].

19. "Florida Faces Political Storm over Draft Building Code," CNN, May 31, 1999 [online], www.cnn.com/US/9905/31/hurricane.building [August 31, 2001]. See also Bob Schildgen, "Unnatural Disasters (areas that suffer repeat flooding yet continue to rebuild)," Sierra, May 1999.

20. C. Rosenzweig and M. L. Parry, "Potential Impact of Climate Change on World Food Supply," Nature 367 (1994): 133–38.

21. Earthquake prediction is controversial. Some European, Japanese, and Chinese teams are working on it, but the United States has largely stepped back.

"Following Loma Prieta, [earthquake] U.S. scientists put earthquake prediction on the back burner." Stephanie Kriner, "Ten Years After Loma Prieta, Mitigation Remains Focus of Earthquake Research," Disaster Relief News Stories, March 9, 2000 [online], www.disasterrelief.org/Disasters/000302quakeresearch/ [August 31, 2001].

Also, according to the United States Geological Service, the answer to the question, "Can You Predict Earthquakes?" is: "No. Neither the USGS nor Caltech nor any other scientists have ever predicted a major earthquake. They do not know how, and they do not expect to know how any time in the foreseeable future. However based on scientific data, probabilities can be calculated for potential future earthquakes." USGS Earthquake Hazards Program, "Frequently Asked Questions, Common Myths About Earthquakes" [online], earthquake.usgs.gov/faq//myths. html#1 [August 31, 2001].

Some minor U.S. research is still going on, for example, at Virginia Tech. See Becky Orfinger, "Taking a New Look at Earthquake Prediction," Disaster Relief News Stories, April 13, 2001 [online], www.disasterrelief.org/Disasters/010412quake study [August 31, 2001]. Andrew Freed, a geophysicist at the Carnegie Institution in Washington, D.C., and Jian Lin of Woods Hole Oceanographic Institution in Woods Hole, Massachusetts, have also published research in the journal *Nature* that contains new data which may lead to improvements in earthquake prediction. See Christina Ward, "Quake Connections? New Study Says Yes," Disaster Relief News Stories, May 15, 2001 [online], www.disasterrelief.org/Disasters/010514 quakestudy/ [August 31, 2001].

22. This was the observational experience of the author while working in wet tropical regions on flood control for many years. The referenced trend was also relayed to the author by Waldemar Boff, former head of the nongovernmental social self-help organization SEOP, in Petropolis RJ Brazil, during our close association working on water recycling technologies in Brazilian favelas between 1994 and 1997. Similar experiences were relayed to the author by scientific colleagues working in the field. However, one of the earliest, and most famous, contemporary references to water as an enemy in the tropics was written by Ernesto "Che" Guevara, who, after conducting numerous military campaigns, wrote, "Rain in tropical countries is continuous during certain months and . . . water is the enemy of all the things that the guerrilla fighter must carry: food, ammunition, medicine, paper, and clothing." Ernesto Guevara, *Guerrilla Warfare* (Cambridge: MIT Press, 1961), p. 53; also online at www.geocities.com/redencyclopedia/guevara.htm [September 5, 2001]. Similar experiences have been relayed by American soldiers who fought in the Philippines in World War II, and later in Vietnam.

23. For a summary of the impact of corruption on such work, see "Corruption and Integrity Improvement Initiatives in Developing Countries," *United Nations Development Programme*, April 1998 [online], magnet.undp.org/docs/efa/corruption. htm [August 9, 2001].

24. "Destructive Storms Drive Insurance Losses Up: Will Taxpayers Have to Bail Out Insurance Industry?" press release, World Watch Institute, Washington, D.C., March 26, 1999.

25. This information was relayed to the author in a series of discussions with Lai Chan, former city engineer, City of Port Louis, Republic of Mauritius, April 1997.

26. Ian Burton et al., *Report from the Adaptation Learning Experiment* (Toronto, Canada: Emergency Preparedness Canada, 1999) [online], www.msc-smc.ec.gc.ca/airg/pubs/alp.pdf [August 1, 2001].

27. Charu Singh, "Growing Public Fury against Government Inaction," Tehelka.com News (Delhi), January 31, 2001 [online], www.tehelka.com/current affairs/jan2001/ca013101gov.htm [August 1, 2001].

28. Richard P. Feynman, "An Outsider's Inside View of the Challenger Inquiry," *Physics Today* 41 (February 1988): 26–37.

29. Roger A. Pielke Jr. and Mary W. Downton, "Precipitation and Damaging Floods: Trends in the United States 1932–97," *Journal of Climate* 13, no. 20 (October 15, 2000): 3625–37; also online at www.esig.ucar.edu/HP_roger/pdf/jc1320.pdf [August 31, 2001].

See also "Societal Changes Increase Flood Costs," *USA Today*, October 24, 2000 [online], www.usatoday.com/weather/clisci/ floods102400.htm [August 31, 2001]. A good summary of how good disaster prevention techniques are ignored in favor of societal inertia is "Flood Policy Reform White Paper," Environmental Defense Fund (2001) [online], www.environmentaldefense.org/programs/Ecosystems/ArmyCorps/floodpolicy.html [August 31, 2001].

30. C. S. Holling, "Engineering Resilience vs. Ecological Resilience," in *Engineering Within Ecological Constraints*, ed. Peter Schultze (Washington, D.C.: National Academy Press, 1996), p. 37; also online at books.nap.edu/books/0309051983/html/37.html [August 15, 2001].

31. David Etkin, "Extreme Events and Natural Disasters in an Era of Increasing Environmental Change," Adaptation and Impacts Research Group, Institute for Environmental Studies, University of Toronto, 1999 [online], www.utoronto.ca/env/em-15-09.pdf [August 1, 2001].

32. "Adapting Urban Areas to Atmospheric Change," Environment Canada Adaptation and Impacts Research Group [online], www.smc-msc.ec.gc.ca/airg/vertical_gardens.htm [August 15, 2001].

33. Roger Pielke Jr. and Daniel Sarewitz, "Breaking the Global Warming Gridlock," *Atlantic Monthly*, July 2000, 54–64; also online at www.theatlantic.com/issues/2000/07/sarewitz.htm [August 1, 2001].

34. "Relevance and Application of the Principle of Precautionary Action to the Caribbean Environment Programme," CEP Technical Report No. 21, United Nations Environment Program (1993).

35. See pp. 53–54.

36. John McLaughlin, "Warning: Quake Coming," *Industry Standard*, March 26, 2001 [online], www.thestandard.com/article/0,1902,22872,00.html#rel_article [August 31, 2001].

37. Ibid.

38. J. Storrs Hall, "Utility Fog: The Stuff That Dreams Are Made Of" and "Utility Fog: A Universal Physical Substance" [online], www.aeiveos.com/~bradbury/Authors/Computing/Hall-JS/UFAUPS.html [July 12, 2001].

39. "Report on AFSC's Hurricane Mitch Relief and Reconstruction Program As of December 31, 1999," American Friends Service Committee, February 15, 2000 [online], www.afsc.org/emap/projects/mich0100.htm [September 5, 2001].

40. "Floyd's Legacy: Record Losses in North Carolina," CNN, September 22, 1999 [online], www.cnn.com/WEATHER/9909/22/ floyd.02/ [September 5, 2001].

41. This concept was put forward by K. Eric Drexler, Chris Peterson, and Gayle Pergamit in *Unbounding the Future: The Nanotechnology Revolution* (New York: Quill William Morrow, 1991), pp. 174–76.

42. Conventional greenhouses aren't as benign as we'd like to think. For a summary of environmental problems associated with their operations, see North Dakota State University's NDSU Extension Service, Greenhouse Pesticide Management, Environmental Protection Issues Web page at www.ag.ndsu.nodak.edu/aginfo/pesticid/publications/GreenH/gpm-2.htm#Pollution [August 31, 2001].

For greenhouse-aesthetics-related items, see the County of Santa Barbara Planning Commission Revised Draft Toro Canyon Plan, "Visual & Aesthetic Resources," February 1, 2001 [online], www.countyofsb.org/plandev/toro/revised_PC_plan/visual_aesthetics.pdf [August 31, 2001].

43. Gayle Pergamit, conversations with the author, October 2000. She is a rocket scientist and coauthor with Eric Drexler and Chris Peterson of *Unbounding the Future*.

44. Michael Braungart, Justus Engelfried, and Douglas Mulhall, "Criteria for Sustainable Development of Products and Production," *Fresenius Environmental Bulletin 2* (Basel, Switzerland: Birkhauser Verlag, 1993), pp. 70–77.

45. IPS is covered in many magazines and books, but a summary is carried on the U.S. Department of State Web site that excerpts portions of a book: Paul Hawken, Amory Lovins, and L. Hunter Lovins, *Natural Capitalism: The Next Industrial Revolution* (New York: Little, Brown & Company, 1999). See the Department of State Web site at usinfo.state.gov/products/pubs/archive/susteco/capital.htm [September 5, 2001].

46. The Virginia-based consultancy McDonough Braungart Design Chemistry markets the methodology. See the company's Web site at www.mbdc.com [September 10, 2001].

47. For more information on Climatex, along with other IPS products and methodologies, see the EPEA Web site at www.epea.com [July 14, 2001].

48. Ibid.

49. Ibid.

50. For a discussion of IFS, see Michael Braungart, Katja Hansen, and Douglas Mulhall, "Biomass Nutrient Recycling," *Water, Environment & Technology* 9, no. 8 (August 1997): 41–45. For work done by Prof. G. L. Chan, see "Integrated Farming," *PROSI*, no. 353 (June 1998) [online], www.prosi.net/mag98/353june/chan353.htm [September 5, 2001].

51. Douglas Mulhall and Katja Hansen, *Guide to Wastewater Recycling in Tropical Regions* (Brussels: European Commission, 1997); see also Web site version at www.hamburger-umweltinst.org/biomass/mnlwide/bnrhmpge.html [July 11, 2001]. This variety of integrated farming system integrates biological processes to achieve high agricultural production and purified water at low cost. Facilities in Brazil and China were designed, built, and operated under my management, the scientific direction of biological engineer Katja Hansen, and the local supervision of Valmir Fachini, Waldemar Boff, and a self-help organization, SEOP. IPS codeveloper Michael Braungart was scientific advisor. The work was supported by the European Commission, Henkel KGaA, and various municipalities. Other projects in Brazil, Fiji, Mauritius, Namibia, and Vietnam were designed and built by G. Lai Chan, an engineer who developed IFS for much of his life and codesigned the Hamburger Umweltinstitut facilities. Funding for his projects came from companies, UN organizations, and ZERI. A separate project in Spain was supervised by Valmir Fachini.

52. The Hamburger Umweltinstitut e.V. (Hamburg Environmental Institute) Web site is at www.hamburger-umweltinst.org/HUI.htm [July 14, 2001].

53. K. Eric Drexler, *Engines of Creation: The Coming Era of Nanotechnology* (New

York: Doubleday, 1986), p. 19. The concept of disassemblers is perhaps one of the most fundamental yet least understood and studied areas of nanotechnology.

54. RoboCup-Rescue Simulation Project Overview [online], robomec.cs.kobe-u.ac.jp/robocup-rescue/simoverview.html [November 24, 2001].

55. See the NASA Small Aircraft Transportation System (SATS) Web site at sats.larc.nasa.gov/main.html. See also an analysis of the Moller SkyCar System and SATS at www.skyaid.org/Skycar/MollerEval_ 10_00.htm#_Toc497038336 [July 10, 2001].

56. "German Train Deal for Shanghai," BBC News, January 21, 2001 [online], news.bbc.co.uk/hi/english/world/europe/newsid_1129000/1129295.stm [September 10, 2001].

CHAPTER 15: LESSONS FROM TOKYO—LEARNING TO PREDICT THE BIG ONE

1. A big barrier to earthquake prediction has been the inability to see down to the depth of tectonic plates, where movements often cause tremors. "Our problem is we're dealing with the Earth and you can't see inside the Earth," said John Filson, coordinator for the USGS earthquake hazards program, quoted by Stephanie Kriner in "Ten Years After Loma Prieta, Mitigation Remains Focus of Earthquake Research," Disaster Relief News Stories, March 9, 2000 [online], www.disasterrelief.org/Disasters/000302quakeresearch [August 31, 2001].

CHAPTER 16: THE LONG VALLEY CALDERA DEFENSE—AVOIDING A DARK AGE

1. Bill McGuire, *Apocalypse: A Natural History of Global Disasters* (London: Cassell & Co., 1999), p. 89.

2. Ibid., p. 61.

3. "Mega Tsunami: Wave of Destruction," transcript, *BBC Horizon,* BBC2, October 12, 2000 [online], www.bbc.co.uk/science/horizon/mega_tsunami.shtml [July 15, 2001].

4. Regarding the Chile landslides, see McGuire, *Apocalypse*, p. 110. Regarding the Cascadia earthquakes, refer to Roy Hyndman, "Giant Earthquakes of the Pacific Northwest," *Scientific American,* December 1995, 68–75; and Robert Yeats, *Living with Earthquakes in the Pacific Northwest* (Corvallis: Oregon State University Press, 1998), pp. 39–62.

5. Tristan Marshall, "The Drowning Wave," *New Scientist* 168, no. 2259 (October 7, 2000): 26.

CHAPTER 17: HOW TO AVERT ARMAGEDDON— HAVE AN ASTEROID FOR LUNCH

1. MIT Lincoln Laboratory, the LINEAR (Lincoln Near Earth Asteroid Research) Project [online], www.ll.mit.edu/linear [August 15, 2001].

2. Arthur C. Clarke and Stephen Baxter, *The Light of Other Days* (New York: Tom Doherty Associates, 2000).

3. Odds are discussed in Carl Sagan, *Pale Blue Dot* (New York: Random House, 1994), p. 313.

4. Carl Sagan and Steve Ostro of the Jet Propulsion Laboratory explained this threat as discussed in ibid., p. 316.

5. See NEAR Web sites at near.jhuapl.edu/ and neat.jpl.nasa.gov/.
See also: "A NEAR Landing on Eros," NASA New Science [online], liftoff. msfc.nasa.gov/News/2001/News-NEAR.asp [August 16, 2001].

CHAPTER 18: THE RIGHT QUESTIONS

1. Howard Bloom, *The Lucifer Principle: A Scientific Expedition into the Forces of History* (New York: Atlantic Monthly Press, 1995), p. 331.

2. Laura Lee, "Really Bad Predictions, Forecasts that Missed by a Mile," *Futurist* 34, no. 5 (September–October 2000): 22.

3. Ibid, p. 20.

4. Ibid.

5. Eric J. Savitz, "Read This, Then Go Back Up Your Data. Corporate America Is Realizing That Every Business Needs a Disaster Plan," *Fortune/CNET Tech Review* (winter 2002): 28–29.

6. K. Eric Drexler, *Engines of Creation: The Coming Era of Nanotechnology* (New York: Doubleday, 1986), p. 94.

7. *Report of the United Nations Conference on Environment and Development,* Rio de Janeiro, June 3–14, 1992, chap. 7, sec. F [online] www.agora21.org/rio92/A21_html/A21en/a21_07.html [August 14, 2001].

8. Sen. Charles E. Schumer, testimony before House Banking Committee hearing on money laundering, March 9, 2000 [online], www.senate.gov/~schumer/html/testimony_on_money_laundering_.html [July 12, 2001].

9. Duncan Graham-Rowe, "Bots Knock the Socks off City Slickers," New Scientist.com, August 11, 2001 [online], www.newscientist.com/hottopics/ai/botsknock.jsp [December 15, 2001]: "Robots can make more cash than people when they trade commodities, according to Jeffrey Kephart at IBM's research centre in Hawthorne, New York. . . . In IBM's test, software-based robotic trading agents–known as "bots"–made seven per cent more cash than people."

10. The United Tribes of the Americas is a nonprofit organization. Its Web site is at www.theuta.com/ [September 5, 2001].

11. The United Tribal Alliance was established as part of the Medicine Wheel

Accord Treaty in 1998 by tribes of indigenous peoples from around the world. It aims to achieve the same sovereign status as national governments have. See www.theuta.com/Pages/history.html [September 5, 2001].

12. See the UTA science and technology Web page, www.theuta.com/Pages/ scien.html [September 5, 2001].

13. Andrew Bolger and Clive Cookson, "Insurers Agree to 5-Year Ban on Genetic Screening," *Financial Times*, October 24, 2001, p. 14.

14. Jeff Karoub, "H.P. Official: Ignorance and Greed could Spoil Nanotech's Credibility," *Small Times*, November 30, 2001 [online], www.smalltimes.com/print_doc.cfm?doc_id=2655 [December 3, 2001].

15. Jill Vardy, "CEOs Post Some Spectacular Departures," *National Post*, December 7, 2001 [online], www.nationalpost.com/search/story.html?f=/stories/20011207/809739.html&qs=spectacular%20departures [December 15, 2001]. Also, Gordon Pitts, "To Every Season, a New CEO?" *Globe and Mail*, October 22, 2001, p. B3; D. C. Denison, "The Increasingly Short-Term CEO," *Financial Post*, November 6, 2001, p. FP12.

16. "Merger of Chrysler and Daimler-Benz was 'a farce that has failed,' " Car Today.com, November 15, 2000, www.cartoday.com/livenews/news/00/11/15.2.asp [September 4, 2001].

17. Signers of the Declaration of Scientists in Support of Agricultural Biotechnology include Nobel prize winners Norman Borlaug, James Watson, Paul Berg, Peter Doherty, and Paul Boyer. The AgBioWorld Foundation is a project of Prof. C. S. Prakash, www.AgBioWorld.org [July 14, 2001].

18. Coates and Jarrett is a Washington, D.C.–based consultancy that, according to its home page, is "one of the few organizations in the world exclusively dedicated to the study of the future. We urge our clients to think long term, 5 to 50 years. We give them a strategic focus, often speaking the unspeakable, giving all the good news and the bad." See www.coatesandjarratt.com/ [September 4, 2001].

19. The Copenhagen Institute for Futures Studies is a Denmark-based consultancy and think tank whose objective is "to strengthen the basis for decision-making in public and private organizations by creating awareness of the future and highlighting its importance to the present." See www.cifs.dk/ [September 4, 2001].

20. Max Glaskin, "Could Your Laptop Do Better Than a Real Estate Agent?" *New Scientist* 172, no. 2319 (December 1, 2001): 22. A prototype software package developed by a team at Glamorgan University in the U.K. has used neural networks to interpret government-housing statistics for forecasting housing prices, with results accurate to 3.8 percent over the last three years.

21. Richard P. Feynman, "There's Plenty of Room at the Bottom: An Invitation to Enter a New Field of Physics" (lecture given to the American Physical Society, California Institute of Technology, December 29, 1959 [online], www.physics.umn. edu/groups/mmc/personnel/pete/There%20is%20plenty%20of%20room%20at%20the%20bottom.htm [July 7, 2001].

22. Searches were done of the Visitor Center Web site on July 10, 2001, and January 25, 2002. Go to the site and search for "nanotechnology" or "artificial intelligence": www.kennedyspacecenter.com/html/site_search.asp [January 25, 2001].

23. The Mars Society Web site, www.marssociety.org [August 16, 2001].
24. Columbia University's Biosphere2 Center [online], www.bio2.edu [January 23, 2002], and interviews with a Biosphere2 official guide during author's visit to Biosphere2, September 1998.
25. "NASA Joins FEMA's Project Impact Effort," press release, FEMA Project Impact, December 7, 2000 [online], www.fema.gov/impact/nasa1207.htm [January 23, 2002].

CHAPTER 19: OVERCOMING CULTURAL AMNESIA

1. Edward Tenner, "The Shock of the Old," *MIT Technology Review* (December 2001): 50–51.
2. Edward Tenner, *Why Things Bite Back: Technology and the Revenge of Unintended Consequences* (New York: Vintage Books, 1996).
3. Frank Waters, *Book of the Hopi* (New York: Penguin Books, 1963), p. ix.
4. Ibid.
5. Thomas E. Mails, *The Hopi Survival Kit: The Prophesies, Instructions and Warnings Revealed by the Last Elders* (New York: Penguin Books, 1997).
6. Ibid., p. 69.
7. Hopi Information Network Web site www.recycles.org/hopi/messages/index.htm [September 4, 2001].
8. Graham Hancock, *Fingerprints of the Gods* (New York: Crown Trade Paperbacks, 1995), p. 502, quoting from the original reference Anthony S. Mercatante, *The Facts on File Encyclopedia of World Mythology and Legend* (London: Facts on File, 1988), p. 26.
9. Graham Hancock extensively catalogues these myths and legends throughout his book *Fingerprints of the Gods.*
10. The Greenpeace organization, for example, was founded on the basis of coastal tribe legends that describe how, someday, "Warriors of the Rainbow" would arise to save Earth from destruction. Some of the Greenpeace founders were anointed as such in a tribal ceremony in the 1970s. This is described in a book by Greenpeace cofounder Robert Hunter, *Warriors of the Rainbow: A Chronicle of the Greenpeace Movement* (New York: Holt Rinehart & Winston, 1979).
11. Lee Brown, "North American Indian Prophecies" (talk given at Continental Indigenous Council, Tanana Valley Fairgrounds, Fairbanks, Alaska, 1986) [online], www.thebearbyte.com/HopiProLB.htm [July 12, 2001].
12. Stephen Bertman, *Cultural Amnesia: America's Future and the Crisis of Memory* (Westport, Conn.: Prager, 2000), also excerpted in the *Futurist* magazine (January/February 2001): 46–51.
13. Ibid., p. 50.
14. Ibid., p. 51.
15. Bruce Lloyd, "The Wisdom of the World: 1,000 Messages for the New Millennium," 2000 [online], www.wfs.org/Q-intro.htm [July 8, 2001].

16. Bruce Lloyd, "Wisdom of the World, Messages for the New Millennium," *Futurist* (May–June 2000).

17. See "Safe keeping," *Economist Technology Quarterly* (September 20, 2001) [online], www.economist.com/science/tq/PrinterFriendly.cfm?story_ID=779564 [September 30, 2001]: "The sands of time may have left intact the stone-chiseled Egyptian hieroglyphics from 2000BC, but a portion of the original census reports of the United States of America for as recent a year as 1960–recorded on UNIVAC type II-A tapes–is now lost forever. Every day, important parts of the world's intellectual record vanish because of failures of the recording systems and media, the recording format becoming obsolete, or publishers who own the material going out of business, as well as the digital rewriting of history and the burning of digital records as political regimes come and go." The article goes on to describe how new digital archive repositories that massively replicate and catalogue data might help to stem the loss of such information.

18. Ray Kurzweil, *PBS News Hour* interview by David Gergen, about his book *The Age of Spiritual Machines: When Computers Exceed Human Intelligence*, September 13, 1999 [online], www.pbs.org/newshour/gergen/september99/gergen_9-13.html [July 14, 2001].

19. Stewart Brand, *The Clock of the Long Now* (New York: Basic Books, 1999).

20. "The Clock of the Long Now, A Talk with Stewart Brand, Interview with John Brockman," *Edge* 46 (August 15, 1998) [online], www.edge.org/documents/archive/edge46.html [July 12, 2001].

21. Ibid.

22. Robert Johnson, "Looking for Floods and Hearing Tales in Unlikely Places," *Wall Street Journal*, November 28, 1999, p. 1.

23. "Disaster that Struck the Ancients," BBC News Online, July 26, 2001, news.bbc.co.uk/hi/english/sci/tech/newsid_1458000/1458327.stm [August 17, 2001].

24. Ibid.

CHAPTER 20: BYPASSING THE ROAD TO HELL

1. Norman Borlaug, "Clash of Trends, Disappearing Water vs. Super Farms," *Futurist* (September–October 2000): 17.

2. Each quote in this paragraph is taken from Odile Nelson, "Science Crosses a Line," *National Post*, November 26, 2001, p. 1.

3. Andy Coghlan, Claire Ainsworth, and David Conar, "Don't Expect Any Miracles," *New Scientist* 172, no. 2319 (December 1, 2001): 4–6.

4. Gary Stix, "What Clones?" *Scientific American*, February 2002, 18–19.

5. For example, every year in America 8 *billion* chickens spend their lives stuffed so tightly together in cages that they can't spread their wings or walk. This is documented in "The Natural History of the Chicken," an independent film documentary produced and directed by Mark Lewis, aired on PBS, July 10, 2001, 8:00 P.M., KCET.

6. Jeremy Bentham, *An Introduction to the Principles of Morals & Legislation* (London, 1789).

7. A good summary of the first group of scientific studies on the emotional lives of animals is given by Susan McCarthy and Jeffrey Moussaieff Masson in their book *When Elephants Weep* (New York: Delta Paperback, 1996).

8. Ibid.

9. One of the most recognized studies on primates using tools was done by Jane Goodall. See Jane Goodall, *In the Shadow of Man* (Boston: Houghton Mifflin, 1988).

10. Margaret Munro, "Smarter Than They Look: Cows Are Much More Sophisticated Than Most People Realize, Researchers Say," *National Post*, November 19, 2001, p. A16.

11. Bob Holmes, "Slaughter of the Innocents," *New Scientist* 172, no. 2319 (December 1, 2001): 34–37.

12. Steven Wise, *Rattling the Cage* (Cambridge, Mass.: Perseus Publishing, 2000). See also "He Speaks for the Speechless," *Discover* (September 1, 2001), 18.

13. Michael Flynn, "A Debt Long Overdue," *Bulletin of the Atomic Scientist* (July/August 2001). This chronicles decades of officially sanctioned lying, deception, and obstruction by the U.S. federal government and major U.S. corporations regarding the cause and extent of illnesses among U.S. nuclear workers; a fact finally admitted officially by the Clinton administration.

14. See the Physicians for Social Responsibility Web site, www.psr.org/, and Computer Professionals for Social Responsibility Web site, www.cpsr.org/ [August 10, 2001].

15. In one of the more remarkable indicators of ambivalence over leadership, an ABC News poll gave former President Bill Clinton one of the highest ratings for job approval–65 percent favorable–among presidents on leaving the office, while 67 percent said he wasn't honest or trustworthy.
See Gary Langer, "Poll: Good Job by the Bad-Boy President Clinton Legacy Shows Wide Split Along Professional, Personal Lines," ABCNews.com, my.abcnews.go.com/PRINTERFRIENDLY?PAGE=abcsource.starwave.com/sections/politics/DailyNews/poll_ clintonlegacy010117.html [September 5, 2001].

16. Daniel S. Greenberg, *Science, Money and Politics: Political Triumph and Ethical Erosion* (Chicago: University of Chicago Press, 2001).

17. Keay Davidson, "Bloated, Whiny and Self-Important," *Scientific American*, September 2001, 98–101.

18. Laurie Garrett, *Betrayal of Trust: The Collapse of Global Public Health* (New York: Hyperion, 2000).

19. Edwin Black, *IBM and the Holocaust: The Strategic Alliance between Nazi Germany and America's Most Powerful Corporation* (New York: Crown, 2001).

20. Seth Shulman, *Owning the Future: Inside the Battles to Control the New Assets: Genes, Databases, and Technological Know-How That Make Up the Lifeblood of the New Economy* (Boston: Houghton Mifflin, 1999).

21. Jill Andresky Fraser, *White-Collar Sweatshop: The Deterioration of Work and Its Rewards in Corporate America* (New York: W. W. Norton & Co., 2001).

22. The meme as a self-replicating cluster of ideas is described in Howard Bloom, *The Lucifer Principle* (New York: Atlantic Monthly Press, 1995). Bloom expounds on the concept of memes as described by Richard Dawkins in *The Selfish Gene* (New York: Oxford University Press, 1976).

23. Bill Joy, "Why the Future Doesn't Need Us," *Wired,* April 2000 [online], www.wired.com/wired/archive/8.04/joy.html [July 7, 2001].

24. *Relevance and Application of the Principle of Precautionary Action to the Caribbean Environment Programme,* CEP Technical Report No. 21, United Nations Environment Program (1993).

25. "Mad Cow Disease: Nicole Fontaine Disappointed and Concerned by the Council's Conclusions," press release, President of the European Parliament, Brussels, November 21, 2000. Application of the precautionary principle led to a Europewide ban on the use of meat and bone meal in animal feed, provoking vast upheavals in European livestock markets. See also "E.U. Bans Meat Products in Animal Feed," Associated Press wire story, December 5, 2000 [online], www.canoe. ca/AllAboutCanoesNewsDec00/05_madcow-ap.html [July 11, 2000].

26. Andrew Pollack, "Farmers Joining State Efforts against Bioengineered Crops," *New York Times National Desk,* March 24, 2001.

27. Many of the discussions referenced in this paragraph took place at or around conferences sponsored by the Foresight Institute regarding the future of nanotechnology, between April and October 2000, in Palo Alto, California, and Washington, D.C.

28. "3M Phasing Out Some of Its Specialty Materials," news release, 3M, May 16, 2000 [online], www.3m.com/profile/pressbox/fluorochem.html [July 11, 2001].

29. "Reducing the Potential Risk of Developing Cancer from Exposure to Gallium Arsenide in the Microelectronics Industry," *NIOSH ALERT DHHS* Publication No. 88-100, October 1987.

30. This conversation occurred at the Eighth Foresight Conference on Molecular Nanotechnology, Bethesda, Maryland, November 2–5, 2000, after a presentation on the properties of materials at the nano scale.

31. A controversy surrounds the impact of energy efficiency on overall consumption. See "California Energy Commission," Nonresidential Electricity Consumption in Silicon Valley Increased from 23,190 Million kWh in 1990 to 26,554 Million kWh in 2000 [online], www.energy.ca.gov/electricity/silicon_ valley_consumption.html [September 4, 2001]. Andrew F. Hamm, "Energy Crisis Scares Silicon Valley," *Silicon Valley/ San Jose Business Journal,* January 19, 2001 [online], sanjose.bcentral.com/sanjose/stories/2001/01/22/story5.html [September 4, 2001]: ". . . Valley [energy] consumption has increased 33 percent since 1994 and demand should grow 11 percent by 2004." James Devitt, "LDEO'S Anderson Assesses National Energy Consumption, Offers Mixed Review of Bush Energy Plan," Earth Institute News, Columbia Earth Institute, June 4, 2001 [online], www.earthinstitute.columbia.edu/news/story6_1_01.html [August 14, 2001]: "...Analyzing previous research, Anderson cites specific surges in California's electricity demand to illustrate how high technology is sapping the state's energy supply:

Electricity consumption in the Silicon Valley is growing three times faster than anywhere else in California. The use of power by California-based Oracle and Sun Microsystems in 2000 amount to a 7-percent increase from 1999 levels. . . ."

32. For example: "Ford and General Motors have both indicated their ability to increase fuel efficiency of light trucks and SUV's by 25 percent by 2004. If the average efficiency in that category alone increased from 20.6 miles per gallon to 25 miles per gallon, within five years the savings would be nearly 1 million barrels of oil per day." "Senator Bingaman Letter Urges Review of Fuel Efficiency for Light Duty Vehicles and Calls for Flexibility in Fuel Delivery Driver's Restrictions," NASEO News, April 9, 2001 [online], www. naseo.org/news/2001_04.htm [September 4, 2001].

33. For an analysis of the economics of sewage treatment, see Katja Hansen and Douglas Mulhall, *Recycling Nutrients through Purification of Municipal Waste Water: Final Report to the European Commission* (Hamburg: Hamburger Umweltinstitut, 1998).

34. "Saving the Environment: A Jobs Engine for the 21st Century," press release, World Watch Institute, September 21, 2000 [online], www.worldwatch.org/alerts/000921.html [September 4, 2001].

35. An analysis of solutions to environmentally destructive subsidies is given by David Malin Roodman, "Reforming Subsidies," in *State of the World 1997*, ed. Lester R. Brown et al. (New York and London: W. W. Norton & Co., 1997), pp. 132–51.

36. John Carman, "'Frontline' Examines SUV's Safety Record," *San Francisco Chronicle*, February 21, 2002, p. D1.

37. John E. Losey, Linda S. Rayor, and Maureen E. Carter, "Transgenic Pollen Harms Monarch Larvae," *Nature* 399, no. 6733 (May 20, 1999): 214. Whether genetically modified maize kills more monarch butterflies than regular crops do has since been contested. See "Butterfly Balls, Genetically Modified Maize is not that Bad for Monarchs," *Economist*, September 22, 2001, 65.

38. "USDA Says U.S. Corn Exports Hurt by StarLink Chaos," CNN.com, November 16, 2000, www.cnn.com/2000/FOOD/news/11/16/biotech.glickman. reut/ [August 15, 2001].

39. "Declaration of Scientists in Support of Agricultural Biotechnology," AgBioworld [online], www.agbioworld.org/ petition.phtml [July 12, 2001].

40. For a list of internal documents released showing how tobacco companies systematically targeted teenagers, see "Secret Tobacco Documents News on the Web" [online], www.tobacco.org/News/98.01 mangini.html [August 15, 2001].

41. For further details on what happens to scientists such as Wigand when they fall afoul of disclosure agreements, see "Inside the Tobacco Deal: Jeffrey Wigand," *PBS Frontline*, aired May 12, 1998 [online], www.pbs.org/wgbh/pages/front line/shows/settlement/timelines/wigand. html [August 12, 2001].

42. Nancy Shute, "Allergy Epidemic," *U.S. News & World Report*, May 8, 2000 [online], www.usnews.com/usnews/issue/000508/allergies.htm [August 12, 2001]; and "More German Children Allergic," *Dispatch Online*, February 19, 1998, www. dispatch.co.za/1998/02/19/features/ALLERGY.HTM [August 12, 2001].

43. "Medicating Kids," *PBS Frontline,* April 10, 2001 [online], www.pbs.org/ wgbh/pages/frontline/shows/medicating/drugs/ [August 12, 2001].

44. Obesity isn't just another irritating health problem: it's an epidemic: "Overweight and Obesity Threaten U.S. Health Gains," press release, U.S. Department of Health and Human Services, December 13, 2001 [online], www.surgeongeneral. gov/todo/pressreleases/pr_obesity.htm [December 15, 2001].

When introducing a report entitled "The Surgeon General's Call to Action to Prevent and Decrease Overweight and Obesity," Surgeon General David Satcher said that "overweight and obesity may soon cause as much preventable disease and death as cigarette smoking." The report says that "approximately 300,000 U.S. deaths a year currently are associated with obesity and overweight (compared to more than 400,000 deaths a year associated with cigarette smoking). The total direct and indirect costs attributed to overweight and obesity amounted to $117 billion in the year 2000. In 1999, an estimated 61 percent of U.S. adults were overweight, along with 13 percent of children and adolescents. Obesity among adults has doubled since 1980, while overweight among adolescents has tripled."

45. Josh Grossberg, "Erin Brockovich–Fact or Fiction?" E! Online News, July 25, 2000 [online], www.eonline.com/News/Items/0,1,6824,00. html [August 10, 2001].

46. Greenberg, *Science, Money and Politics,* p. 222. Greenberg documents the contrast between a chronically poor public understanding of science, and historically steady increases in funding for science. See chap. 13, "Public Understanding of Science," pp. 205–34.

47. "Truth or Consequences," *Economist,* September 15, 2001, 70–71.

48. "The Real Scandal: America's Capital Markets Are Not the Paragons They Were Cracked Up to Be," *Economist,* January 19, 2002 [online], web1.infotrac. galegroup.com/itw/infomark/706/673/20271539w1/13!help_InfoMark [January 20, 2002].

49. Kelly Hearn, "Army Looks to Nanotechnology, Robotics," July 1, 2001 [online], www.vny.com/cf/News/upidetail.cfm?QID=198956 [August 8, 2000]; "Congressional Panel Hears Plans for Nanotechnology in the Military," June 27, 2001 [online], www.smalltimes.com/print_doc.cfm?doc_id=1546 [August 8, 2001].

50. See the Web site "Will Spiritual Robots Replace Humanity by 2100?" for comments by these technologists on risks and benefits of molecular technologies, www.stanford.edu/dept/symbol/Hofstadter-event.html [August 7, 2001].

51. K. Eric Drexler, *Engines of Creation: The Coming Era of Nanotechnology* (New York: Doubleday, 1986). Nanotechnology risks are described in part 3, "Dangers and Hopes," pp. 171–231.

52. Doug Brown, "Army to Invest in Cyberuniform Center," ZDNet, June 21, 2001 [online], www.zdnet.com/intweek/stories/news/0,4164,2779752,00.html [August 8, 2001].

53. Thomas McCarthy, "Molecular Nanotechnology and The World System" [online], www.mccarthy.cx/WorldSystem/nature.htm [August 8, 2001].

54. Greg Jaffe, "Tug of War; In the New Military, Technology May Alter the Chain of Command," *Wall Street Journal,* March 30, 2001, p. A1.

55. "Watching You," *Economist*, September 22, 2001, 64. Part of this article discusses the pros and cons of remote control antihijacking systems that may themselves be taken over by malicious hackers.

56. National Institutes of Health, *Guidelines for Research Involving Recombinant DNA Molecules*, January 2001 (effective June 24, 1994, with several amendments in the 1990s) [online], www4.od.nih.gov/oba/rac/guidelines/guidejan01.htm [July 7, 2001].

CHAPTER 21: USING OPEN SOURCE

1. John Markoff, "News Analysis, Judging a Moving Target," *New York Times*, June 29, 2001, p. C1.

2. The private company Celera and the National Institutes of Health fought with each other for years over decoding the human genome, then started to cooperate, but still continued fighting. Justin Gillis, "Celera wins NIH Grant," WashingtonPost.com, March 1, 2001, www.washtech.com/news/biotech/7931-1.html [August 15, 2001].

3. Andy Coghlan, "Huge Chunk of Our Genome is Set to be Privatised," *New Scientist* 172, no. 2321 (December 15, 2001): 6.

4. The struggle over patents and generic drugs intensified in September 2001, when Brazil threatened to produce its own version of the AIDS drug Viracept. At the time of writing it had negotiated a 40 percent price reduction with the drug's manufacturer. See Jennifer L. Rich, "Roche Reaches Accord on Drug With Brazil," *New York Times*, September 1, 2001 [online], www.nytimes.com/2001/09/01/business/worldbusiness/01DRUG.html?searchpv=day05 [September 1, 2001].

For an example of how the battle over genetic resources began, see Shulman, *Owning the Future: Inside the Battles to Control the New Assets: Genes, Databases, and Technological Know-How That Make Up the Lifeblood of the New Economy* (Boston: Houghton Mifflin, 1999), pp. 131–35.

5. "Study Examines Data Withholding in Academic Genetics. Many Genetic Researchers Denied Access to Resources Related to Published Studies," press release, Massachusetts General Hospital, January 22, 2002 [online], www.mgh.harvard.edu/DEPTS/pubaffairs/Releases/012202datawithholding.htm.

6. Shulman, *Owning the Future*, p. 81.

7. For more details on definition of open source, see "The Open Source Definition," OpenSource.org, www.opensource.org/docs/definition_plain.html [August 8, 2001].

8. Lee Gomes, "Is Microsoft Secretly Using Open Source?" *Wall Street Journal*, June 18, 2001 [online], www.zdnet.com/zdnn/stories/news/0,4586,2776342,00.html ?chkpt=zdnn_tp_ [July 15, 2001].

9. Ibid.

10. See the "Open Source Disclosure Project" [online], www.foresight.org/priorart/index.html [July 14, 2001].

11. Wade Roush, "Web Tolls Ahead?" *MIT Technology Review* (January/February 2002): 20–21.

12. Charles Babcock, "Torvalds blasts Microsoft's Mundie on open source," ZDNet News, May 12, 2001 [online], www.zdnetindia.com/news/international/stories/22289.html [July 15, 2001].

CHAPTER 22: REDESIGNING DEMOCRACY FOR ARTIFICIAL INTELLIGENCE

1. Bruce Lloyd, "The Wisdom of the World: 1,000 messages for the New Millennium" [online], www.wfs.org/Q-intro.html [September 5, 2001].

2. Theodore Kaczynski, "The Unabomber Manifesto, Industrial Society and Its Future," published jointly by the *New York Times* and the *Washington Post*, September 19, 1995.

3. Bill Joy, "Why the Future Doesn't Need Us," *Wired*, April 2000 [online] www.wired.com/wired/archive/8.04/joy.html [July 7, 2001].

4. Ibid.

5. Thomas McCarthy, "Molecular Nanotechnology and the World System" (1999–2000) [online], www.mccarthy.cx/worldsystem/nature.htm [August 8, 2001].

6. David W. Moore, "Americans Support Teaching Creationism As Well As Evolution in Public Schools Divided on Origins of Human Species," Gallup News Service, August 30, 1999 [online], www.gallup.com/poll/releases/pr990830.asp [September 7, 2001].

7. Kim Stanley Robinson, *Red Mars, Green Mars,* and *Blue Mars* (New York: Bantam, 1993–97).

8. "The Mars Declaration," Mars Society Web site, www.mars society.org [July 14, 2001].

CHAPTER 23: LIBERATING EACH ONE OF US

1. Richard Eckersley, "Is Life Really Getting Better?" *Futurist* (January 1999): 23.

2. Ibid., p. 24.

3. United Nations Development Program, *Human Development Report* 1999 [online], www.undp.org/hdro/report.html [August 16, 2001], pp. 128, 138. This statistic is often quoted, and appears, for example, in Jean Bertrand Aristide, *Eyes of the Heart, Seeking a Path for the Poor in the Age of Globalization* (Monroe, Maine: Common Courage Press, 2000), p. 5.

4. Jill Andresky Fraser, *White-Collar Sweatshop: The Deterioriation of Work and Its Rewards in Corporate America* (New York: W. W. Norton & Co., 2001), p. 3.

5. Some leading business writers and economists have documented the effects of cybereconomy work on the individual. Richard Sennett, *The Corrosion of Character: The Personal Consequences of Work in the New Capitalism* (New York: W.W. Norton & Co., 1998); and Joanne B. Ciulla, *The Working Life: The Promise and Betrayal of Modern Work* (New York: Times Business Books, 2000).

6. Mortimer Zuckerman, "A Nation Divided," *U.S. News & World Report,* October 18 1999 [online], www.usnews.com/usnews/issue/991018/18edit.htm [July 12, 2001].

7. "Youth Trends in Brief," *Futurist* (March/April 2001): 7.

8. As measured by the so-called National Income and Product Accounts (NIPA), the personal saving rate (PSR) is the fraction of personal income that is not consumed for taxes, Social Security and Medicare, and personal expenses such as mortgage, the grocery bill, and clothes. From 1990 to 2001, the U.S. personal savings rate declined from about 7 percent to −1 percent, then rose slightly in early 2002 as consumer spending rates tapered off. There has been an extensive discussion over the validity of this measure. Many economists argue that it isn't an accurate measure of real savings, because stock market and real estate gains are excluded from the savings rate, and the markets produced great gains for a decade. However, these arguments took on a different meaning after the NASDAQ dropped 60 percent in 2000, effectively reversing the "wealth effect" referenced frequently by Federal Reserve Chairman Alan Greenspan. When added to the negative savings rate, such asset losses exacerbated the real decline rather than negating it. For a discussion of the issue, see K. C. Swanson, "Why Is the Personal Savings Rate Negative for the First Time Since 1933?" The Street.com, February 28, 2001, www.thestreet.com/funds/investing/1322929.html [August 16, 2001].

9. "Employers to Face Double Digit Health Care Cost Increases For Third Consecutive Year," Businesswire, October 23, 2000 [online], www.mostchoice.com/Business/Benefits/Health/B_Ben_Health_ health_care_costs_rise.cfm [September 1, 2001]; "As Insurance Companies Jack Up Premiums, Employers and Workers Face Higher Costs for Health Insurance Coverage," Labor Research Association, December 28, 2000 [online], www.laborresearch.org/ health_care/ins_costs_2001.htm [September 1, 2001].

10. This phenomenon is nothing new. In a *New York Times* piece, Jerome Segal, a chronicler of the General Progress Index developed by Redefining Progress in Oakland, California, depicts how most families still spend about four-fifths of their core budget on core needs, often in the same ways as they did a hundred years ago. Food and clothing costs have dropped while housing costs have gone up as a percentage of that core. "In 1901, 80 percent of spending went for food, housing and clothing. In 1999, 81 percent of spending went for food, housing, clothing, transportation and health care. A century of economic growth has certainly brought some genuine progress, but it is less clear-cut than we sometimes think." Jerome M. Segal, "What We Work for Now," *New York Times,* September 3, 2001 [online], www.rprogress. org/media/opeds/010903_ca.html [September 30, 2001].

11. *High School and Youth Trends, National Institute on Drug Abuse,* Monitoring the Future Study, 2000 [online], www.drugabuse.gov/Infofax/HSYouthtrends.html [July 12, 2001]; Donna Leinwand, "Club Drugs Sending More Young People to Hospitals," *USA Today,* July 25, 2001, p. 3A [online], www.usatoday.com/usatonline/2001 0725/3508804s.htm [July 25, 2001]. Moreover, a multiyear study released in 2001 concludes that the aim of Congress to make American schools drug free by the year 2000 has been a "complete failure." See "Malignant Neglect: Substance Abuse and

America's Schools," press release, The National Center on Addiction and Substance Abuse at Columbia University, September 5, 2001 [online], www.casacolumbia.org/ newsletter1457/newsletter_show.htm?doc_id=80623 [September 5, 2001].

12. Daniel Q. Haney, "Obesity Growing as World Problem," *Los Angeles Times,* February 16, 2002 [online], www.latimes.com/news/nationworld/wire/sns-ap-world-obesity0216feb16.story [February 16, 2002].

13. Alvin Toffler, *Future Shock* (New York: Bantam Books, 1974), p. 259.

14. Marvin Cetron and Owen Davies, "Trends Now Changing the World," *Futurist* (March–April 2001): 27–35.

15. "National Health Care Expenditures, 1999," National Health Care Administration [online], www.hcfa.gov/stats/nhe-oact/ [July 12, 2001].

16. "National Health Care Expenditures Projections, 2000–2010," National Health Care Administration [online], www.hcfa.gov/stats/nhe-proj/proj2000/ default.htm [July 12, 2001].

17. Kate Zernike, "Anti-Drug Program Says It Will Adopt a New Strategy," *New York Times,* February 15, 2001 [online], www.nytimes.com/2001/02/15/national/ 15DARE.html [July 12, 2001].

18. This decline is documented by Laurie Garrett, *Betrayal of Trust: The Collapse of Global Public Health* (New York: Hyperion, 2000), chap. 4.

19. David Shaw, "Inhale, Lie, Exhale, Lie," *Los Angeles Times,* February 13, 2001, p. A1. This was part of a four part series regarding deceptive practices in Hollywood.

20. Ibid.

21. Greg Krikorian, "Federal and State Prison Populations Soared Under Clinton, Report Finds," *Los Angeles Times,* February 19, 2001 [online], www.latimes. com/news/state/20010219/t000015042.html [July 14, 2001].

22. "The United States is the country with the highest prison population rate in the world–just over 700 per 100,000 of the national population, or five times the overall world rate. In announcing these figures recently the U.S. Department of Justice reported that this means that 1 in every 142 United States residents is being held in a penal institution." Roy Walmsley, "An Overview of World Imprisonment: Global Prison Populations, Trends and Solutions" (paper presented at the United Nations Programme Network Institutes Technical Assistance Workshop, Vienna, Austria, May 10, 2001) [online], www.kcl.ac.uk/depsta/rel/icps/world_imprison ment. doc [September 10, 2001].

23. Angela Antonelli, "Regulation: Demanding Accountability and Common Sense," *Issues 2000, The Candidates' Briefing Book* (Washington, D.C.: Heritage Foundation [online], www.heritage.org/issues/chap4.html [July 12, 2001]. Since this statistic was published, thousands of new pages were added to the Federal Register, with a big surge at the end of the Clinton administration. Some of these were rolled back by the Bush administration, but a net surplus of pages remains on the books.

24. *Human Development Report 1999,* United Nations Development Program [online], www.undp.org/hdro/report.html [August 16, 2001], pp. 128, 138.

25. Ibid.

26. Ellen E. Schultz, "Companies Quietly Use Mergers and Spinoffs to Cut Workers Benefits," *Wall Street Journal,* December 27, 2000, p. 1.

27. Ibid.

28. George Soros, "The Capitalist Threat," *Atlantic Monthly*, February 1997, 45–58; *Open Society, Reforming Global Capitalism Reconsidered* (New York: Public Affairs, 2000).

29. Robert Pool, *Fat: Fighting the Obesity Epidemic* (Oxford: Oxford University Press, 2000), conclusion.

30. "More Teen Girls Smoking in U.S.," CNN, March 28, 2001 [online], www.cnn.com/2001/HEALTH/03/27/women.smoking.02/index.html [July 14, 2001].

31. Lloyd Steven Sieden, *Buckminster Fuller's Universe* (Cambridge, Mass.: Perseus Publishing, 1989), p. 33.

32. Robert Freitas Jr. has given an overview of potential nanomedicine applications with his *Nanomedicine, Volume 1: Basic Capabilities* (Georgetown, Tex.: Landes Bioscience, 1999). See also online www.nanomedicine.com/ [August 16, 2001].

33. Kim Stanley Robinson, *Red Mars, Green Mars*, and *Blue Mars* (New York: Bantam, 1993–97).

34. "15. Antenna Criteria: Antenna on or above a structure shall be subject to the following: . . . f. The structure must be architecturally and visually (color, size, bulk) compatible with surrounding existing buildings, structures, vegetation, and uses. Such facilities will be considered architecturally and visually compatible if they are camouflaged to disguise the facility. . . ."

"An Ordinance of the City of University Place, Washington, Pertaining to Personal Wireless Telecommunications Facilities," University Place, Washington, Ordinance No. 152, June 1997 [online], www.mrsc.org/ords/t-z/u54-152.htm [September 1, 2001].

35. Melissa Mertl, "An E-Nose for Trouble," *Discover*, September 2001, 20.

36. See "Nano Bar Codes–Bio-Nanotech in Action," *Scientific American*, September 2001, 71, sidebar.

37. Roger Highfield, "'Electric Trees' Envisioned as Power Source," *Vancouver Sun*, October 25, 2001, p. A12. Prof. Bernard Witholt, chairman of the Institute of Biotechnology in Zurich, forecasts that plants may produce electricity from sugars or photosynthesis.

38. Smart dust, robotics.eecs.berkeley.edu/~pister/SmartDust/ [July 14, 2001]. See also DARPA MEMS Web site, www.darpa.mil/mto/mems/summaries/Projects/index.html [July 14, 2001].

39. Ibid.

40. Ray Kurzweil, "The Age of Spiritual Machines" (sermon to the First Unitarian Universalist Church of San Diego, January 23, 2000) [online], firstuusandiego.org/public/sermons/012300Kurzweil.ram [July 11, 2001].

41. Michael Flynn, "A Debt Long Overdue," *Bulletin of the Atomic Scientist* (July/August 2001).

42. Paul Kurtz, "Humanist Manifesto 2000. A Call for New Planetary Humanism," *Free Inquiry*, fall 1999, p. 11.

43. "Leading Causes of Death in the Developing World," *Newsweek*, February 4, 2002, 47. The graph depicts heart disease, lower respiratory infections, and stroke as the three leading causes of death in the developing world, and cites the World Health Organization as the source of statistics.

370

44. In 1998 more than 300 million persons were displaced by natural disasters, according to the World Resources Institute's State of the World Report (as reported by CNN, January 16, 1999 [online], www.cnn.com/US/9901/16/calamity). A large body of broadly circulated news reports show what happens to human rights when a huge natural disaster strikes. Often those rights disappear, at least temporarily. Residents are subject to the whims of distant government authorities far from the scene. The 2001 Gujarat earthquake in India was an example, where months after the quake hundreds of thousands were still living in subhuman conditions while central governments dallied over what to do. An attempt has been made to establish rights for persons displaces by natural disasters. See "Guiding Principles on Internal Displacement," United Nations Office for the Coordination of Humanitarian Affairs, first submitted in 1998, by Francis Deng, UN special representative for internally displaced persons [online], www.reliefweb.int/ocha_ol/pub/idp_gp/ idp.html [July 12, 2001].

45. Dava Sobel, *Longitude: The True Story of a Lone Genius Who Solved the Greatest Scientific Problem of His Time* (London: Fourth Estate Ltd., 1998), p. 53.

APPENDIX A

1. The three laws of robotics are summarized by Isaac Asimov in *Robot Visions* (New York: Byron Preiss, 1991), p. 423. These are taken from his *Handbook of Robotics, 56th Edition, 2058 A.D.*, as quoted in Isaac Asimov, *I, Robot* (New York: Gnome Press, 1950). In Isaac Asimov, *Robots and Empire* (New York: Doubleday, 1985), chap. 63, the "Zeroth Law" is extrapolated, and the other three laws modified accordingly.

APPENDIX B

1. Foresight Guidelines on Molecular Nanotechnology, Foresight Institute [online], www.foresight.org/guidelines/current.html [December 15, 2001].

A BRIEF SAMPLING FOR THE SCIENTIFICALLY INCLINED READER

NANOTECHNOLOGY

(Note: This selection focuses on the theoretical possibilities of nanotechnology.)

Drexler, K. E. "Building Molecular Machine Systems." *Trends in Biotechnology* 17, no. 1 (1999): 5–7; [online] www.imm. org/Reports/Rep008.html.

———. "Molecular Manufacturing: Perspectives on the Ultimate Limits of Fabrication." *Phil. Trans. R. Soc. London A* 353 (1995): 323–31.

———. "Molecular Machines: Physical Principles and Implementation Strategies." *Annual Review of Biophysics and Biomolecular Structure* 23 (1994): 337–405.

Freitas, Robert A., Jr. "Tangible Nanomoney." *Nanotechnology Industries Newsletter* 2 (July 2000): 2–11; [online] www.zyvex.com/Publica tions/papers/Nanomoney.html.

———. "Respirocytes in Nanomedicine." *Graft: Organ and Cell Transplantation* 52 (May 2050): 148–54 (special "Future Issue," published May 2000); [online] www. eurekah.com/nanomedicine/index.html.

———. "Some Limits to Global Ecophagy by Biovorous Nanoreplicators, with Public Policy Recommendations." *Zyvex* (April 2000); [online] www.foresight.org/NanoRev/Ecophagy.html.

———. "Exploratory Design in Medical Nanotechnology: A Mechanical Artificial Red Cell." Artificial Cells, Blood Substitutes, and Immobil. *Biotech.* 26 (1998): 411–30; [online] www.foresight.org/Nanomedicine/Respirocytes. html.

Minsky, Marvin. "Virtual Molecular Reality." In *Prospects in Nanotechnology: Toward Molecular Manufacturing*, edited by Markus Krumenacker and James Lewis. Wiley, 1995.

ROBOTICS

Moravec, Hans. "Robots: Re-evolving Minds at 10^7 Times Nature's Speed." *Cerebrum* 3, no. 2 (spring 2001): 34–49; [online] www.frc.ri.cmu.edu/~hpm/project.archive/robot. papers/2000/Cerebrum.html.

———. "Robust Navigation by Probabilistic Volumetric Sensing: Mass-Market Mobile Utility Robot Development." *Progress Report 4 (Photorealism!?)*, March 20, 2001 (DARPA Mobile Autonomous Robot Software program support) Carnegie Mellon University. [online] www.ri.cmu.edu/~hpm/project.archive/robot. papers/2001/ARPA.MARS/Report.0103.html.

ARTIFICIAL INTELLIGENCE AND ENABLING TECHNOLOGY

Gershenfeld, Neil, and Isaac L. Chuang. "Quantum Computing with Molecules." *Scientific American*, June 1998, 66–71; [online] www.media.mit.edu/physics/publications/papers/98.06.sciam/0698gershenfeld.html.

Kurzweil, Raymond. "Live Forever–Uploading The Human Brain . . . Closer Than You Think." February 1, 2000; [online] www.kurzweilai.net/meme/frame. html?main=/articles/art0157.html.

———. "The Twenty-First Century: A Confluence of Accelerating Revolutions." Keynote speech at the Foresight 8th Tech. Conference, November 2000; [online] www.kurzweilai.net/meme/frame.html?main=/articles/art0184.html. (A complete archive of works written by Raymond C. Kurzweil and a directory of selected articles about Kurzweil or the Kurzweil companies can be found online at www.kurzweilai.net/meme/frame.html?m=10.)

Kymisis, John, Clyde Kendall, Joseph Paradiso, and Neil Gershenfeld. "Parasitic Power Harvesting in Shoes." *Proc. of the Second IEEE International Conference on Wearable Computing* (October 1998): 132–39; [online] www.media.mit.edu/physics/publications/papers/98.08.PP_wearcon_final.pdf.

Maguire, Yael, Edward Boyden, and Neil Gershenfeld. "Towards a Table-Top Quantum Computer." *IBM Systems Journal* 39, no. 3/4 (2000): 823–39; [online] www.media.mit.edu/physics/publications/papers/00.11.maguire.pdf.

Minsky, Marvin. "Why People Think Computers Can't." *AI* 3, no. 4 (fall 1982). Reprinted in *Technology Review* (November/December 1983) and in Denis P. Donnelly, ed., *The Computer Culture* (Cranbury, N.J.: Associated University Presses, 1985).

GENETIC PROGRAMMING, DNA COMPUTING, AND POTENTIALLY RELATED GENETIC TECHNOLOGIES

Bloom, Howard. "Instant Evolution–The Influence of the City on Human Genes: A Speculative Case." Visiting Scholar New York University Presented at the Fifth Center for Human Evolution Workshop "Cultural Evolution" Seattle, Washington, May 11, 2000; [online] www.howardbloom.net.

———. "Sociobiology and Politics Beyond the Supercomputer: Social Groups as Self-Invention Machines." Paper presented before a joint session of the European Sociobiological Society, the International Political Science Association, and The Association for Politics and the Life Sciences. Later appeared in Albert Somlt and Steven A. Peterson, eds., *Research in Biopolitics*, vol. 6, Sociobiology and Biopolitics (Greenwich, Conn.: JAI Press, 1998), pp. 43–64; [online] www.howardbloom.net.

Koza, John R., Forrest H. Bennett III, David Andre, and Martin A. Keane. "Genetic Programming: Biologically Inspired Computation That Creatively Solves Non-Trivial Problems." In *Evolution as Computation, DIMACS Workshop, Princeton, January 1999*, edited by Laura F. Landweber and Erik Winfree (Heidelberg: Springer-Verlag, 2001), pp. 15–44; [online] www.genetic-programming.com/eac2001chapter.pdf.

Koza, John R., Martin A. Keane, Jessen Yu, Forrest H Bennett III, and William Mydlowec. "Automatic Creation of Human-Competitive Programs and Controllers by Means of Genetic Programming." *Genetic Programming and Evolvable Machines* 1, no. 1/2 (2000): 121–64; [online] www.genetic-programming.com/gpem control.pdf.

Myers, Eugene W., Granger G. Sutton, Art L. Delcher, Ian M. Dew, Dan P. Fasulo, Michael J. Flanigan, Saul A. Kravitz, Clark M. Mobarry, Knut H. J. Reinert, Karin A. Remington, Eric L. Anson, Randall A. Bolanos, Hui-Hsien Chou, Catherine M. Jordan, Aaron L. Halpern, Stefano Lonardi, Ellen M. Beasley, Rhonda C. Brandon, Lin Chen, Patrick J. Dunn, Zhongwu Lai, Yong Liang, Deborah R. Nusskern, Ming Zhan, Qing Zhang, Xiangqun Zheng, Gerald M. Rubin, Mark D. Adams, and J. Craig Venter. "A Whole-Genome Assembly of *Drosophila*." *Science* 287 (March 24, 2000): 2196–2204.

Rubin, Gerald M., Mark D. Yandell, Jennifer R. Wortman, George L. Gabor Miklos, Catherine R. Nelson, Iswar K. Hariharan, Mark E. Fortini, Peter W. Li, Rolf Apweiler, Wolfgang Fleischmann, J. Michael Cherry, Steven Henikoff, Marian P. Skupski, Sima Misra, Michael Ashburner, Ewan Birney, Mark S. Boguski, Thomas Brody, Peter Brokstein, Susan E. Celniker, Stephen A. Chervitz, David Coates, Anibal Cravchik, Andrei Gabrielian, Richard F. Galle, William M. Gelbart, Reed A. George, Lawrence S. B. Goldstein, Fangcheng Gong, Ping Guan, Nomi L. Harris, Bruce A. Hay, Roger A. Hoskins, Jiayin Li, Zhenya Li, Richard O. Hynes, S. J. M. Jones, Peter M. Kuehl, Bruno Lemaitre, J. Troy Littleton, Deborah K. Morrison, Chris Mungall, Patrick H. O'Farrell, Oxana K. Pickeral,

Chris Shue, Leslie B. Vosshall, Jiong Zhang, Qi Zhao, Xiangqun H. Zheng, Fei Zhong, Wenyan Zhong, Richard Gibbs, J. Craig Venter, Mark D. Adams, and Suzanna Lewis. "Comparative Genomics of the Eukaryotes." *Science* 287 (March 24, 2000): 2204–15.

Yurke, Bernard, Andrew J. Turberfield, Allen P. Mills Jr., Friedrich C. Simmel, and Jennifer L. Neumann. "A DNA-fuelled molecular machine made of DNA." *Nature* 406 (August 10, 2000): 650–58.

NATURAL DISASTERS, CLIMATE CHANGES, AND SOCIOECONOMIC IMPACTS

Alley, R. B., et al. "Holocene Climatic Instability: Aprominent, Widespread even 8200 Years Ago." *Geology* 25, no. 6 (1997): 483–86.

Alley, R. B., et al. "Abrupt Increase in Snow Accumulation at the End of the Younger Dryas Event." *Nature* 362 (1993): 527–29.

Bass, B., T. Byers, and N.-M. Lister. "Integrating Ecohydrological Research into Land-Use Planning." *Hydrological Processes* (December 1998).

Bostrom, Nick. "The Doomsday Argument Is Alive and Kicking." *Mind* 108, no. 431 (1999): 539–50.

Dore, M., and David A. Etkin. "The Importance of Measuring the Social Costs of Natural Disasters At a Time of Climate Change." *Australian Journal of Emergency Management* 15, no. 3 (spring 2000): 46–48; [online] www.msc-smc.ec.gc.ca/airg/Social_costs_Disasters_Australian.doc.

Etkin, D. A. "Extreme Events and Natural Disasters in an Era of Increasing Environmental Change." *Emerging Environmental Issues in Ontario, Report of a May 14, 1999, Workshop.* Sponsored by the Institute for Environmental Studies and the Integrated Environmental Planning Division, Ontario Ministry of the Environment. (Toronto: University of Toronto, Institute for Environmental Studies) I.E.S. Environmental Monograph No. 15 (1999), pp. 19–22.

———. "Natural Hazards." *Human Activity and the Environment* (Ottawa: Statistics Canada, 1999), pp. 187–92.

———. "Risk Transference and Related Trends: Driving Forces Towards More Mega-Disasters." *Environmental Hazards* 1 (1999): 69–75; [online] www.msc-smc.ec.gc.ca/airg/pubs/risk_transference.pdf.

———. "Socio-Economic Impacts of Extreme Events, and Issues on the Effectiveness of Warnings." Proceedings: Sixth Workshop on Operational Meteorology. Nov. 29–Dec. 3, 1999. Halifax, N.S., pp. 189–97.

Firth, C. R., and W. J. McGuire. "Volcanoes in the Quaternary." *Geol. Soc. Lond. Spec. Public.* Bath: Geological Society Publishing House, 1999.

McGuire, W. J. "A Risky Business." *Geographical* 71, no. 2 (1999): 10–11.

———. "Island Edifice Failure and Associated Tsunami Hazards." *Pure and Applied Geophysics* 157 (2000): 899–955.

———. 1999. "Cooling the Inferno." *Reinsurance* 29, no. 11 (1999): 24.

McGuire, W. J., D. R. Griffiths, P. L. Hancock, and I. S. Stewart, eds. *The Archaeology of Geological Catastrophes.* Bath: Geological Society Publishing House, 2000.

Moss, J. L., W. J. McGuire, and D. Page. "Ground Deformation Monitoring of a Potential Landslide at La Palma, Canary Islands." *Journal of Volcanology and Geothermal Research* 94 (2000): 251–65.

Talley, L. D., G. J. Fryer, and R. Lumpkin. "Oceanography." In *The Pacific Islands: Environment & Society,* edited by M. Rapaport, pp. 19–32. Honolulu: Bess Press, 1999.

Xia, J., G. Huang, and B. Bass. "Combination of a Differentiated Prediction Approach and Interval Analysis for the Prediction of Weather Variables under Uncertainty." *Journal of Environmental Management* 49 (1997): 93–106.

Zebrowski, E. J. *Perils of a Restless Planet: Scientific Perspectives on Natural Disasters.* Cambridge: Cambridge University Press, 1999.

MASS EXTINCTIONS

Ward, P. "The Cretaceous/Tertiary boundary in the marine realm: a 1990 perspective." *GSA Special Paper* 247 (1990): 425–31.

———. *On Methuselah's Trail: Living Fossils and the Great Extinctions.* New York: W. H. Freeman and Company, 1991.

———. *Rivers in Time: The Search for Clues to Earth's Mass Extinctions.* New York: Columbia University Press, 2001.

INDEX

A.I. (film), 26, 104
ABB, 66
aboriginal peoples, 242
Academy of Sciences, China, 134
Adams, Douglas, 23, 86, 209
adaptation, 191; and cultural roadblocks, 202; and Impacts
  Research Group, En-
  vironment Canada, 194, 205;
  and the Internet, 206; and mitigation, differences between,
  193; and molecular assembler,
  207–15; and protection from
  planetary extremes, 234–54;
  and public support, 203; and
  redundancy, 205; and utility fog,
  209–10; and vertical gardens, 205
Adaptation Learning Experiment, 202
adaptation strategies: bear losses, 194;
  change use, 198; glossary of, 194;
  modify the threat, 195; move, 198;
  prevent effects, 197; research new,
  199; share losses, 195
adaptive technologies, examples of,
  207–15, 236–54
adhesives, 89
Advanced Digital Communications, 182
aerocars, 65, 89, 122, 216
aerospace companies and molecular technologies, 251–53
Age of the New Individual, 216
Agenda 21, 177; rewriting, 240
aging. *See* immortal memory; medicine;
  nanobacteria; robotic companions
agricultural production, substituting, 199

agriculture: and adaptation,
  235–56; and genetics, 73,
  235
agroindustry, 235
AIBO, 126
AIDS, 54, 73–75, 83, 270, 297,
  307
air traffic control, 122
alien species, 306
allergy, 273
Alley, Richard, 171
Alvarez, Luis and Walter, 163
American Association for the Advancement of Science, 35
American Chemical Society, 137
American foreign policy and health,
  76–77
ancient civilizations, 181–85; investigations of, 184
ancient structures and Black Sea, 165
animal emotions, 264
animal experiments, 265
animal rights movement, 264–65, 312
animal testing, 265
animals, ethical treatment of, 265
antibiotics, resistance to, 74–75
AOL/Time Warner, 129, 135
Arkwright Mutual Insurance Company, 174
Armageddon, 158
Army Corps of Engineers, 195–96
art and culture. *See* cultural diversity; Hollywood; philosopher king; special effects
  artists
artificial atom, 36
artificial environments, 131, 301

artificial habitat, 132
artificial intelligence, 30, 52, 93, 128, 287,
   311–14; and space exploration, 253;
   impacts of, 234–54; inevitability of, 312;
   right to life, 311
artificial silicon retina, 103, 128, 130
artificially intelligent computers and globalized
   decision-making, 290
artificially intelligent machine, 109
Ashby, W. Ross, 193
ASIMO, 21, 126
Asimov, Isaac, 52, 104, 118, 136, 280
assembler, definition of, 38–39
assembly, skepticism about, 39
asteroids, 226–29; Eros, 228–29; hit, 148;
   mining and extraction bots, 227;
   munchers, 227; planet-killing, 226–29, 292
astrobiology, 161
astronomy. See asteroid; comet Shoemaker-Levy
   9; Hubble
AT&T, 58
AT&T Labs, 130
Athabasca oil sands, 64
athereosclerotic plaque, 56
Atlantis, 183; in Cuba, 158; skeptics, 180
atomic force microscope, 32
atomic microscopy, 42
augmented reality, 299; definition of, 58
Ausubel, Jesse, 145
automotive coatings, 65; manufacturers, 45,
   64–65

backing up human memory, 260
bacteria. See nanobacteria
Baillie, Mike, 170
Bass, Brad, 194
Bauval, Robert, 180
Baxter, Stephen, 226
Bayh-Dole Act, 131
Beer, Stafford, 193
Bekaert ECD Solar Systems, 114
Bell Labs, 32, 130
Bell Lucent Labs, 33
BellSouth, 130
Benfield Greig Hazard Research Center, 166
Bentham, Jeremy, 264
Bertelsmann, 135
Bertman, Stephen, 259
Biela's Comet, 160
Bill and Melinda Gates Foundation, 76
Binnig, Gerd K., 32

biocide nanocoatings, 114
biodegradability, 213
biodegradable pigments, 65
biodegradable plastics, 125
biological weapons, 248
Biosphere 2, 253
biotechnology, 41, 272
Black, Edwin, 266
blood nanobacteria. See Nanobacterium san-
   guineum
Bloom, Howard, 232
Bloomberg School of Public Health, 78
Boies, David, 133
Boies, Schiller, and Flexner, 133
books. See e-books; e-ink; electronic paper
Borg, 100, 110, 287
Borlaug, Norman, 263
Bostrom, Nick, 145, 147
Bourke, Mary, 164, 261
brain pills, 105
Brand, Stewart, 261
Braungart, Michael, 211–12
Brazil, 144
broadband, 58; satellites, 59
Brockovich, Erin, 275
Brownlee, Donald, 142
Buckminsterfullerenes, 32
Burns, Marshall, 45, 48
bush robot, 92
business. See investment
business management, 246–47; revolution, 97

calcification, 55–56
California Institute of Technology, 41, 208
California NanoSystems Institute, 36, 131
Canadarm, 127, 131
cancer cell, 73–74
Capek, Karel, 100
car. See aerocar; automotive manufacturers;
   skycar
Carbon Nanotechnologies, Inc., 131
carbon nanotube, 33–35, 44, 94, 96; armor, 223,
   302
Carnegie Mellon University, 125
Carson, Rachel, 268
Cascadia subduction zone, 221
cascading failures, 191
cataclysms, past: Toba Sumatra, 173; Yellowstone Park, 173
cataracts, 56
catastrophe: defense against, 189; part of evolu-

tion, 161; scientific resistance to, 256; and sustainable lifestyle, 189
catastrophic leaps as catalysts for evolutionary changes, 158
catastrophic threats to society, 163
catastrophism: definition of, 161; skepticism about, 161, 181
Catastrophists, 162
CAVE, 58
Celera Genomics, 125
cell ecology, 118
Center for Science, Policy and Outcomes, Columbia University, 206
Challenger (space shuttle), 31, 203
chameleon coatings, 121, 301
Changchun, China, 114
Charlie Rose Show, 135
Chemical and Engineering News (magazine), 42, 135, 137
chemical nanotechnology, 66
chemicals industry, role of, 236
Cherpitel, Didier, 175
China, 195
Chinese Society of Materials, 134
Chokkagata earthquake, 152
Chow, Alan and Vincent, 103
Ciftcioglu, Neva, 56
Ciprofloxacin, 75
Cisco, 58, 129
civilizations, lost, 180. *See also* ancient civilizations
Clarke, Arthur C., 104, 136, 226
climate adaptation and food fabrication, 210
climate adaptation researchers, 236–37
climate catastrophes, 132; skepticism of, 179
climate change: and economic activity, 198; and ice cores, 171; and Irish oak rings, 169; preventing, 193
climate change handbook, 193, 200
climate modeling, 237
climate upheavals, 172
climate, worst recorded, China, Middle East, South America, 171
Clock of the Long Now, 261
cloning, 54, 267, 303; and human embryos, 263
closed system, 23. *See also* Earth, as a closed system
closed-loop systems, 212–14
clothing. *See* smart clothing
Club of Rome, 116
CNET, 135

Coates and Jarratt, 249
coatings, 65, 118, 269; impact-proof, 67
Code of Federal Regulations, 297
Cognitive Enhancement Research Institute, 105
Cold War, 54, 74–75, 184, 304
collision-avoidance system, 122
colon polyps, 73
color at nano level, 65, 269
Columbia University, 36, 134
comet bombardment, 160
Comet Shoemaker-Levy 9, 156–57, 159–60, 185, 226, 280, 305
communities, floating, 96
Computer Motion, (company), 127
Computer Professionals for Social Responsibility, 265
computer, flexible, 48
computing: and corporate disaster recovery, 237–38; and logic circuits, 33, 45; power, growth in, 29, 97; speed, 51
concatenation algorithm, 130
confluence of technologies, 73
consciousness: defining, 306; surviving indefinitely, 262
constructed environment, 131, 253
construction bots, 215
construction, reinforcing buildings, 94
contamination from destruction of human technologies, 113
convergence of robotics, genetics, nanotechnology, and artificial intelligence, 72
convergence. *See* technology
Conversay, 130
Copenhagen Institute for Futures Studies, 249
Cornell University, 36, 134
coronary artery plaque, 56
corporate concentration, 247
corruption, as barrier to disaster prevention, 201
cosmic commons, 149
credibility gap, 268–77
Cremo, Michael, 184
crisis management systems, redesigning, 214
critter-cam, 67
Cuban Academy of Sciences, 182
cultural amnesia, 255–62; and New Madrid fault, 203
cultural diversity, 100
Cumings, John, 69,71
Curl, Robert F., Jr., 32
currencies, universal units, 84, 241
cyberbartering, 132

cybercash, 132
cybermoney, 132
cyberserfdom, 294
cybersex, 58, 218
cybersuit, 96–97
cybertraveling, 88
cyborg, 104

DARPA. *See* Defense Advanced Research Projects Agency
Darwinism. *See* catastrophism; evolution; elephant in the room of environmentalism
David, Saul, 136
Davis, James, 208
Davos forum, 54
DDT, 76
death. *See* endless human; immortal memory; Robo sapiens
Defense Advanced Research Projects Agency (DARPA), 45, 79, 125, 128, 134, 301
degradation pathways, 273
Dekker, Cees, 32
Delft University of Technology, 32–33
democracy and artificial intelligence, 286–92, 314
dendrochronology, 169
dengue fever, 74
dental plaque, 56
dentistry. *See* nanobacteria; nanotoothpaste
Department of Energy, 113
designer drugs, 77, 290
desktop digital fabricator, 36,
desktop fabricator, 16, 97, 236
desktop factory, 19
desktop manufacturing, 43, 46, 213, 216, 300; and hearing aids, 44
diet industry, 274
digital divide, 78
digital fabrication, 131, 299; types of, 43
digital fabricators, 37, 43, 45
digital matter control, 91
digital terrorist, 248
digital workplace, 294
Dillon, William, 172
disassembly, 119, 212–14, 228
disaster agencies, 238
disaster and human rights, 308
disaster porn, 158
disaster preparedness: and nanotechnologists, 248; societal resistance to, 204; strategies, 191–202. *See also* adaptation

disaster prevention, 175, 218–20, 235–54. *See also* adaptation
disaster recovery: and energy resilience, 214–15; and systems redundancy, 214–15; and space repair, 214–15; and self-disassembly, 214–15
disaster relief, worsening future disasters, 175
disaster researchers in developing nations, 200
disaster resistance and superstrong materials, 94
disaster response, 203; and World Trade Center, 214; and construction bots, 214; and repair bots, 214
disaster risks, exposure of West Coast to, 204
disaster-adaptation shortfalls, 204
disaster-resilient industrial base, 247
disasters, complacency about, 233
Discovery Channel, and promoting space based molecular technologies, 135, 252
diseases: cures for, 54, 74, 93; impacts of, 54
Disney, 129, 135
disruptive technologies: definition of, 22, 54; examples of, 64; and nanobacteria, 56
DNA, 41, 70, 105, 179; computers, 32; labeling, 223; and molecular manufacturing, 74; motors, 32, 108; research, regulation and development of, 272
Dow Jones News Service, 135
Drake, Frank, 138
DRAM. *See* dynamic random access memory
DreamWorks, 129
Drexler, K. Eric, 27, 32, 38, 39–40, 46, 50, 114, 136, 213, 238
drought, as a cause of political instability, 144
drugs, relative costs of, 75
Dunn, Christopher, 181
dust cloud of 536 C.E.; famine, 171
dynamic random access memory, 51
dysgenic pressures, 148

E*TRADE, 132
early warning system, 222
Earth: as a closed system, 23, 159, 160–61, 240; isolated from space, 159; protective magnetic field, 159; not what it seems, 159; as an open ecosystem, 159, 310, 314
Earth Summit 1992, Rio de Janeiro, 177
earthquakes, 183; America's largest, 168; Gujarat, India, 203, 216; and health-care system, 153; insurance, 176, 216; and just-in-time supply lines, 153; and liquefaction, 224; Missouri, 154; New York, 153; predic-

tion, and Tokyo, 217–20; St. Louis, 153; and stock market crash, 153; and toxic pollution, 152

earthquake-resistant design, former Soviet Union, San Francisco, 197

earthquake-warning system, 208

ebay, 132

e-book, 84

Eckersley, Richard, 293

ecocommunities, 118

ecofascism, 287

ecoimperialism, 201; and disaster mitigation, 201

ecological implications of nanotechnology, 269

ecological rebirth, 120

ecologists and environmentalists, 239–41

economics: of molecular assembly, 309–10; of superdisaster preparedness, 310

Economist (magazine), 135, 276

economy and artificial intelligence, 241; splintering of, 102

ecosystems, historic breakdowns, 183

education. *See* enhanced intelligence; neural implants, *Homo provectus*, robotic companion. *See also names of individual educational institutions and Web site URLs (in endnotes).*

Educational Foundation for Nuclear Science, 265

Eigler, Donald M., 32

E-Ink, 47

El Niño, 206

electric power. *See* solar power; electromagnetic transport; energy consumption

electromagnetic transport, 90, 216; vehicles, 95

electronic ink, 49

electronic newspaper, 47

electronic paper, 21, 47, 49, 86, 116

electronic pen, 48

electronic store display, 47

elephant in the room of environmentalism, 176

Ellenbogen, James, 131

emergency response center, New York, 233

EMI (company), 135

emotion chip, 109

emotional refugees, 273

employment security, 294–96. *See also* digital workplace; economy; Internet; investors; virtual reality

encryption, 80, 97, 130–31

endless human, 262

energy: consumption by molecular technologies, 269; deficit, 114–15; efficiency, 64, 269; infrastructure and natural disasters, 239; paradox, 60; payback, 114; resilience, 215

Engelfried, Justus, 212

enhanced intelligence, 59

Ennex Corporation, 45, 48

Enron, 54, 276, 290

entertainment. *See* Hollywood; virtual reality

environment: global regulation, 286; natural stability, 160; and persistent materials, 269

environmental cleanup and nanotechnology, 96

environmental doctrine, 159

environmental engineering and spirituality, 303

environmental impacts of new materials, 57, 65

environmental implications of feedstock, 50

environmentalism, 159; and solar cells, 61; and technologism, war between, 177; *See also* sustainable development

environmentalists, and blame for holding back science, 271

environmentally friendly GMO, 272

environmental movement: and GMOs, 268; and space colonization, 301

environmental organizations, 240; and artificial intelligence, 241; and disaster preparedness, 240; and redefining sustainability, 240

environmental performance, 125

environmental pressure groups, 118

environmental problem, definition of, 270

environmental refugees, 175

environmental restoration, and Florida Everglades, 196

environmental risks, perception of, 270,307

environmental theory, deficiencies, 176,189

e-paper. *See* electronic paper

Ericsson, 48, 129–30

eternal life. *See* endless human; immortal memory; Robo sapiens

ethical programming, 109

ethics, 138, 247, 264–77, 302, 306; research as amoral, 270

Etkin, David, 205

European Commission, 134

European Space Agency, 131

evolution: catastrophic version, 162; as a gradual process, 158; irregular process, 159

Excite@home, 54

exoskeleton, 70, 224, 301–302

extinction: dinosaurs, 163; risks, 145–49

extraterrestrial civilization, 149
extreme engineering, 238
extreme environments, 231
extreme-environment clothing, 69
Extropian philosophy, 108
Extropy Institute, 106, 108, 112
eyeglass lenses, 67
eyesight. *See* artificial silicon retina

Fåhraeus, Christer, 48
farming. *See* integrated farming; agriculture; food fabrication
fashion. *See* smart clothing
Fauci, Anthony, 77
Federal Emergency Management Agency (FEMA), 199, 240, 249, 253
Federal Reserve, 298
feedstock, 50, 116, 119, 225; regulatory controls, 119
Fenwick & West, 133
Fermi, Enrico, 111
Feynman Prize in Nanotechnology, 139
Feynman, Richard P., 31, 62, 67, 204
fiber-optic networks, 53
Final Fantasy: The Spirit Within (film), 121
finance and molecular technology, 79
financial bubble, 267. *See also* personal savings rate
financial markets and memes, 267. *See also* investors
Financial Times, 135
fire-resistant fiber reinforcing, 224
First Nations, 242
fiscal policy and molecular assembly, 241
Fiske, John, 188
flatwire, 66
Fleischer, Richard, 136
Flem-Ath, Rand and Rose, 181
flood evidence, Australia, 159
flood: Bangladesh, 175; Huang He river, 196; Mississippi River Valley, 196, 199; Yangtze River, 175
Foley & Lardner, 134
Food and Drug Administration (FDA), 105, 235
food fabrication: and synthesis, 210; software, 235
food synthesis and animal rights, 306
food-toxic environment, 274, 296
foot-and-mouth disease, 265
Foresight! Guidelines for Molecular Nanotechnology, 214

Foresight Institute, 38, 112, 249, 280, 285, 308
fossil fuel industry, 60, 239
fractal robots, 90–92
Frankenstein science, 104, 280
Frankfurter Allgemeine Zeitung, 135
Fraser, Jill Andresky, 267, 294
freedom of information, 283
Fu, Ping, 45
fuel cell, 60, 64
Fuller, Buckminster, 299
Future Shock, 11, 137

GRAIN., definition of, 30
Galileo, 156, 160
Galileo (satellite), 160
Galileo (spacecraft), 156
gallium arsenide as an environmental problem, 269
Garrett, Laurie, 153, 266
General Electric, 135
genetic design, 100
genetic modification, transparency of information, 273
genetic programming, 98, 100, 105; and patents, 284
genetic warfare, 125
genetically engineered biological agent, 147
genetically modified organisms (GMO), 268–69, 271; and U.S. corn exports, 271
genetically modified products, processes used in development of, 272
genetics, 30; industry, 54; impacts, overview, 73. *See also* cloning; DNA; DNA motors
genomics, impacts, overview, 73
geology. *See* earthquake, volcano
geomorphologist, 159
geomorphology: and sediment core sampling, 159; and sediment samples, 164
Gershenfeld, Neil, 46, 97, 137
Global Alliance for Vaccinations and Immunization, 76
global positioning systems, 77,122
global warming, runaway, 148
globalization, 298
Goodyear, 64
government: inability to regulate technology, 242, 308; repressive totalitarian global regime, 149
Great Depression, 144, 150
Great Kanto earthquake, 150–51
green backlash, 273

green stress, 275
Greenberg, Dan, 266, 276
greenhouse gas reduction, 193
guidance systems for flying, 65, 216
guidelines for development of technology. *See Foresight Guidelines*; intelligent product system; Laws of Robotics
Gunn, Joel D., 171
Gyricon Media, 47

habitat: destruction, causes of, 298; encroachment, 118; preservation, 118
Hadfield, Peter, 151, 154
Halas, Naomi, 131, 137
Hall, John Storrs, 209
Hamburger Umweltinstitut, 124, 212
Hancock, Graham, 180, 182, 184, 257
Hapgood, Charles, 181
happiness. *See* human rights; technology liberating the individual
Harrison, John, 309
Harvard University, 36
Harvard University Medical School, 134, 283
Hassan, Fekri, 261
health care, 54, 72, 74, 76–77, 93, 247–48, 299; failure of, 296–97; preventative, 266. *See also* heart disease; nanobacteria; robotic care
health implications of nano-sized materials, 119
heart disease, 72, 74, 307
Hebrew University, Jerusalem, 134
Herrenknecht AG, 92
Hewat, Marcus, 30
Hewlett-Packard, 128, 132, 244
Hillis, Danny, 261
Hinzmann, Brock, 45
Hitachi, 58
Hitchhiker's Guide to the Galaxy, The, 23, 86, 209
Hofstadter, Douglas, 137–38
Holland, John, 139
Holling, C. S., 205
Hollywood, 129, 136, 158; prop makers, 45; special effects and video game designers, influence of 99
Holocaust, 266
home medical clinic, 299
home medical diagnosis, 299
Homo provectus, 85, 102, 110, 311
Homo sapiens, 83, 102, 107, 110, 146, 171, 179, 184, 312; decline of dominance, 312; replacing, 138
Honda Motor Company, 126

Hopi Indians, 256–58
housing. *See* adaptation; carbon nanotubes; construction
Hu, Evelyn, 131
Hubble Space Telescope, 156
Hudson River, 116
human consciousness, religious definitions of, 303
human genome, 39; defective, 179; patenting of, 283
Human Genome Project, 179
human rights: and artificial intelligence, 308; and liberating the individual, 216, 314; and mass survival, 304–307
human technology, as part of nature, 146
human vulnerability to the natural environment, 145
human-computer relationship, 78
hurricane, 122; losses, 174
Hurricane Andrew, 175, 195, 206, 216
Hurricane Floyd, 164, 166, 169
Hurricane Mitch, 175, 195, 216
Hutchison, Dan, 51

IBM, 33, 128, 130, 132, 266; Almaden Laboratories, 32; Zurich Research Laboratory, 32
ice age, 171, 182
ice storm, 197
idea economy, 282
Iijima, Sumio, 32
Ikonos 2 (satellite), 184
Immersadesk, 58
immortal memory, 258–59
immune system, 73
incarceration rate, 297
indigenous peoples, 242
Industrial Technology Research Institute, Taiwan, 134
infection. *See* health care; medicine; nanobacteria
infrastructure planning, 238
inkjet printer, 44
Institute for Soldier Nanotechnologies, 278
insurance, 195, 243; and genetic screening, 243; and natural disasters, 202, 216, 243
integrated farming system, 212
Intel, 128
intellectual freedom, 283
intellectual property, 282–85, 314
intelligent machines, 285; societal dependence on, 287

intelligent product system, 211–14; as guideline for development of nanotechnology 211–14; and consumable goods, 212; and closed loop systems, 212; and disassembly, 212; and integrated farming, 212; and recycling, 212

interdisciplinary communication, 269

Intergovernmental Panel on Climate Change (IPCC), 192, 194

International Red Cross (Federation of), 175, 249, 308

International Space Station, 52

Internet, 45, 54, 88, 117, 129; and climate adaptation, 211; and disaster recovery, 237; and drug manufacturing, 242; and earthquake warning, 208; and food fabrication, 211; voting, 290

Internet Architecture Board, 129

Internet Engineering Task Force, 129

Internet Software Consortium, 129

Intuitive Surgical Inc., 127

investment, 53; 244–45; and disaster preparedness, 238; breakout industries, 244; liabilities, 244–45; sector index funds, 244; seed capital companies, 132. See also adaptation; desktop manufacturing; digital fabrication; digital workplace; economy; employment; guidelines; health care; insurance; molecular assembly; robotics

investors, 38, 244–45

IPCC. See Intergovernmental Panel on Climate Change

ispo Gmbh, 69

Jacobson, Joseph, 46, 131

James, Peter, 181

Japan Today, 135

Japan, 126. See also Tokyo, Kobe

Jet Propulsion Lab, NASA, 128

Joy, Bill, 111, 138, 268, 288

Jupiter, 156–57, 160–61, 280

JX Crystals (company), 63

Kaczynski, Theodore, 288, 305

Kajander, Olavi, 56

Karl Storz Endoskope, 127

Kasparov, Garry, 17, 26

Kellogg (company), 81

Kelly, Kevin, 139

Kennedy Space Center, 252

Keys, David, 171

kidney stones and nanobacteria, 56

Kleiner, Harry, 136

Kobe, 150–51, 153, 175, 203, 214, 217

Koza, John R., 27, 100, 137, 139

Kroto, Harold W., 32

Kubrick, Stanley, 104, 136

Kurzweil, Ray, 27, 111, 137–38, 146, 303

La Niña, 206

labor unions, 246

landslide: and Alaska, 165; and Canary Islands, 166, 306; cause of tsunami, 165, 222; and Mediterranean, 165; and methane hydrate, 173

language translation software, 86. See also translation software

Laser Dynamics Laboratories, Georgia Institute of Technology, 65

Law of Requisite Variety, 193–94, 198–99, 300

Laws of Robotics, 109, 280

lawyers, 133

Leakey, Richard, 184

legends, validity of, 242

leisure and molecular technologies, 246

Lernout & Hauspie, 130

Levy, David, 156, 185

liability and molecular technologies, 79

life, eternal. See endless human; immortal memory; Robo sapiens

life insurance, 243

life sciences, 124

lifestyle, temporary, 296

Linux (operating system), 285

liquid crystal display, 119

lithography, soft, 42

Lloyd, Bruce, 260

Lockheed Martin, 131

Long Now Foundation, 261

Long Valley Caldera, 221

Los Alamos Laboratory, 129

Los Angeles Times, 135,297

Lost City, Cuba, 182

Lotus effect, 67,69

Lucasfilms, 130

Lucent Technologies, 32,47

MacDonald Dettwiler and Associates, 131

machines, asking for rights, 303

Maddox, Sir John, 179

MaestroSign system, 50

magma domes, 221

Mails, Thomas E., 257
malaria, 54, 74–75, 307–308
malaria, cure for: and Canberra, Australia, 76; and Gambia, 76; and Gates Foundation, 76; and Glaxo Wellcome, 76; and Johns Hopkins University, 76; and NIH, 76
manufacturing: food, 117; production principles, 117. *See also* digital fabrication; desktop manufacturing; molecular assembler
Mars, 53, 137, 291–92, 301
Mars Society, 131, 253, 291
Marsden, Brian, 156
Marshall, Tristan, 165
Martin, James, 111
mass extinction, 161; as a way to control populations, 305
mass survival, 304–307
Massachusetts General Hospital, 283
Massachusetts Institute of Technology (MIT), 36–37, 128
Masse, Marcel, 286
material abundance, 117
materials characterization, 133
Max Planck Institute, 112, 134
McCarthy, Thomas, 278, 289
McGuire, Bill, 166
mechatronics, 72
media, and nanotechnology, 135. See also names of individual publications, networks, and companies
medical robots, 94
medicine, 247–48. *See also* AIDS; artificial silicon retina; health care; heart disease; malaria; organ replacement; surgery
megacomputing, 51, 302
megaterraforming projects, 199
megatsunami, 177
memes: in financial markets, 267; guilt by association, 273; in scientific ethics, 267; shifting of, 272
memory: collective loss of, 259–62; transfer into robotic body, 262
MEMS. *See* micro-electro-mechanical systems
Merkle, Ralph, 139
methane hydrate, 172
micro-electro-mechanical systems (MEMS), 41, 43, 72
microsensors into tires, 64
Microsoft, 129, 133, 283, 285
Midnight Sun Heating Stove, 63
military, 53; participation in natural-disaster mit-

igation, 248–49, 278; security, 248, 280; and terrorism, 302. *See also* Defense Advanced Research Projects Agency; Institute for Soldier Nanotechnologies; molecular deterrence; National Security Agency; Smart Dust; stealth technology; U.K. Defense Ministry
millions of instructions per second (MIPS), 29
Minatec, France, 134
mining bots, 228, 250
Minsky, Marvin, 27
MIPS. *See* millions of instructions per second
Miramax, 129
Mirkin, Chad, 42
MIT Media Lab, 46, 131
MIT Technology Review (magazine), 47, 135
MIT. *See* Massachusetts Institute of Technology
mitigation. *See* adaptation and mitigation
Mitre Corporation, Virginia, 131
Mitsui & Co., 34
molecular age and government, 241–42
Molecular Alley, 35
molecular assembler, 36, 39, 84, 117–19, 123, 207–15, 229, 308–309; motivation for inventing, 227–29; prize for inventing, 309; and resource extraction, 250; skepticism about, 308–309
molecular assembly: and globalized delivery systems, 214; and greenhouses, 211; impacts of, 234–54; preconditions for, 38
molecular biosynthesis, 210
molecular chemistry and nanotechnology, 41, 212
molecular computer, 225
molecular computing, 34, 51
molecular construction, 119
molecular defenses and natural disasters, 181, 229
molecular economy, 116–17, 293–95
molecular machines, 41
molecular manufacturing, 74; and local independence, 291; and mass survival, 211, 304; side effects of, 214
molecular nanotechnology, 31, 38; and military deterrence, 278–79; and nation states, 289; pathways to, 40
molecular revolution, impacts of, 55
molecular technologies: benefits for consumers, 236; dangers of, 278, 288; definition of, 22; as defense against natural calamity, 234; implications of, 24, 235–54; and mass survival, 314; and megachanges, 55

Molecular Electronics Corporation, 128
molecular future: as ecological question, 240;
    contradictions of, 59; definition of, 11
Molecular Manufacturing Enterprises, 132
molecular technology and megaimpacts, 117
molecules as feedstock, 50
Moller, Paul S., 89
monetary system and molecular assembly, 241
money. *See* economy; financial markets;
    investors; new economy; personal savings
    rate
money trail, 74
Montreal Protocol, 207
Moore, Jim, 165
Moore, Patrick, 146
Moore's Law, 45
Moravec, Hans, 27, 92, 137–38
Morrison, David, 162
Motorola Corporation, 81
mountain pine beetle, 172
Munich Re, 174

nanites and excavation, 223
nano, as a cliché, 38
Nano (forename), 38
Nano (television show), 135
Nanobaclabs, 56
nanobacteria, 35,56,307. *See also Nanobacterium
    sanguineum*
nanobacterial infections, 57
Nanobacterium sanguineum, 56
NanobacTX, 56
nanobes, 56
nanobots, 93, 116, 119
NanoBusiness Alliance, 131
nanocell, 45
nanoceramics, 114
nanocoatings, 60, 66, 70, 114
nanocomputer, 33, 35, 51
nanocrystalline: cell, 62; powder, 67
nanodrugs, 77
nanoecology, 113, 118, 161
nanoeconomy, 119
nanoelectronics, 35
nanofiber particles, 66
nanofuel, 89
Nanogate Technologies GmbH, 66
nanogel, 114
nanoglue, 114
Nanoinvestor News (Web site), 135
nanolabel, 223

Nanomagazine.com, 235
nanometals, 232
nanometer, definition of, 30
nanoparticles, effects on human metabolism, 66,
    211
nanopigments, 114
nanoproducts, 211
nanoreplicator, rogue, 147
nanorods, 65
nanosatellites, 227
nano-scale: bearing, 69, 71; coatings, 114;
    mechanical-biological mites, 223; particles,
    118; products, definition of, 38
nanosensor, 73, 219, 222
nanoshells, 66, 131
nanospies, 88, 302
nanostocks, 244–45
nanostructure, 38
nanostructured film, 65
nanosurface, 62, 73
Nanotech Investor (magazine), 132
Nanotech Planet (Web site), 135
nanotech skiwear, 69
Nanotechnews.com, 135
nanotechnology, 30, 32, 43, 50, 118, 244–45;
    business and, 36, 131; chronology of, 32;
    definition of, 38; development guidelines,
    280; ecological implications of, 118; and
    environmental risks, 270; and fluores-
    cence, 36; and food fabrication, 210; and
    friction, 69; and lubricants, 67; and manu-
    facturing cost reduction, 72; and near-
    Earth object sensing, 208; venture capital,
    244–45
Nanotechnology (magazine), 135
nanotools, 40
nanotoothpaste, 35
nanotube structural reinforcement, 210
nanotube-reinforced skins, 122
nanowar, 98
nanowasher, 38
nanowires, 66, 80
nanoWorkbench, 44
Naples, 176
Napster, 133, 283
NASA, 65, 89, 95, 131, 203, 216, 228, 252
NASA: and remote sensing for disaster pre-
    paredness, 253; Goddard Institute of Space
    Studies, 170
NASDAQ, 53
National Bureau of Standards, 133

National Center for Atmospheric Research, 206
National Center of Supercomputing, 58
National Institute for Materials Science (Japan), 40
National Institute of Allergy and Infectious Diseases, 76
National Institutes of Health (NIH), 76, 125, 134, 280, 283
National Institute of Standards and Technology (NIST), 133
National Nanotechnology Initiative, 36
National Oceanic and Atmospheric Administration (NOAA), 253
National Science Foundation, 38, 134, 154
National Security Agency (NSA), 78–79, 128, 134
national security and ethics, 304
natural catastrophe, 139, 150
natural disaster prevention: and Canada, 202; and Florida, 202; and Mauritius, 202
natural disasters: and environmental organizations, 177; exploitation of victims, 176, 201; and human rights organizations, 177; impacts on technology, 54; insurance, 174; and public awareness, 201; refugees, 175; societal impacts of, 153; and toxic pollution, 213; and UN organizations, 192
natural disaster zones, development in, 174
natural extremes, defenses against, 234–54
natural habitat, 117
natural resources and molecular assembly, 250
nature and molecular assembly, 38
Nature (journal), 111, 135, 170, 179
nature's galactic clearance projects, 189
nature's time bombs, 24, 144, 180
Near Earth Asteroid Rendezvous satellite (NEAR), 228–29, 252
near-Earth objects, 157–58, 179, 226–30, 251, 308–10; collision, frequency of, 162–63; and molecular assembly, 226–30; sensors, 208, 227
near-term catastrophic threats, 163
NEC Research Labs, 32
NEMAX, 53
neural chip, 110
neural implants, 262, 287
neural network, 219
neutrinos, 111, 118, 160
New Century Nanometric Technology Research, 114
new economy, 53, 267, 295; and stress, 296

new individual, 299–300, 302
New Madrid fault, 168
New Scientist (magazine), 135, 152, 263
New York Stock Exchange, 21
New York Times, 135, 282
News Corp., 135
night vision technology, 54
NIH. *See* National Institutes of Health
NIST. *See* National Institute of Standards and Technology
NOAA. *See* National Oceanic and Atmospheric Administration
Nokia, 129–30
nongovernmental organizations, role of environmentalists, 239–40. *See also individual organization names*
nonpoint source, 274
Nortel, 57, 129–30
Northridge earthquake, 153–54, 176, 195
Northwestern University Institute for Nanotechnology, 36, 42
NSA. *See* National Security Agency
Nuance (company), 130
nuclear weapons control, compared to nanotechnology controls, 280
nyms, 121

O Instituto Ambiental, 212
obesity, 296, 298; as an environmental problem, 274
oceanography, 182. *See also* National Oceanic and Atmospheric Administration; tsunami
oil and gas, 60. *See also* fossil fuel industry
open source, 47, 98, 282–86; hardware, 47; software, movement, 60, 266; definition of, 285
Optobionics Corporation, 103, 128
Oracle (company), 129
organ replacement, 61
organized labor, 246
orphan diseases, 75, 307
overabundance and molecular assembly, 116, 118–19
overconsumption and molecular assembly, 116
overproduction and molecular assembly, 116
Oxford GlycoScience, 283
Oxford University, U.K., 32, 134

pace of natural genetic change, 146
Pacific Tsunami Warning Center, 143
paper. *See* electronic paper

paperless office, 116
paradigm shift in industry, 51
paradoxes and possible futures, 19
Patent Policy Working Group, 285
patents, 106, 282–86; invalid, 266, 284; and pharmaceuticals, 283; and secrecy in science, 283
pathways to molecular technology, 40
PBS, 135
Pergamit, Gayle, 40,137,211
personal computer operating systems, replacing, 54
personal debt, 294
personal fabricator, 46
personal privacy, 80, 93. *See also* stealth technology; stealth surveillance, Smart Dust
personal savings rate, 295
personalized fabrication on demand, 44
pets. *See* AIBO
Peterson, Chris, 21, 32, 40, 136–37
petroleum. *See* fossil fuel industry
pharmaceutical companies, 125, 185, 296
philanthropists, 250
Philips (company), 58
philosopher-king, 99
photolithography, 42
photo-optical computers, 62
photovoltaic cells, 61
Physicians for Social Responsibility, 265, 309
physics disaster, 248
Pielke, Roger, 206
pilotless drones, 67
pincers, to manipulate atoms, 39
planet-altering collision, 156
Polaroid, 54
politics. *See* government; Internet voting; drought, as a cause of political instability
Pompeii, 176
Ponting, Clive, 183
Pool, Robert, 298
positioning for molecular assembly, 38
posthumans, 102, 107
precautionary principle: and Montreal Protocol, 207, 268; and relinquishment, 207, 268; rethinking the concept of, 306–307
prediction, dangers of, 53, 200
prehistory, investigation into, 182
printable bicycle, 46
privacy. *See* personal privacy
probability theory and risk assessment, and existential risk, 146

Procter and Gamble, 125
Program for the Human Environment, Rockefeller University, 145
programmable chip, 78
programmable matter, 36
programmable voice, 30
programmed nanoparticles, 70
Project Impact, FEMA, 199
prophesies: differentiation from legends, 256; skepticism about, 256
prosthetics, 45, 73
proteins, 125; self-assembling, 43
publishing. *See* e-books; e-ink; electronic paper
Pugwash Conferences, 265
pyramids, 181

quantum computing, 79
Queen's University, Belfast, 170

Raindrop Geomagic, 45
random-access memory, 130
rapid prototyping, 43–44, 46
real estate and molecular technologies, 251
RealSpeak software, 130
real-time intelligence imagery, 68
reassembly, 50
recombinant DNA, 272
reconstructive surgery, 44
recreation. *See* cybersex; robotic companions; virtual reality
recycling: chain, 213–14; economy, 214
Red Herring (magazine), 135
redesigning Earth, 195
religion. *See* cloning; Galileo; Kurzweil, Ray
relinquishment of technology, 206, 268, 280
remediation, 116
remote sensing, and forecasting of natural disasters, 237, 253
renewable energy, 60, 239
Rennie, John, 263
repair bots, 215
reproduction via artificial intelligence, 108
rescue bots, 214. *See also* RoboRescue
research, and national security, 184
resilience, definition of, 204–205
resilient communities, 243
resistance: to environmental extremes, 197; forest fires and, 205
resource depletion, 148
retinal implant. *See* artificial silicon retina
Rice University, 35, 46, 66, 128, 131

Risk Management Systems, 152
risk management and reduction, 175. *See also* adaptation; disaster preparedness; disaster prevention; September 11; World Trade Center
risks, categories of, 147, 307
Robinett, Warren, 32
Robinson, Kim Stanley, 132, 136, 291, 301
Robo sapiens, 85, 101, 109–10, 118, 242, 308, 311
Robo servers, 85, 94, 101, 110, 308, 311; and democratic control, 287
RoboCup, 70, 126, 214
RoboRescue, 214
robot, definition of, 100
robot wars, 71
robotic care of the elderly, 94
robotic companions, 94
robotic health care, 248
robotic medicine, 72
robotic organs, 102
robotic replenishment and climate adaptation, 210
robotic surgery, 127
robotics, 30, 100, 125; pervasiveness of, 109; and surveillance, 80
Robotics Institute: Carnegie Mellon University, 125; University of South Florida, 126
robots, capabilities of, 109
Roddenberry, Gene, 136
Rohrer, Heinrich, 32
Ross, Dr. Aki (fictional character), 121
Roukes, Michael, 41

Sagan, Carl, 227
Saint Thomas, 120–24
San Francisco Chronicle, 135
Sandia National Laboratory, 127
Santa Fe Institute, 112
Sarewitz, Daniel, 206
Saxl, Ottilia, 137
scanning tunneling microscope, 32, 41
scanning-probe microscope (SPM), 44
Scansoft, 130
Schweizer Optik GmbH, 68
Schweizer, Erhard K., 32
science: and animal rights, 264; public attitude toward funding, 276; and public trust, 264,290; public understanding of, 276; and science fiction 16, 21. *See also* medicine; molecular assembly; nanotechnology; technology; environmentalism

Science (journal), 135
Scientific American (magazine), 135, 263–64, 266
scientific discovery, rate of, 39
Seagram (company), 135
Search for Extraterrestrial Intelligence (SETI) Institute, 138
Securities and Exchanges Commission, 244
security of the individual, 281,299
Sega (company), 58
self-assembling nanobots, 228
self-assembly, 62, 228; proteins, 43
self-aware computers, 105, 265
self-disassembly, 213, 215
self-repairing satellites, 111
self-replicating, intelligent machines, control by corporations, 139
self-replication, 38
seismic activity. *See* earthquake; volcano; asteroid
Sellers, Jane, 181
sensors, 57
September 11, 17, 79, 154, 210, 237, 256, 304
Seven of Nine (fictional character), 104
sewage treatment, 270
sex. *See* cybersex; virtual sex
sexually transmitted diseases and virtual sex, 60
shatterproof glass, 66
Shelley, Mary, 104
Sheridon, Nicholas K., 47
Shoemaker, Carolyn, 156, 185
Shoemaker, Eugene, 156, 185
Shoemaker-Levy Comet. *See* Comet Shoemaker-Levy 9
Shulman, Seth, 266
Silicon Valley, 35
Singularity, 27–29, 83, 102, 106; definition of, 104
Skeptical Inquirer (magazine), 180
skepticism. *See* Atlantis; catastrophism; climate catastrophe; molecular assembler; prophesies
sky bots, 225
skycar, 89
Small Times (Web site), 135
Smalley, Richard E., 32, 131
smart card, 58, 290
smart clothing, 87
smart dust, 80, 302
SmartPaper, 50
smokestack technologies, 207
social backlash, 296

social disaffection, as a path to terrorism, 280
social extremes and contradictions, 302
social workers, and impacts of molecular technologies, 251
societal implications of technologies, 137, 293–99
societal problems, fear, 299
software, 282–86, 299. *See also* computing; food fabrication software; Internet software consortium; nanotechnology development guidelines; open source; translation software
solar cell, 61–62, 114, 218
solar coating, 215
solar economy, 115
solar grid, 223
solar panels, flexible, 115
solar power station in orbit, 52
solar sails, 227
solar wind, 159
Sony Electronics Inc., 58, 126, 129, 135
Soros, George, 298
space agencies: of Brazil, 131, 240; of Canada, 131, 240; of China, 131, 240; and molecular technologies, 251–53; and near-Earth object detection, 227; of Russia, 131, 240
space cable, 94
space colonization and ethics, 306
Space Daily, 135
space elevator, 94–95, 302
Space Imaging, Inc., 184
space organizations and molecular technologies, 251-3
space suit, 69
Space.com, 135
Spaceguard Foundation, 157, 227
special interest tactics, 267, 276, 291. *See also* ethics; subsidies
special-effects artists, 120
species extinction, 118
species: definition of, 108; roulette, 189
Speechworks, 130
Spielberg, Steven, 104, 136
SPM. *See* scanning-probe microscope
sport. *See* RoboCup; virtual reality
Sports Utility Vehicles, risk perception, 271
Sprint (company), 58
SRI Consulting Business Intelligence, 45
standards, 131
Stanford University, 100
startups. *See* investment

Star Trek, 100, 109, 287, 301
stealth surveillance, 67
stealth technology, 66
stock markets: and artificial intelligence, 241; and earthquake 153. *See also* investment; investors; nanostocks
storm. *See* adaptation; climate; hurricane
Stothers, Richard B., 170
Strong, Maurice, 177
subsidies, 190; to fossil fuel industry, 53; to polluting industries, 270; to wasteful industries, 240, 267
suburbanization, 117
Sun Microsystems, 36, 58, 129
sun, energy from, 159
superhuman intelligence, 102
superintelligence, 147; flawed, 149
superorganism, 301
supersonic tunnel, 84
surgery, 72. *See also* medicine
sustainability, redefining, 189–90, 240, 306
sustainable development: definition of, 176, 178; and disaster preparedness, 176, 213
sustainable manufacturing, 213
sustainable society. *See* sustainable development
systems redundancy, 215. *See also* Internet

Taniguchi, Norio, 32
taxation and molecular assembly, 241
teaching. *See* education
Technanogy (company), 132
Technion Institute of Technology, Haifa, 134
technological arrest, 148
technological change, rate of, 53
technological singularity, 105
technologies and businesses to track, examples of, 244-5
technologism, 177
technology: abuse, 267; backlash, 67; bubble, 53; as a continuation of natural processes, 146; convergence of, 55; development as exponential, 28; and evolution, link, 146; liberating the individual, 293; using to improve life, 299–300; variables affecting progress of, 55; wealth and happiness, 293
technology-evolution continuum, 146
teen pregnancy and virtual sex, 60
Tel Aviv University, 134
telementoring and medicine, 127
television, future of. *See* augmented reality; virtual reality

Tenner, Edward, 255
teraflops, 100
terraforming, 117, 278, 286, 300; as adaptive strategy, 195–96; Mississippi Basin, 205; opposition, 197; technologies, 190; Three Gorges Dam, 195
terrestrial impact frequency, 164
terrorism: and the Internet, 60; low tech vs. high tech, 279. *See also* September 11; World Trade Center
text-to-speech, 130
thermal photovoltaics, 62–63
Thompson, Richard, 184
Thorpe, Nick, 181
threats to progress, 145–49
three-dimensional viewing, 72
3M Corporation, 269
tidal wave. *See* tsunami
time, relaying messages over extended spans of, 262
tobacco industry, 273
Toffler, Alvin, 11, 137, 295
Tokai Bank, 152
Tokai earthquake, 152
Tokyo, 150–51, 154, 217–20
Tokyo Science University, 32
Toppan Printing Company, 47
tornado protection. *See* adaptive technologies, examples of
Torvalds, Linus, 285
Toshiba, 58
Tour, James M., 128, 131
toxic industrial processes, 115, 119, 211
toxic interactions, 117
Toyota Technological Institute, 114
tranquilizer prescriptions for kids, 274
transcending upload, 148
transhuman rights movement, 308, 314
transhumanism, 106–107
translation software, 60, 100, 219
transportation, 64; and molecular technologies, 253. *See also* electromagnetic vehicles; skycar; space elevator
travel. *See* aerocar; electromagnetic transport; guidance systems; transportation; virtual reality
tropics, risks of, 74
tsunami, 143, 163, 222; areas at risk of, 165, 191; barrier, 223–25; damage, Kodiak Alaska, 178
Tunguska, 158, 162

tunnels for transportation. *See* transportation; supersonic tunnel
typhoon. *See* hurricane

U.K. Defense Ministry, 135
U.S. Air Force, 68
U.S. Environmental Protection Agency (EPA), 113, 116, 249, 270
U.S. Geological Survey, 172, 208
U.S. National Academy of Sciences, 192
U.S. National Science and Technology Council, 177
U.S. Navy Center for Applied Research in Artificial Intelligence, 111
U.S. Patent Office, 284
U.S. Tsunami Warning Center, 143
UCLA, 36
UCSB, 36
Ulrich, Howard, 165
Unabomber, 288, 305
Uniformatarians, 162
uniformity doctrine, 162
Union of Concerned Scientists, 265, 308–309
UNI-SOLAR, 115
United Nations, 192
United Nations Environment Program, 192, 207
United Nations Framework Convention on Climate Change, 193
United Tribal Alliance, 243
United Tribes of the Americas, 243
Unitika, 114
University of Alberta, 134
University of Berkeley, Department of Physics, 69, 71
University of California, 32, 134
University of California, Davis, 113
University of Edinburgh, 137
University of Freiburg, 126
University of Illinois at Chicago Medical Center, 58
University of Melbourne, 134
University of North Carolina, 32, 44, 58, 129, 171
University of Rochester, New York, 129
University of San Diego, 105
University of Tokyo, 134
University of Toronto, 134
University of Washington, 134
University of Windsor, 159
urban sprawl, 117
utility fog, 89, 209, 214, 223

variety, environmental concept of, 193
Varmus, Harold, 76
Venter, Craig, 125
venture capital investment bubble, 244–45
venture capitalists, 132, 244–45. *See also* startups; investment
Vesuvius, Mount, 176
Viacom, 132
Vinge, Vernor, 16, 26–27, 105
virtual conference, 88
virtual junkies, 88
virtual personalities, 121
virtual reality, 23, 87; booth, 58; gloves, 58; interface, 44; suit, 218
virtual sex, 60, 87
virtual workplace, 59
VLSI Research, 51
voice activation, 130
voice recognition, 130
voice synthesis, 130
voice-enable text-based Web sites, 130
volcanic dust, 124
volcano: Krakatau, 173, 221. *See also* Long Valley Caldera
volcano collapse: and El Hiero, 166; La Palma, effects on North America, 166; and Marquesas, 166; and Réunion, 166; and Tristan da Cunha, 166
volcanoes: and climate patterns, 221; and nanosensors, 222
Von Däniken, Erich, 184
Von Ehr, James, 40, 131–32, 137
Von Neumann, John, 233

Wachowski, Larry and Andy, 136
Wall Street Journal, 52, 285
war: by remote control, 278–79; earthquake as a cause of, 150. *See also* genetic warfare; military; terrorism; Internet
War on Drugs, 242, 296
Ward, Peter D., 142
waste recyclers and nanotechnology, 210
wastewater treatment, 212. *See also* intelligent product system

water, 117; as an enemy in the tropics, 201; industry, 200; synthesizers and nanotechnology, 210; systems and nanotechnology, 210
Waters, Frank, 257
wave, highest recorded, 165
wealth, polarization and concentration of, 297
wealth effect, 296
weather, as cause of societal stress, 274. *See also* climate; hurricane
Weizmann Institute, 32
Wells, H. G., 103
"What Is Nano?" (investor guide), 244
Wigand, Jeffrey, 273
wildlife. *See* animal rights movement; species, definition of
Williams, R. Stanley, 244
Williams, Stanley R., 32
Williams, Tad, 136
Wired (magazine), 135
wireless licenses, cost of, 53
women in molecular science, 137, 249
World Commission on Environment and Development, 176
World Future Society, 249, 308
World Health Organization (WHO), 75
World Technology Evaluation Center, 38
World Trade Center, 113, 126, 163, 208, 214–15, 233, 237, 279, 305
World Transhumanist Association, 106, 249
world war, 150
World Wide Web Consortium, 285

Xerox Corporation, 47, 132

Yale University, 145
Yokohama, 151

Z Corporation, 37, 44
zero pollution, 116
Zerrl, Alex, 71
Zhejiang University, 134
Zubrin, Robert, 132, 137, 291
Zyvex, 131–32, 310